PREDICTION OF TURBULENT FLOWS

The prediction of turbulent flows is of paramount importance in the development of complex engineering systems involving flow, heat and mass transfer, and chemical reactions. Arising from a programme held at the Isaac Newton Institute in Cambridge, England, this volume reviews the current situation regarding the prediction of such flows through the use of modern computational fluid dynamics techniques, and attempts to address the inherent problem of modelling turbulence. In particular, the current physical understanding of such flows is summarised and the resulting implications for simulation discussed. The volume continues by surveying current approximation methods whilst discussing their applicability to industrial problems. This major work concludes by providing a specific set of guidelines for selecting the most appropriate model for a given problem. Unique in its breadth and critical approach, the book will be of immense value to experienced practitioners and researchers, continuing the UK's strong tradition in fluid dynamics.

PREDICTION OF TURBULENT FLOWS

Edited by

G. F. HEWITT AND J. C. VASSILICOS
Imperial College, London

CAMBRIDGE UNIVERSITY PRESS
Cambridge, New York, Melbourne, Madrid, Cape Town, Singapore, São Paulo

Cambridge University Press
The Edinburgh Building, Cambridge, CB2 2RU, UK

Published in the United States of America by Cambridge University Press, New York

www.cambridge.org
Information on this title: www.cambridge.org/9780521838993

First published 2005

Printed in the United Kingdom at the University Press, Cambridge

Typeface Times 11/14 pt.   *System* LATEX 2$_\varepsilon$ [TB]

A catalogue record for this publication is available from the British Library

Library of Congress Cataloguing in Publication data
Prediction of turbulent flows / edited by G. F. Hewitt and J. C. Vassilicos.
p.   cm.
Includes bibliographical references and index.
ISBN 0 521 83899 1 (alk. paper)
1. Turbulence – Mathematical models.   2. Unsteady flow (Fluid dynamics) – Mathematical models.
I. Hewitt, G. F. (Geoffrey Frederick)   II. Vassilicos, J. C.

TA357.5.T87P74   2005
532′0527′015118 – dc22   2004054471

ISBN-13 978-0-521-83899-3 hardback
ISBN 0 521 83899 1 hardback

Contents

1

Introduction

G. F. Hewitt and J. C. Vassilicos

1 Background

In 1999, a major programme on turbulence was held at the Isaac Newton Institute (INI) at Cambridge, England, which was aimed at taking an overview of the current situation on turbulent flows with particular reference to the prediction of such flows in engineering systems. Though the programme spanned the range from the very fundamental to the applied, a very important feature was the involvement and support (through the UK Royal Academy of Engineering) of key players from industry. This volume, which has evolved from the INI programme, aims to address the needs of people in industry and academia who carry out calculations on turbulent systems.

It should be recognised that the prediction of turbulent flows is now of paramount importance in the development of complex engineering systems involving flow, heat and mass transfer and chemical reactions (including combustion). Whereas, in the past, the developer had to rely on experimental studies, based usually on small scale model systems, more and more emphasis is being placed nowadays on the use of computation, often through the use of commercial computational fluid dynamics (CFD) codes. Superficially, the use of such computational methods seems ideal; they allow painless extension to large scale and can often give information on fine details of the flow that are not economically accessible to experimental measurement. Furthermore, the results can be presented in an easily accessible and attractive form using the sophisticated computer graphics now generally available. Such methods have become big business!

Unfortunately, there is a major problem in the application of CFD techniques in predicting industrial turbulent flow systems, namely the inherent modelling of the turbulence itself. Though low Reynolds number turbulent flows (close to the

Prediction of Turbulent Flows, eds. G. F. Hewitt and J. C. Vassilicos.
Published by Cambridge University Press. © Cambridge University Press 2005.

transition from laminar flow) can be modelled in a reasonably fundamental way using the techniques of direct numerical simulation (DNS), such methods are of little direct relevance to industry. Firstly, the Reynolds numbers for which such calculations can be performed are very limited, and certainly well below those of prime interest to industry. Secondly, the computing resources required are enormous; whilst it is true that computing is getting ever faster and cheaper, it seems unlikely that the usual industrial range of Reynolds numbers will be accessible to DNS-type methods in the near future.

In order to achieve closure of turbulent flow predictions, therefore, it is necessary to invoke some form of *turbulence model*. There are a bewildering variety of such models available in the literature, with the two main classes being the *Reynolds averaged Navier–Stokes (RANS)* models and the *large eddy simulation (LES)* models. The first (and most numerous) class has its origin in the classical averaging of the Navier–Stokes equations due to Osborne Reynolds; here, no attempt is made to deal with the detailed structure of the turbulence. Rather, statistical quantities are obtained and models derived for their prediction based on *modelling hypotheses* that often have several (sometimes many!) adjustable constants that are optimised by comparison with experimental data. A number of distinct types of RANS model have been developed; perhaps the most widely used are based on the assumption of an effective, isotropic, turbulent viscosity whose value can be determined from the local averaged turbulence quantities. Archetypal amongst this sub-class is the k–ε model in which the turbulent viscosity is related to the turbulent kinetic energy (k) and the turbulence dissipation rate (ε). Despite the fact that the k–ε model is contradicted by a wide range of experimental data (for instance by the fact that the planes of zero shear and maximum velocity are different in channels with one rough and one smooth wall), it is still used almost universally (and perhaps often unthinkingly) in industrial CFD predictions. More advanced (though more complex) RANS models (such as the *Reynolds stress transport models*, RSTM) are available which are not limited by the assumption of small scale isotropy.

The second general class of models is the *large eddy simulation (LES)* models. Here, whilst it is recognised that the prediction of the fine details of the turbulent flow is infeasible, an attempt is made to model the temporal and spatial characteristics of the larger eddies. The smaller eddies are dealt with using *sub-grid models*. This class of models is beginning to be used in prediction of industrial systems, particularly where an understanding of the local fluctuating behaviour is important. However, the computational requirements are very large compared to the RANS models. A problem with LES models is the representation of the region near the wall.

2 Objectives

The above brief background will give an indication of the importance and difficulties of modelling turbulent systems. As a result of the intensive interactions between industrial practitioners and academic specialists which were brought about by the INI programme, the idea emerged of generating an overview of turbulent flow prediction methods and the project to produce the present volume was launched. The objectives were:

(1) To summarise current understanding of the physics of turbulent flow and implications for modelling.
(2) To review prediction methods and their applicability to various industrial prediction problems.
(3) To provide a specific set of guidelines (a 'route map') for the choice of model for a given problem.

3 Structure of the volume

The volume is structured into eight chapters (including this present one). The remaining chapters are as follows:

Chapter 2: Developments in the understanding and modelling of turbulence. This chapter essentially provides a summary of what is known about the nature of turbulent flows, particularly in the light of the work of the INI programme. The features common to all turbulent systems are discussed and those features specific to particular manifestations of turbulence are reviewed.

Chapter 3: RANS modelling of turbulent flows affected by buoyancy or stratification. Flows driven wholly or partly by buoyancy (arising from density variations caused by temperature or concentration gradients) are common in practice. Stratification arising from density gradients is also important in many systems. These effects present significant challenges in turbulence modelling and these challenges are specifically addressed in this chapter. Moreover, this chapter presents generic material on turbulence models (for instance the k–ε and second order closures for the RANS models) which can also be applied in the more general case where buoyancy is less significant.

Chapter 4: Turbulent flames. In modelling turbulent flames, it is necessary not only to model the turbulence but also to represent the interaction between the turbulent fluctuations and the chemical reaction. This provides a particular challenge since mixing by small scale turbulence has a strong effect on local chemical reaction. This chapter begins with a discussion of the two main types of combustion (non-premixed and premixed respectively). The various approaches are summarised (chemical equilibrium, flamelets, pdf methods and models, eddy breakup models etc.). Not all combustion systems fall into premixed and non-premixed types; the premixing can be partial. These cases are discussed.

Chapter 5: Boundary layers under strong distortion: an experimentalist's view. Turbulent boundary layers represent a severe challenge to prediction methods. Not only are they regions of strong distortion and change, but their interactions with shocks and vortices give rise to additional problems. This chapter reviews prediction methods, observations and measurements available for boundary layers under strong distortion and attempts to identify areas in which future research should be concentrated.

Chapter 6: Turbulence simulation. As was mentioned above, LES is being increasingly used to address industrial problems. This chapter reviews the current situation on this type of modelling, evaluating the barriers to its more widespread use. The chapter also discusses direct numerical simulation (DNS); DNS can provide useful physical insights into turbulence processes which can be used to develop closures applicable at the higher Reynolds numbers.

Chapter 7: Computational modelling of multi-phase flows. Multi-phase flows (i.e. flows involving two or more phases – gas, liquid and solid) are not only usually turbulent but also have the additional complication of having moving interfaces within the flow. In the case of flows having two or more fluid phases, these interfaces may be deformable. Nevertheless, computational modelling is playing an increasing role in the prediction of multi-phase systems. For instance, industrial systems involving dispersed flows (sewage treatment plant, agitated vessel reactors, etc.) are now widely predicted using the commercial CFD codes. This chapter reviews the background to the prediction of such dispersed flows. For flows in which interface distortion is significant, predictions are more difficult; however, there is a growing range of methods for modelling the interfacial behaviour and these are reviewed and examples of their application presented.

Chapter 8: Guidelines and criteria for the use of turbulence models in complex flows. As was stated above, the practitioner is faced with a bewildering array of turbulence models. When applied to a given system, the models can give different (and sometimes very different) results. Which is the correct model to use? Though, because of the basic limitations discussed above, there is no truly 'correct' model for most engineering turbulent systems, it is undoubtedly true that some models perform better than others for particular classes of problem. Recognising this fact, this chapter attempts to provide a reference guide to the choice of model to be used in particular circumstances.

2

Developments in the understanding and modelling of turbulence

J. C. R. Hunt, N. D. Sandham, J. C. Vassilicos, B. E. Launder,
P. A. Monkewitz and G. F. Hewitt

Abstract

Recent research is making progress in framing more precisely the basic dynamical and statistical questions about turbulence and in answering them. It is helping
both to define the likely limits to current methods for modelling industrial and
environmental turbulent flows, and to suggest new approaches to overcome these
limitations. This chapter had its basis in the new results that emerged from more
than 300 presentations during the programme held in 1999 at the Isaac Newton
Institute, Cambridge, UK, and on research reported elsewhere. The objective of
including this material (which is a revised form of an article which appeared in the
Journal of Fluid Mechanics – Hunt *et al.*, 2001) in the present volume is to give a
background to the current state of the art. The emphasis is on the *physics* of turbulence and on how this relates to modelling. A general conclusion is that, although
turbulence is not a universal state of nature, there are certain statistical measures and
kinematic features of the small-scale flow field that occur in most turbulent flows,
while the large-scale eddy motions have qualitative similarities within particular
types of turbulence defined by the mean flow, initial or boundary conditions, and
in some cases, the range of Reynolds numbers involved. The forced transition to
turbulence of laminar flows caused by strong external disturbances was shown to
be highly dependent on their amplitude, location, and the type of flow. Global and
elliptical instabilities explain much of the three-dimensional and sudden nature of
the transition phenomena. A review of experimental results shows how the structure
of turbulence, especially in shear flows, continues to change as the Reynolds number of the turbulence increases well above about 10^4 in ways that current numerical
simulations cannot reproduce. Studies of the dynamics of small eddy structures and

Prediction of Turbulent Flows, eds. G. F. Hewitt and J. C. Vassilicos.
Published by Cambridge University Press. © Cambridge University Press 2005.

their mutual interactions indicate that there is a set of characteristic mechanisms in which vortices develop (vortex stretching, roll-up of instability sheets, formation of vortex tubes) and another set in which they break up (through instabilities and self-destructive interactions). Numerical simulations and theoretical arguments suggest that these often occur sequentially in randomly occurring cycles. The factors that determine the overall spectrum of turbulence are reviewed. For a narrow distribution of eddy scales, the form of the spectrum can be defined by characteristic forms of individual eddies. However, if the distribution covers a wide range of scales (as in elongated eddies in the 'wall' layer of turbulent boundary layers), they collectively determine the spectra (as assumed in classical theory). Mathematical analyses of the Navier–Stokes and Euler equations applied to eddy structures lead to certain limits being defined regarding the tendencies of the vorticity field to become infinitely large locally. Approximate solutions for eigen modes and Fourier components reveal striking features of the temporal, near-wall structure such as bursting, and of the very elongated, spatial spectra of sheared inhomogeneous turbulence; but other kinds of eddy concepts are needed in less structured parts of the turbulence. Renormalized perturbation methods can now calculate consistently, and in good agreement with experiment, the evolution of second- and third-order spectra of homogeneous and isotropic turbulence. The fact that these calculations do not explicitly include high-order moments and extreme events suggests that they may play a minor role in some aspects of the basic dynamics. New methods of approximate numerical simulations of the larger scales of turbulence or 'very large eddy simulation' (VLES) based on using statistical models for the smaller scales (as is common in meteorological modelling) enable some turbulent flows with a non-local and non-equilibrium structure, such as impinging or convective flows, to be calculated more efficiently than by using large eddy simulation (LES), and more accurately than by using 'engineering' models for statistics at a single point. Generally it is shown that where the turbulence in a fluid volume is changing rapidly and is very inhomogeneous there are flows where even the most complex 'engineering' Reynolds stress transport models are only satisfactory with some special adaptation; this may entail the use of transport equations for the third moments or non-universal modelling methods designed explicitly for particular types of flow. LES methods may also need flow-specific corrections for accurate modelling of different types of very high Reynolds number turbulent flow including those near rigid surfaces.

This chapter is dedicated to the memory of George Batchelor who was the inspiration of so much research in turbulence and who died on 30th March 2000. These results were presented at the last fluid mechanics seminar in DAMTP Cambridge that he attended in November 1999.

1 Introduction

'The problem of turbulence' has been seen as one of the great challenges of mathematics, physics and engineering for more than 100 years, by Lamb, Einstein, Sommerfeld, Ishlinski and others. Much of the interest in meeting this challenge is because of its practical value; the solution of many technical, industrial and environmental problems increasingly requires improvements, both in our fundamental understanding of turbulence, and in the utilization of advances in computation to calculate, at appropriate levels of accuracy and speed, the characteristic features and statistical properties of these flows (e.g. Hunt 1995; Holmes, Lumley & Berkooz 1996).

Major centres for mathematical science and theoretical physics are holding intensive programmes on turbulence (examples being at Ascona, Monte Verita 2nd Symposium on Turbulence, Switzerland (Gyr, Kinzelbach & Tsinober 1999) and the Institute for Theoretical Physics, Santa Barbara in 2000) to complement regular summer schools and conferences, such as the European Turbulence Conference and Turbulent Shear Flow Symposia. In this chapter we draw some general conclusions about current questions and developments in research on turbulence and its practical applications, resulting from the programme at the Isaac Newton Institute at Cambridge (UK) between January and June 1999. This involved more than 400 participants, visiting for various periods, and about 300 presentations by academic and governmental researchers, and those working on problems in industrial and environmental organizations, some of which combined with the Royal Academy of Engineering to provide generous support for the programme. All three disciplines of mathematics, physics and engineering were well represented. We also refer here to other recent research developments reported in the scientific literature and at the International Congress on Industrial and Applied Mathematics held at Edinburgh in July 1999. Detailed reports on various aspects of the programme have been published by Voke, Sandham & Kleiser (1999); Launder & Sandham (2001); Vassilicos (2000b); Hunt & Vassilicos (2000).

This chapter is aimed at a broad fluid mechanical readership. It focuses, inevitably somewhat selectively and subjectively, on progress in research towards the major questions of the subject and certain practical objectives, both of which provided a framework for the programme. Although these were formulated well before the programme began, they evolved by progressive adjustment and addition during the six-month period. They essentially finally became the following.

(i) To consider broadly and in depth whether fluid turbulence in its different manifestations has some common features (in some defined statistical sense) that are universal to all kinds of fully turbulent flow, or whether any commonality only exists within certain types of turbulence (such as those driven by mean shear, or natural convection). In other words is there one 'problem of turbulence' or several?

(ii) To explore the promising directions for tackling the fundamental problems of turbulence dynamics, some of which go back to the 1930s (see Constantin 2000; Frisch 1995). Within this fell the following specific questions.

(a) Is Taylor's (1938) conjecture about turbulence correct? It is that the normalized mean rate of energy dissipation, $\hat{\varepsilon} = \varepsilon/(u_0^3/L_x)$ (where ε is the dimensional dissipation rate, u_0 is a typical r.m.s. velocity, and L_x is a typical integral length scale) of a turbulent flow field (away from a boundary) is independent of the turbulent Reynolds number $Re = u_0 L_x/v$, if the Reynolds number is sufficiently large, i.e.

$$\hat{\varepsilon} \to \text{const} \quad \text{as} \quad Re \to \infty. \tag{2.1}$$

If this is true (as is generally assumed in statistical models), what are the implications for the structure of the velocity field? If it is not, as some investigations suggest, what is the asymptotic relation between the rate of energy dissipation and the Reynolds number? These questions are central to turbulence theory and modelling: for example, Taylor's conjecture is part and parcel of such turbulence modelling, such as k–ε.

(b) Turbulence forces on mean flows are due to Reynolds stresses and arise from correlation between vorticity and velocity components. A fundamental understanding of Reynolds stresses requires, therefore, an understanding of the turbulent velocity fluctuation field. What is the nature of the 'wiggliness' and 'smoothness' of the velocity field as $Re \to \infty$, a question first raised by Richardson (1926) who wondered whether the velocity, even though its magnitude is finite, might be so 'wiggly' that it is not effectively differentiable anywhere (as with a Weierstrass function or some other fields with a non-integral Hausdorff fractal dimension). Without this wiggliness, the velocity field would not have the gradients necessary for energy dissipation to remain finite as $Re \to \infty$, Eq. (2.1). An alternative concept is that as $Re \to \infty$, turbulence is fundamentally intermittent with a finite number of distinct points where the derivatives are singular, separated by smooth regions in between. Some combinations of such distributions of near-singularities (defined as singularities in the limit as the Reynolds number tends to infinity) are necessary if Taylor's conjecture (2.1) is to be valid. Furthermore, how are such distributions consistent with the idea that velocity fields at the small scales may be self-similar over an increasing range of length scales as Re increases, a concept essential to large eddy simulations? How can deviations from self-similarity be considered in the context of multiple-scale velocity fields?

(c) Can even stronger singularities occur in which the velocity and vorticity at points in the flow tend to infinitely large values in a finite time t^*, after a finite-amplitude turbulent flow field has been initiated at $t = 0$? Although this phenomenon has never been observed, some special mathematical solutions to the Euler and the

Navier–Stokes equations suggest that it may be possible (Leray 1933; Kerr 1999; Moffatt 1999; Ohkitani & Gibbon 2000; Doering & Gibbon 2000). Are near-singularities of Navier–Stokes turbulence the remnants of finite-time singularities of the Euler equations? Does the tendency for such singular events to occur determine the 'tail' of the probability distribution of the turbulent flows and if so, how?

(d) What is the nature of the eddy transfer or 'cascade' process, in which when $Re \gg 1$ (if (2.1) is correct) the velocity fluctuations right down to the smallest scales reach a quasi-equilibrium state in the 'Lagrangian' or 'turn-over' time scale of order L_x/u_0? Also, to what extent are small-scale processes (depending on the precise definition) independent of the large-scale motions? Some physical models have suggested an infinite cascade involving vortical events at each 'eddy' scale (Tennekes & Lumley 1971; Frisch 1995), whereas others have suggested that relatively few complex events are needed (e.g. Lundgren 1982). The upscale energy transfer equally needs better understanding through study of the large-scale dynamics, which depends on how these eddy motions are correlated over large distances (Davidson 2004).

(e) To what extent do the large-scale motions of the turbulence tend to become independent of initial and boundary conditions, or, if the flow was initially laminar, of the particular process of transition to turbulence (George 1999): is this by means of internal self-organization or by chaotic interactions or both? Landau & Lifshitz (1959): 'We have seen that, whatever the initial phases β_j, over a sufficiently long interval of time the fluid passes through states arbitrarily close to any given state, defined by any possible choice of simultaneous values of the phase ϕ_j. Hence it follows that, in the consideration of turbulent flow, the actual initial conditions cease to have any effect after sufficiently long intervals of time. This shows that the theory of turbulent flows must be a statistical theory.' Batchelor's (1953) view was more conditional: '. . . we put our faith in the tendency for dynamic systems with a large number of degrees of freedom, and with coupling between these degrees of freedom, to approach a statistical state which is *independent* (partially, if not wholly) of the initial conditions. With this general property of dynamical systems in mind, rather than investigate the motion consequent upon a particular set of initial conditions, we explore the existence of solutions which are asymptotic in the sense that the further passage of time changes them in some simple way only.' This and the other fundamental questions provide a context for considering the appropriate future directions for the statistical computational models of turbulence needed for practical purposes.

(f) How are fully developed turbulent velocity fields related to their sources of energy, whether from initial conditions, continuing instabilities within a flow, or from boundary conditions such as a rigid wall?

(iii) Are certain statistical properties of fully developed inhomogeneous turbulence near plane rigid surfaces independent of the upstream or outer flow conditions and what is their form? This question refers to flows with and without a significant velocity $\bar{U}$ greater than the typical fluctuating velocity u_*; firstly, what is the mean velocity profile

$\bar{U}(x_3 u_*/v)$, whose mathematical form may be determined by the dependence on the Reynolds number of the outer flow (Barenblatt & Chorin 1998)? Secondly, under what conditions are the velocity spectra $\Phi_{11}(k_1)$ and $\Phi_{22}(k_1)$ along the streamwise direction given by

$$\Phi_{11}(k_1), \ \Phi_{22}(k_1) \propto u_*^2 k_1^{-1}, \tag{2.2}$$

when $\Lambda^{-1} < k_1 < x_3^{-1}$, for $x_3 \ll h$, where h is the thickness of the boundary layer/pipe and Λ is an outer length scale much greater than h (Marušić & Perry 1995)? Thirdly, for turbulent flows with or without a mean velocity component, how general is the self-similar form of the two-point velocity correlation of the normal components

$$R_{33}(x_3, x_3') = \overline{u_3(x_3)u_3(x_3')}/\overline{U_3^2}(x_3) = f(x_3/x_3') \quad \text{for} \quad x_3 < x_3' \tag{2.3}$$

(Hunt *et al.* 1989)?

(iv) To what extent do the asymptotic forms as $Re \to \infty$ for the statistics and characteristic eddy structures differ from those found when Re is finite? Are there distinct sub-classes of turbulence corresponding to different ranges of Re (or of Rayleigh number for natural convection (cf. Castaing *et al.* 1989))?

(v) To consider how fundamental research on turbulence might lead to improvements in turbulence-simulation methods and statistical models. The deficiencies of current models, as pointed out by industrial participants, tend to become apparent when they are applied to turbulent flows that are highly inhomogeneous and rapidly changing (over the length and time scales of the large eddies), which is to be expected since these 'non-conforming' situations do not correspond with the assumptions that underpin the models, e.g. Launder & Spalding (1972), Lumley (1978). Because industry is now familiar with the use of such models, it was requested that their rationale and limitations should be defined and explained using recent research, such as that on inhomogeneous turbulence. Since the models are often applied to 'non-conforming' flows, interest was expressed in interpreting the often puzzling results of the computations in these situations. Moreover, significant modifications are being proposed to existing modelling methods and these need to be evaluated and understood.

Questions (reviewed by Geurts 1999; see also Geurts & Leonard 1999) about the limitations of large eddy simulation methods are closely linked to those on the fundamental dynamics and statistics, since the methods involve computing the 'resolved' velocity field above a certain 'filter' scale l_f that is greater than that of the smallest 'Kolmogorov' eddies of the turbulence l_k. (Only if the Reynolds number of the turbulence is small enough, typically $Re < 10^3$, is it possible to avoid this approximation and compute the turbulence directly, e.g. Moin & Mahesh 1999.) Discussions were mainly focused on constant-density flows, though the importance of turbulence in two-phase flows (Hewitt 1999; Reeks 1999), buoyancy-dominated flows (Banerjee 1999; Launder 1999), and compressible flows (Bonnet 1999; Gatski 1999) was reviewed. There are many detailed questions about this filtering approximation; for example what happens when very small-scale, highly anisotropic and often non-Gaussian motions are generated near boundaries, or how is the predictability of a simulated flow affected by randomness of the unresolved small scales, a problem of interest for forecasting environmental flows and controlling engineering flows (Lesieur 1999).

2 Origins

Turbulent flows are generated in different ways. Laminar flows (i.e. flows that in any one realization in a fluid with simple boundaries are exactly predictable for all time given a finite amount of data about the flow) can become unstable when small fluctuations are caused by boundaries with complex (fractal) shapes (Queiros-Conde & Vassilicos 2000) and complex movements (Warhaft 2000) or by the effects of body forces, e.g. electromagnetic forces. A fully developed turbulence is reached when one or more of these processes has generated velocity fields that are chaotic in space and time, having smooth spectra and smooth probability distributions. Once this state is reached, which requires that the Reynolds number is large enough, these general qualitative properties are observed not to change even when quite substantial perturbations are introduced, say in relation to u_0 and L_x, such as mean-flow distortions or damping via body forces or suspended particles. In flows with certain interaction body forces (e.g. electromagnetic or gravitational) the turbulence can generate resonances with local singularities (Kerr 1999; McGrath, Fernando & Hunt 1997). A significant mechanism, discussed later, for increasing the level of chaotic behaviour in turbulence at high values of Re is the continuous growth of instabilities from infinitesimal initial amplitudes even when the turbulence is fully developed.

A key question of recent research has been to identify and describe the different mechanisms affecting the evolution of unstable fluctuations from perturbations on laminar flow into fully developed turbulence. The first of two types of transition proposed by C. C. Lin (Ffowcs Williams, Roseblat & Stuart 1969) was a 'slow' evolution (as in wakes, jets and curved shear flows) when the initial shear flow is unstable to a single mode and there are distinct bifurcations as subsequent modes develop by nonlinear interaction. The second type is a fast evolution (as in pipe flow) when nonlinear growth of a single transition rapidly produces a spectrum of velocity fluctuations comparable in width to that of fully developed turbulence. In this case there is a 'fast' transition or breakdown and a sudden change from smooth laminar flow to turbulence or patches of turbulence.

These categories can be understood physically by considering how these insta-bilities develop in particular flows. For example, in boundary layers, disturbances grow in the streamwise direction via a 'slow' transition process with distinct modes, typically evolving from linear to three-dimensional weakly nonlinear form (Smith 1999). Depending on whether the streamwise extent of the boundary layer flow is short or long, the disturbances may or may not develop into turbulence for a given value of *Re*. In other flows, for example with recirculation, such as Taylor–Couette or certain wake flows, the disturbances stay at the same amplitude everywhere in the flow and do not necessarily generate turbulence, so that the slow evolution of nonlinearities implies slow transition. As Huerre & Monkewitz (1990) first clarified, in the former case of the boundary layer the 'slowly' evolving instabilities are 'convective', and may or may not lead to transition anywhere in the flow. In the latter type of flow absolute instabilities fill the domain, and are slowly evolving. In most cases where there is a fast evolution of nonlinear disturbances, whether the instabilities are convective or absolute, there is fast transition to turbulence.

Recent research has provided some insights into the questions raised by this framework, but the framework itself has not seriously been questioned. Whereas many features of slow instabilities have been analysed, the key fundamental question is to understand the breakdown process, which can also reappear in a fully turbulent flow as a 'transition' from one form of turbulence structure to another – see below. The most thoroughly studied case is that of global instability of rotational flows with locally closed elliptical streamlines (and particle paths), whose general significance for turbulence was pointed out by Gledzer *et al.* (1975), Malkus & Waleffe (1991) and others, and whose theoretical understanding was developed by Bayly (1986). The discussions showed that both global (referring to a non-local classical linear analysis in terms of normal modes) and local (referring to WKB short-wave asymptotics along individual trajectories – see Lifschitz & Hameiri 1991) solutions along streamlines (see Cambon & Scott 1999; Leonard 1999) lead to the exponential growth of three-dimensional, wave-like disturbances. Disturbances grow over times that scale on the inverse of the strain rate and not on the inverse of vorticity. These two time scales can be comparable but they can also be very different, as in the vortex interaction experiments of Leweke & Williamson (1998). Nonlinear interactions develop over a period of the order of the rotation time (and not significantly less), leading to a fully developed, multi-length-scale, turbulent flow, i.e. a breakdown. Recent laboratory experiments on a laminar vortex distorted by an adjacent vortex by Leweke & Williamson (1998) and on a fully turbulent vortex undergoing compression by Borée *et al.* (1999) demonstrate how this basic inviscid instability (Lundgren & Mansour 1997) causes rapid transition of laminar flows and a rapid change in the structure of a turbulent flow. This is consistent with the integral of helicity, $H = \int u \cdot \omega \, dV$, being conserved for these isolated structures or contained flows. Note that the local helicity $h = u \cdot \omega$ increases from zero in the base flow to positive and negative values of order u_0^2/L_x, showing that increasing $|h|$ is not necessarily an indicator of a slower cascade of vorticity.

With an analysis employing global instability methods, Le Dizes (1999) showed that the mechanisms involved in the elliptic instability are essentially equivalent to the growth of three-dimensional perturbations within vortices explained as 'vortex core dynamics' by Hussain (1999).

New qualitative evidence presented by Durbin (1999) from the numerical simulations of Wu *et al.* (1999) was consistent with the analysis of Malkus & Waleffe (1991), in that these global or elliptical instabilities are the cause of the rapid evolution of boundary layer instabilities when the imposed disturbances (whether at the wall or from external velocity fluctuations) have a significant amplitude u_e. These produce sufficiently large closed streamline regions around the critical layer for rapid three-dimensional instabilities to grow locally in the form of the very low amplitude disturbances. 'Bottom up' (i.e. forward pointing) triangular spots are generated at the wall while 'top down' (backward pointing) spots are generally formed from very low amplitude disturbances imposed at the top of the boundary layer. Typically, the former rapid evolution and transition occurs if u_e is comparable to the fluctuations in the fully turbulent boundary layer, i.e. u_*, where u_* is the friction velocity, typically $U_0/20$, where U_0 is the mean velocity outside the layer.

A qualitative understanding of this spot transition is currently applied in the design of turbomachinery blades. As it passes downstream, the entrainment flow into the growing spot affects the mean velocity profile, so as to reduce the deficit in the momentum flux of the boundary layer profile and its tendency to thicken in the adverse pressure gradient towards the trailing edge of the aerofoils (Hodson 1999).

The effects of different kinds of external turbulence on the transition to turbulence of laminar boundary layers on isolated bodies can only be analysed theoretically in the initial stages of transition for the idealized case of a very thin flat plate and when the external fluctuations have a vanishingly small amplitude (e.g. Wundrow & Goldstein 2001). For practical situations, various approximate theoretical methods have been developed for different types of laminar flow and external turbulence (Atkin 1999). Some concepts were presented about how velocity fluctuations of free-stream turbulence outside a boundary layer affect fluctuations within it and cause 'bypass' of the transition process via a sequence of instabilities. The first mechanism is that fluctuations are 'convected' along the streamlines and enter the growing boundary layer where its velocity gradient amplifies the small-scale fluctuations algebraically (Voke 1999; see also Trefethen *et al.* 1993) and where larger-scale fluctuations may amplify the eigen modes of the layer. Different methods for calculating the amplification of disturbances in shear flows where the streamlines are slightly non-parallel were discussed (Lingwood 1999), in particular the relative merits of the widely used parabolized stability equations (PSE) and the rigorous triple-deck asymptotic theory (Lucchini 1999; Healey 1999). The second mechanism is caused by the action of external fluctuations travelling over the layer as a localized disturbance, such as a moving wake. Numerical simulations confirm the theory that if they travel at the same speed as the free-stream speed U_0, the fluctuations they induce tend to be maximal at the top of the layer because the flow within the layer is 'sheltered' from the larger scales of the external turbulence; the smaller scales can induce fluctuations and diffuse downwards. However, the upper fluctuations rapidly grow as Kelvin–Helmholtz billows and spread the resulting small-scale turbulence downwards. Both effects can cause transition. In engineering models they are usually represented as a diffusion-like transport process in one-point statistical models (Savill 1999, 2002). Durbin (1999) argued that such approximations might be incorrect because they imply that the external turbulence would have a greater effect on the boundary layer when its length scale is increased.

The third major area of controversy about the initiation and persistence of turbulence concerns the cycle of growth and decay of fluctuations in turbulent boundary layers near a rigid surface (or 'wall'). There are minor differences between the mechanisms proposed for the lower range of the turbulent Reynolds number (Re), but a major difference in the proposed self-generation mechanisms when Re becomes very large. When $Re \lesssim 10^4$, as many numerical simulation and laboratory measurements have demonstrated, instabilities in initially laminar boundary layers become nonlinear, and then develop into longitudinal vortices. These deflect upwards and downwards the mean spanwise vorticity of the boundary layer, causing low-speed and higher-speed strips with associated elongated vorticity fields. These become unstable, grow, and disrupt the local flow structure; significant velocity fluctuations are generated that may extend into the outer layer; following their decay, the streamwise vortical regions re-form once more (Jiménez 1999; Sandham 1999; Hussain 1999). In order to describe the possible mechanisms, elements of the flow field have been studied in isolation, e.g. by conditional sampling (Hussain 1986), or by proper orthogonal decomposition of the measured two-point velocity correlations (Holmes *et al.* 1996). Different techniques have then been used to analyse these fields, for example by local stability analysis or by calculating the temporal evolution of a few low-order modes.

While the main picture, as described above, is common to all investigations, there are some significant differences among the models, particularly regarding whether the near-wall dynamics is a pure instability mechanism quite independent of the fluctuation velocity field in the outer part of the boundary layer (or channel flow) (Schoppa & Hussain 2002), or whether random fluctuations in this outer field stimulate resonant modes near the wall. This interaction may be essential to sustain the cycle of growth and decay of the 'near wall' vortical structures (Sandham 1999). Apparently no experimental evidence yet exists that leads to a clear distinction between the validity of these concepts.

In turbulent boundary layers the continual generation of small-scale instabilities or resonances near the wall at moderate Re means that the flow is susceptible to being controlled, for example by adjusting small panels up and down to modify the growth of instabilities (Hussain 1999; Holmes *et al.* 1996; Carpenter 1999). If the eddy structure undergoes substantial change at very high Reynolds number, as indicated by experiments, will the effectiveness of such wall techniques be reduced? Another turbulence-control technique is the introduction of long-chain molecules which reduce the frictional drag of liquid flows in large pipelines; since the mechanism here is through 'damping' of eddy straining, it should not change qualitatively at high Reynolds number, as is observed (Sreenivasan 1999, Ptasinski *et al.* 2003).

In most turbulent free shear flows, the effect of instabilities on the mean flow and turbulence is surprisingly not generally considered to be so significant, because they occur on the edges of the free shear layer. In these flows the energy of the turbulence is mostly generated in the interior of the flow where there is a strong local interaction between the gradient of the mean velocity ∇U and the Reynolds stresses $\overline{u_i u_j}$. The role of instabilities may be particularly significant on the edges of clouds and plumes when influenced by body forces and external turbulence (e.g. Baht & Narasimha 1996).

3 New measurements and simulations

The most universal and fundamental aspects of turbulence, namely the small-scale statistical structure when *Re* is large enough for a significant inertial, spectral sub-range to exist, can still only be studied in detail through measurements, because numerical simulations and theoretical models are only approximate. Furthermore, no facilities yet exist in which it is possible to mount controlled experiments with a well-designed velocity field at very high Reynolds numbers (Nieuwstadt 1999); the measurements still have to be made in artificial turbulent flows constructed for other purposes, such as in one part of an aeronautical wind tunnel where $Re \sim 10^4$–10^5 (Arneodo *et al.* 1999; Saddoughi & Veeravalli 1994), or in strong jets (Van Atta 1991), or else in the atmospheric boundary layer where $Re \geq 10^5$.

Recently, large wind tunnel and atmospheric measurements have related the small-scale motions to the velocity field as a whole, $u(x, t)$ for example, by calculating the conditional *n*th-order moments of components of $\Delta u(r)$ at each value of $|u|$, denoted by $\overline{(\Delta u^n; |u|)}$ (Praskovsky *et al.* 1993; Tsinober 2000; Sreenivasan 1999). These results showed some dependence of the amplitude and even the structure of small-scale eddy motion on the large-scale eddy motion. In this context we note that Nie & Tanveer (1999) rigorously derived, from the Navier–Stokes equation, Kolmogorov's four-fifths law for the third-order structure function without Kolmogorov's assumption of local isotropy. Hence, some elements of the small-scale structure can persist even in the presence of large-scale effects. We conclude that the mechanisms for direct connections between large and small scales in different types of turbulence will only be better understood with more detailed and different types of statistical measurements.

These new experimental and theoretical results are nevertheless consistent with earlier studies showing significant amplitudes of the non-isotropic component of the second- and higher-order moments of small-scale turbulence that were dependent on the large-scale mean shearing motion or the non-isotropic, non-Gaussian eddies of natural convection (Saddoughi & Veeravalli 1994; Hunt, Kaimal & Gaynor 1988). Other kinds of local conditional statistics taken effectively at two points were reported that were designed to elucidate the detailed structures of small-scale eddies at high *Re*. Wavelet analysis of these measurements is a natural generalization of the structure-function analysis of these measurements providing increased information both in physical and scale spaces (Arneodo *et al.* 1999; Brasseur 1999). For example, the wavelet transform was applied to the study of the dynamics of the Burgers equation leading to a clear demonstration of how a single localized characteristic flow structure in the field (in this case in the form of a shock) can, all by itself, determine the high-wavenumber energy spectrum, and how energy transfer can be studied concurrently in both scale and physical spaces (Brasseur 1999). Furthermore, by applying wavelet methods to experimental, one-point turbulence velocity data, Arneodo *et al.* (1999) were able to show that if the turbulence is a multiplicative cascading process (which it may well not be), then this process is not self-similar. From their atmospheric measurements of low-order and higher statistics, Kholmyansky & Tsinober (2000) and Sreenivasan (1999) deduced that small-scale turbulence may not be completely self-similar at the Reynolds numbers currently attainable. However, flow visualization studies in the laboratory (Schwarz 1990; Douady, Couder & Brachet 1991), casual observations in the environment of dust, bubbles, clouds etc., and numerical simulations at lower *Re* (Passot *et al.* 1995; Jiménez *et al.* 1993; Ohkitani 1999) have shown that characteristic structures exist in the form of rolled up vortical layers and elongated vortices. However, there is still no conditioned experimental data for these structures that are sufficiently detailed for their precise analysis.

Although flow visualization provides multi-point qualitative information this can only be provided systematically by making measurements at three or more points simultaneously. Warhaft (2000) showed how three-point velocity and temperature measurements in active grid turbulence (which produces enough energy to simulate many features of high-*Re* turbulence) could demonstrate quite clearly the existence of 'scalar/vortical fronts' on thin surfaces across which there are intense scalar gradients (see Mydlarski & Warhaft 1998; Chertkov, Pumir & Shraiman 2000). This is consistent with numerical simulations and theoretical concepts about the first stage of intense scalar mixing, the second stage being the rolling up of these fronts leading to intense local mixing (Nieuwstadt 1999).

A complementary approach to understanding turbulence structure is to consider how the relative velocity $\Delta u(t)$ and the distance Δ between pairs of fluid particles vary with time t (and the initial spacing Δ_0 at $t = 0$). Richardson's (1926) atmospheric measurements, that helped stimulate the Obukhov–Kolmogorov theory, suggest that

$$\Delta^2 = G_\Delta \varepsilon t^3, \tag{2.4}$$

where Δ is smaller than the turbulence length scale L. This statistical relation, which has great practical value for estimating concentration fluctuations (Derbyshire, Thomson & Woods 1999), can now be understood better in terms of the eddy motions in the turbulence: an approach of practical value on the scale of synoptic storms. (A collaboration between industrialists and academics was initiated during the programme to use this approach for establishing the limitations of Richardson's law.) Laboratory experiments (albeit at quite low $Re \leq 10^2$) of Tabeling (1999) using the new techniques of simultaneously measuring Δ and the pattern of the flow field (with particle imaging), showed how $\overline{\Delta^2}$ increases mostly because of sudden separation events between a minority of particle trajectories, the majority of them remaining close to each other for a very long time. This is consistent with simulation results and the approximate theory of Fung *et al.* (1992) who argued that these rare, sudden and intense separation events occuring in saddle-point regions where streamlines converge and diverge most rapidly (where the 'scalar fronts' described above tend to form). It may be because of the scarcity of these particle separation bursts that, in some flow fields, the Richardson constant G_Δ in Eq. (2.4) is of the order 0.1 or smaller (see also Section 4.2). There remains much uncertainty about this fundamental constant that can only be settled by experiments at high enough Re for there to be a wide self-similar 'inertial range' (see Voth, Satyanarayan & Bodenschatz 1998).

Turbulent flows in practice are inhomogeneous and bounded either by a rigid wall or by a region of non-turbulent flow in which there might be some kind of laminar motion or none at all. In the latter case, there is a transition between the turbulent and laminar flow, with a randomly moving 'interface' separating the rotational velocity fluctuations of the turbulence from irrotational motions, which decay to zero over a distance L_x from the interface. Recent experiments and simulations for both these types of boundary are more detailed than earlier studies and suggest that new concepts and models are necessary for these critical boundary regions of turbulent flows.

Although the structure of turbulence near a 'wall' in the absence of mean flow has only been studied in detail over the past 20 years, its main features (for convective turbulence or mechânically generated turbulence) have now been established through similar findings in experiments, e.g. Kit, Strang & Fernando (1997), numerical simulations, e.g. Perot & Moin (1995a, b), Banerjee (1999), and approximate models, e.g. Craft & Launder (1996): namely that the length scales of eddy motions parallel to the wall, are largely determined by the length scales of eddy motion away from the wall, L_x (e.g. as in thermal convection, Castaing *et al.* 1989), while the length scales normal to the wall are determined by the distance (x_3) to the wall. The normal velocity decreases towards the wall and parallel components increase (by up to 30%) until they are within a fluctuating shear layer of thickness $\mathcal{L} \ll L_x$. For *Re* below about 10^3, the surface shear layer produced by the energy-containing eddies is laminar and its thickness is of order $Re^{-1/2}$. As *Re* increases above about 10^4, this overall structure, including the form of the spectra, does not change, but the shear layer at the wall changes character and its thickness $\mathcal{L}$ becomes approximately proportional to $(\log Re)^{-3}$ as opposed to $Re^{-1/2}$. For a smooth wall there is a very thin inner viscous layer of thickness $h_* \sim v/u_*$, where u_* is the friction velocity of the energy-containing eddies. (See reviews in Plate *et al.* 1998.)

By contrast, when there is a mean shear flow parallel to the wall the eddy structure is quite different and has a greater qualitative change as *Re* increases, especially in its relation to the eddy structure far above the wall. Measurements have been recently reported of the spectra $\Phi_{11}(k_1)$, $\Phi_{22}(k_1)$ of the streamwise and spanwise velocity fluctuations in turbulent pipe flows of radius h for *Re* up to 10^4 (Kim & Adrian 1999) and in the atmospheric boundary layer of thickness h for ($Re > 10^5$) (e.g. Hoxey & Richards 1992; Fuehrer & Friehe 1999). These confirmed over different ranges of x_3 the main result of the earlier studies of Marušić & Perry (1995) and others; in pipes the range was close to the 'wall' (i.e. $h_* < x_3 < h_s \sim 0.2h$), while in the atmosphere the range was very close to the ground ($z_0 < x_3 < h_s \sim 0.01h$), where $h_*/h \sim 10^{-3}$ and $z_0/h \sim 10^{-4}$. In both cases the large-scale spectra had an invariant self-similar structure for eddy scales (k_1^{-1}) greater than the distance (x_3) from the wall, but less than a long streamwise length scale (Λ) i.e. $(2\pi/\Lambda) \geq k_1 \geq (2\pi/x_3)$. It was found that the spectra $(\Phi_{11}(k_1), \Phi_{22}(k_1)) = (C_{*11}, C_{*22})u_*^2 k_1^{-1}$, where u_* is the surface friction velocity, and C_{*11}, C_{*22} are approximately independent of x_3/h.

Not only is the depth h_s of the surface layer, where these self-similar spectra are observed, sensitive to the value of *Re*, but so is the value of Λ/h. In both cases the maximum value of Λ is significantly greater than both h and the scale of the eddy structures in the outer part of the flow (which can be explained in terms of the formation of 'streaks' by vertical fluctuations interacting with the mean shear, Jiménez 1999). At lower *Re*, Λ varies with x_3/h quite rapidly in a pipe (Kim & Adrian 1999) and less so in a boundary layer, but in both cases reaches about $18h$ when $x_3/h \sim 0.2h$, whereas in the very high-*Re* range of the atmospheric boundary layer Λ is approximately equal to $3h$–$5h$ and does not vary significantly with z. (This formula, proposed by Davenport (1961), has been used by wind engineers ever since!) These experiments confirm the theoretical model of Townsend (1976) and Perry (1999) that at very high *Re* the variances of the parallel components $\overline{u_1^2}$, $\overline{u_2^2}$, obtained from integrating the spectra, vary in proportion to $u_*^2[\ln(\Lambda/z) + \text{const}]$. Most measurements now agree that in this range of wavenumber the spectra for the normal velocity component $\Phi_{33}(k_1)$ and the co-spectra of the shear stress $\Phi_{13}(k_1)$ are constant with wavenumber, i.e. $\Phi_{33}, \Phi_{13} \cong (C_{*33}, C_{*13}) u_*^2$. So on integration these variances are proportional to u_*^2. This is consistent with the definition $u_*^2 = -\overline{u_1 u_3}$ and with the general result for high-Re shear flows that $\overline{u_3^2}/(-\overline{u_1 u_3})$ is of order unity.

These statistical results are also broadly consistent with the main features of the eddy structure, namely the elongated contours of instantaneous high and low streamwise velocity, found in numerical simulation (Jiménez 1999) and atmospheric observations of elongated streamwise vortices, and with the sloping eddy structures expanding in diameter as the distance from the wall increases, as seen in laboratory experiments (Perry 1999).

The difference between the magnitude of Λ/h and its variation with the normal distance (x_3/h) as Re increases much above 10^4 is consistent with the possible change of the eddy structure. A more vigorous vertical exchange of large eddies from the outer region of boundary layers towards the wall is seen in the atmosphere in the form of moving 'cat's paws' on water surfaces or cornfields, an increased value of the cross-correlation $\hat{R}_{33}(x_3, x_3)$, defined in (2.3) (Brown & Thomas 1977; Hunt & Morrison 2000), and an increase in the vertical turbulence $(\overline{u_3^2})$ with height (z) in the surface layer (Högström 1990).

The turbulence in the outer regions of a boundary layer and throughout the whole thickness of free shear flows is dominated by the interface with the exterior non-turbulent flow. An industrial participant regretted that despite its importance for aeronautical applications this aspect of inhomogeneous turbulence has received far less research attention than that near the wall. As Lumley (1999) remarked, the dynamics of these interfaces also determine how local regions of intense vortical motions evolve within a general turbulent flow when Re is very large. Recent analysis by Bisset *et al.* (1998) of previously published numerical simulations of wakes (Moser, Rogers & Ewing, 1998), has shown that turbulence statistics have a local structure when expressed in terms of the normal distance n_I from such interfaces. Even though the Reynolds numbers of these simulations were not large $(Re \sim 10^2)$ it was found that the conditional profiles of the variables as a function of n_I vary sharply near $n_I = 0$ because of the very active small-scale motions at the interface; the vorticity variance $\overline{\omega^2}(n_I)$ and dissipation $\bar{\varepsilon}(n_I)$ were approximately constant for $n_I/h \lesssim -0.01$, and vanished for $n_I/h \gtrsim -0.01h$, showing that the interface is even thinner than the expected scale of order $Re^{-3/4}$. Even the large-scale variables such as the conditional mean velocity $\bar{U}$ and temperature $\bar{T}$ also have sharp jumps at the interface. The computed flow fields show how large scales bring 'fresh' fluid from the interior of the region to the interface where it is mixed both at saddle-point regions (defined with respect to the moving surface) and at engulfing nodal regions at the back of the large folds in accordance with the experimental results of Ferré *et al.* (1990) and Gartshore (1966).

These results should help explain and improve some of the *ad hoc* steps taken in numerical calculations of interface processes which largely ignore the intermittency of the turbulence. The rate of boundary entrainment E_b or slow movement outward of the mean interface position, i.e. $E_b = d\overline{x_{3I}}/dt$, is approximated by a diffusion-like process in the models (Turner 1986). But since the eddy viscosity v_e outside the interface is zero, this would mean $E_b = 0$. Therefore, as explained by modellers (Leszchiner 1999), a small non-zero value of v_e has to be assumed, although its magnitude has only a small effect on E_b (cf. Cazalbou, Spalart & Bradshaw 1994). Further studies are needed to resolve the uncertainty in the value of E_b for turbulent layers, which experimenters and modellers find is of the order of u_*, the r.m.s. velocity fluctuation. George (1999) on the other hand argued that E_b is determined by a weak diffusive process and is much smaller, being of the order $u_*(u_*/U_0)$ (George & Castillo 1997).

These discussions about the structure of the fluctuating velocity field at the wall and at the outer interface in turbulent boundary layers are related to the current controversies about the form of the mean velocity profile normalized on the surface friction velocity, $U(n_*)/u_*$, where $n_* = x_3 u_*/v_*$. Since its discovery by von Kármán (1930) the profile has generally been accepted as having a logarithmic form, i.e.

$$U/u_* = A \ln(n_*) + B, \tag{2.5}$$

where $A \cong 2.5$ and $B \cong 5.6$ are experimental coefficients that were assumed to be effectively invariant with Reynolds number for $Re \lesssim 10^3$. There is a similar log profile over rough surfaces. The form (2.5) is now questioned, firstly by close examination of new measurements in high-Re turbulent boundary layers (e.g. those of Zagarola & Smits 1998), and secondly by reconsidering similarity theory, which leads Barenblatt & Chorin (1998) to propose that the data are better described by a 'power law' profile of the form

$$U/u_* = A'n_*^{\alpha} + B', \tag{2.6}$$

where α is a function of Re. The largest differences between (2.5) and (2.6) (which are of the order of 10–20%) appear at locations where $n_* \leq 10^2$. Perry (1999) and other experimenters have commented that this is where the measurements are most uncertain because, at very high values of Re, the measurement points are so close to the wall that the accuracy of the measurement is not great enough to distinguish between the formulae. However, these differences matter because even small changes of, say, 3% in the pressure drop along pipes or in the skin friction of aircraft are economically significant. Establishing the form of this universal near-wall profile is also considered essential as a boundary condition for many widely used statistical models applied to this kind of turbulent flow (see Section 5).

4 Eddy structures

4.1 Dynamics

We review here and in the following section the various ways in which progress in dynamical and statistical calculations are contributing to the basic problems of turbulence, set out in Section 1. Because the full flow field at high Re can neither be calculated analytically nor simulated numerically, various idealizations and approximations are made in constructing theoretical models. Note that even where complete simulations are possible (as they are for $Re \leq 10^3$), theoretical models are still being actively developed to understand the flow, to extend the statistical results to higher values of Re and to provide methods for faster practical calculations.

There are two main theoretical approaches. One, described in this section, is to focus on the internal dynamics and external interactions of typical observed forms of eddy structure. In some studies the eddy flow fields are idealized in order to simplify the analysis. The other approach, described in Section 5, is to calculate in some simplified way the dynamics of an approximation of the overall flow field, usually in terms of its representation by a set of defined functions, e.g. its Fourier coefficients. The objective of either type of calculation is usually to derive or explain certain statistics of the whole flow, e.g. spectra, dissipation and transfer of energy and the probability distribution of the velocity field.

Studies of eddy structures are assisted by experiments and numerical simulations of the interactions between particular isolated vortical motions and surrounding flow fields that are characteristic of larger-scale structures within a turbulent flow (e.g. Couder 1999). The intrinsic assumptions involved in overall dynamic models (see Section 5), can be examined by studying how the turbulence responds to a narrow band of forcing frequencies or step perturbations (i.e. the relaxation process). Revealing experiments of this kind were conducted by Kellog & Corrsin (1980) and are now recommended as an essential element in improving models of turbulence spectra (Adrian & Moser 2000).

Much research continues to be based on the analysis of how small perturbations with random velocity and vorticity fields, $u(x, t)$, $\omega(x, t)$, with an integral scale l develop with time (t) in a more energetic velocity and vorticity field U, Ω with length $L \gg l$ (Hunt 1999; Cambon 1999; Leonard 1999). Initially, the strain rate of the large-scale fields is greater than that of the small-scale field u, i.e. $(u/l)/S = \mu \ll 1$, where $S = U/L$ and μ is a small parameter. Studies along these lines show firstly how structures in the small-scale field evolve, secondly how they may react back on the large-scale field and thirdly how they affect the overall dynamics and statistics of turbulence (e.g. Cambon & Scott 1999).

These stages can be explained in terms of the contributions of the large- and small-scale fields in the vorticity equation. The linear terms, caused by the direct interaction of the fields, whose magnitude can be characterized as $\gamma_L \mu S^2$, are $\nabla \Lambda (U \Lambda \omega + u \Lambda \Omega)$, and the nonlinear term caused by self-interaction of the small-scale fields, whose magnitude is $\gamma_{NL} \mu^2 S^2$, is $\nabla \Lambda (u \Lambda \omega - \langle u \Lambda \omega \rangle)$, where the average operation is denoted by $\langle \rangle$ as it must be taken over the large scale L. Here γ_L and γ_{NL} are coefficients of order unity that depend on the nature of the interaction.

The basic form of the large-scale straining may be characterized by its second-order invariant II, where II is normalized on the squares of the symmetric strain ratio and the vorticity, i.e.

$$II = \frac{(\partial U_i / \partial x_j)(\partial U_j / \partial x_i)}{\Sigma^2 + \frac{1}{2}\Omega^2} = \frac{\Sigma^2 - \frac{1}{2}\Omega^2}{\Sigma^2 + \frac{1}{2}\Omega^2}, \tag{2.7}$$

where

$$\Sigma_{ij} = \frac{\partial U_i}{\partial x_j} + \frac{\partial U_j}{\partial x_i}, \qquad \Sigma^2 = \Sigma_{ij}\Sigma_{ji}, \qquad \Omega^2 = |\Omega|^2.$$

It is also convenient to define the normalized third invariant

$$III = \frac{(\partial U_i / \partial x_j)(\partial U_j / \partial x_k)(\partial U_k / \partial x_i)}{\left(\Sigma^2 + \frac{1}{2}\Omega^2\right)^{2/3}}. \tag{2.8}$$

Consider the case where II has a significant component, i.e. $II + 1 > \mu$, then it is found that (except if $\partial U_i / \partial x_j$ is perfectly axisymmetric) the linear amplification of the non-uniform vorticity ω of the small scales leads to the formation of distinct thin layers, or sheets, parallel to ω and aligned in the direction of the strain, which rotate if $\Omega \neq 0$ (e.g. Betchov 1956). This increases the magnitude of the nonlinear term and in general γ_{NL} increases faster with γ_L. Since 'sheets' of small thickness tend to have a finite width, they begin to roll up and distort in other ways by self-induction (through the term $(u \cdot \nabla)\omega$) (Kida & Tanaka 1994; Passot *et al.* 1995; Kevlahan & Hunt 1997). Scalar fields are distorted by these motions into similar patterns of planar and rolled-up sheets (Brethouwer *et al.* 2002).

The straining produced by the rolling-up weakens the part of the sheet that is feeding into the roll-up, which therefore tends to become an isolated vortex structure (e.g. Pullin & Saffman 1998). In some circumstances if the initial small-scale velocity field is a coherent structure with a particular orientation and symmetry with respect to the axis of an irrotational strain, the nonlinear term does not grow as fast as the linear term (Gibbon 1999; see also Lundgren 1982) (i.e. $\gamma_{NL} \ll \gamma_L$), for example if wide vortex sheets or tubes are formed. In this general case the vorticity cannot grow without limit because the form of the small-scale flow is generally unstable to even smaller-scale levels of fluctuation which can grow exponentially on this distorted but slowly changing structure, e.g. as Kelvin–Helmholtz-like billows, provided viscous effects are small enough locally (Passot *et al.* 1995). Both these types of nonlinearity, operating on the time scale S^{-1}, tend to limit the growth of $\overline{\omega^2}$ and to amplify those components of the velocity fluctuations which the linear distortion tends to suppress (e.g. normal to the mean velocity in a shear flow). The reduction of anisotropy by direct nonlinear mechanisms in straining flows leads to different results than statistical modelling based on the 'scrambling' process of Rotta (1951). For example, the components amplified by the linear process are not correspondingly reduced by the nonlinear process.

The growth of $\overline{\omega^2}$ can also be limited by another mechanism: when the stretching of the vorticity extends beyond the scale L over which the large-scale strain is correlated, then the effective strain is weaker because it has a random orientation and magnitude (Leonard 1999). This limiting process also takes place on the strain time scale S^{-1}. The numerical study of Ohkitani (1998) shows that the mean-square growth of vorticity is less than that of the length of fluid elements $\overline{l^2}$, which is consistent with the existence of these self-limiting mechanisms.

The growth of the small-scale turbulence affects the non-uniformity of the large-scale strain field (i.e. $\nabla \nabla U$). This provides another mechanism for the limitation of the growth of small scales. The nonlinear self-induced terms $\nabla \wedge (u \wedge w)$ at the large scale L affect the vorticity on this scale, which can grow until a significant perturbation in the large structure U develops, such as a set of closed streamlines. This greatly limits the straining of the small scales (Kerr & Dold 1994; Nazarenko, Kevlahan & Dubrulle 1999). This example of an upscale process, which has been verified experimentally (Couder 1999), requires a stable large-scale flow with a significant amplitude for the large-scale perturbation to develop driven by the small-scale turbulence. (See also Sulem *et al.* 1989.)

Where the large-scale straining is purely rotational, e.g. with vorticity Ω_3, then $II \approx -1$. This motion has no direct stretching effect on the vorticity of the small-scale turbulence on a time scale S^{-1}; it merely rotates the vorticity and velocity fields which, for example, leads to oscillations in the ratios of the moments $(\overline{u_1^2}/\overline{u_2^2})$ if the turbulence is initially anisotropic. On a longer time scale $T_L = L_x/u_0$, the small-scale turbulence increases the separation of fluid elements by a distance Δ_3 in the direction of Ω. Since this nonlinear process amplifies the vorticity component ω_3, in the direction of Ω, in regions where $\Delta_3 > 0$ and reduces ω_3 in regions where $\Delta_3 < 0$, vortical structures emerge parallel to Ω. Furthermore, those having the same sense of vorticity as Ω are stabilized against small fluctuations by the large-scale rotation, while those with the other sense do not tend to form and are unstable (Cambon 1999). Where these structures are formed very close to each other, they tend to rotate around each other and to merge into large structures (e.g. Hopfinger, Browand & Gagne 1982). This is one of several examples of where, as certain eddy structures form, they tend to merge with others nearby and suppress other types, both effects tending to amplify the local gradients of vorticity at the edges of the structures.

In simulations of the development of an initial distribution of vortices at $Re \approx 150$, Ohkitani (1999) showed how the vorticity was amplified in the form of sheets which then rolled up quite rapidly and, through viscous diffusion, turned into a distribution of elongated vortices or 'worms'. These persisted until the turbulence finally decayed. Despite the relative brevity of the sheet-roll-up phase, his calculations showed that this mechanism provided more of the transfer of energy to small scales than the longer lasting 'worm'-like phase. Perhaps this explains why the rolled-up vortex sheet in an extended straining flow (recently reviewed by Pullin & Saffman 1998) can only persist and be continually regenerated at very high values of Re (such as seen by traces in atmospheric and oceanic turbulence) and not at the lower values of Re that can be directly simulated numerically.

Recent theory supported by numerical simulations and experiments has revealed more about the dynamics of these small but mature vortices, especially how they contribute locally to the role of dissipation of energy ε and how, outside them, they induce helical streamlines and straining fields which affect other vortices (e.g. Ohkitani 1999; Kida, Miura & Adachi 2000; Douady *et al.* 1991; Vassilicos 2000a). It is found that these vortices have a finite length (which can be as large as the integral scale at low Re), are not very curved, and have a finite lifetime; these effects may be caused more by the growth of instabilities within the vortices stimulated by the random fluctuations in the surrounding flow than by strong mutual interactions between vortices (Verzicco & Jiménez 1999; Melander & Hussain 1993; Miyazaki & Hunt 2000). At high enough values of Re, this breakup stage could presumably be followed by a new cycle of sheet formation-roll-up-vortices-breakup; this could occur everywhere and at random places through the flow. The role of 'background' fluctuations outside the structure can be significant during all these stages (Hunt 1999; Tsinober 2000). Chertkov *et al.* (2000) and Ooi *et al.* (1999) have shown how different stages of this cycle can be mapped on to a graph of invariants of the velocity strain field ('*Q–R*' or *II–III* plots) from numerical simulations (see also Tsinober 2000). Hunt (1999) suggested that this provides a semi-deterministic model for the natural time scale of the small-scale eddy motions.

Understanding how large or intense the structures grow and the different time scales at which they break-up affects how they contribute to the overall dynamics and statistics of the turbulence. Some investigators have suggested on theoretical grounds (Hunt 1999), others on the basis of studying vortices in two-dimensional flows (Kiya, Ohyama & Hunt 1986) and low-order model behaviour of boundary layer eddies (Holmes *et al.* 1996), that most vortical structures move round each other, so that their interactions are long range and on average not very strong (see Section 5). On the other hand Pradeep & Hussain (2002) and Moffatt (2004) suggest that strong interactions (when the helicity integral of the structure changes) might occur sufficiently often to affect the dynamics and especially the extreme values in the probability distribution. A new analysis (see Moffatt (2004)) of flow vortices grouped in two anti-parallel pairs and orientated so that they collide with each other at right angles showed how, even when viscosity is included in the calculation, the velocity and vorticity tend locally to an infinitely large value at a finite time t_∞ that is independent of viscosity even when Re is finite. This specific calculation of a realizable flow (assuming it remains stable and the critical symmetry of the vortices is exact) is consistent with some earlier theory that such singularities could exist (Leray 1933; Pumir & Siggia 1990). Numerical simulations conducted during the programme by Ohkitani & Gibbon (2000) showed that a class of stretched solutions identified by Gibbon, Fokas & Doering (1999) leads to a finite-time singularity.

Mathematical studies are helping to define bounds and general properties of such singularities that form in a finite time. Doering & Gibbon (2000) proved during their stay at the INI that there cannot be a finite-time singularity of the Navier–Stokes equation if the ratios between a set of statistically defined microscales, all smaller than the Taylor microscale, are increasing fast enough with time. Constantin (2000) obtained results on the inviscid Euler equations using the Cauchy–Weber relations (for the vorticity and velocity of distorted fluid line elements) which may point to the absence of finite-time, geometrically regular, self-similar singularities. The study of singularities and near-singularities is inherently linked to the pivotal dependence of dissipation on Reynolds number (Taylor's conjectured equation (2.1)) and to the way that this dependence is determined by the flow geometry of the underlying velocity field (see Doering & Constantin 1998; Flohr & Vassilicos 1997; Angilella & Vassilicos 1999; Kerswell 1999).

The mechanisms for the growth and interaction of the largest-scale eddy structures are necessarily studied in the context of particular types of turbulent flows because these structures always retain some influence or 'memory' from initial or boundary conditions (see Section 5). Nevertheless, where the large-scale eddies are free to move and interact with each other, and whether produced homogeneously at some initial time or in some local region (as in a boundary layer), it is found that they have some general features in common. In both two- and three-dimensional turbulence there is a greater tendency for large vortical eddies to grow by the mechanism of boundary entrainment (discussed in Section 3) and by engulfing small eddies, than to be diminished by breakup caused by occasional collisions with other large eddies. There are stronger tendencies for three-dimensional as compared to two-dimensional vortices to become unstable and to interact (even at a distance) with other structures. The net growth rates of these structures is much less than for two-dimensional vortices, which is consistent with statistical results (e.g. Lesieur 1990). These mechanisms shed some light on the unresolved questions (Herring 1999; Davidson 2004) about the nature of long-range effects in two- and three-dimensional turbulence and the convergence of volume integrals $I_k = \int_0^\infty r^k R(r) dr$ of the cross-correlations $R(r) = \overline{u(x)u(x+r)}$. Davidson (2004) considers the angular momentum of large but finite volumes of isotropic turbulence, whose large eddies have small enough initial momentum that their energy spectrum $E_{(k)}$ is much less than $O(k^2)$ – Saffman (1967). He gives a new argument for why Loitsyanskii's (1939) integral I_4 is finite, and approximately constant. This integral plays a critical role in statistical models discussed in Section 5.

4.2 Kinematics and statistics of eddy structures

The objective identification of these 'structures' and the assessment of their contributions to the overall statistics of a flow are as important as their dynamics. There are three main approaches to the identification problem, according to the sampling method used and the type of eddy being analysed (Bonnet & Glauser 1993). The first uses statistics, such as two-point Eulerian correlations, to extract the forms of modes defined in fixed coordinate systems (e.g. Devenport's 1999 study of eddies in the near wake). The second is based on measurements available at high Re, well-focused identification methods using conditionally sampled data are necessary; new developments in the measurement and analysis of the multiple-scale properties and individual events using wavelet analysis were reported by Arneodo *et al.* (1999) and Nicolleau & Vassilicos (1999) (see also Silverman & Vassilicos 1999). When complete data of flow fields are available (which are only obtainable from numerical simulations and therefore at moderate values of Re), the forms of eddy structures can be evaluated in terms of tensorial invariants of the velocity gradient field (Section 4.1). For example, vortices near the wall in boundary layers have been identified using as the threshold criteria various combinations of II, Σ^2, Ω^2 (Hussain 1999; Perry 1999; Lesieur 1990). Small-scale vortices in homogeneous turbulence were identified, for example as regions of low pressure, by Kida *et al.* (2000).

The third 'optimal' type of approach is to combine the first two, for example by using data at a point from a single realization, in combination with correlations from an ensemble, to infer the local eddy structure. To test the conjecture by Tabeling

imposed distortion of the turbulence, such as the time for elements to leave a turbulent boundary layer and enter the wake behind an obstacle, or be compressed in a shock wave – both being examples of aeronautical turbulent flows that are considered to be unsatisfactorily calculated by the present generation of practical models (Hills & Gould 1999).

(iii) Turbulence dynamics are intrinsically inhomogeneous when the integral scale L_x is of the order of or greater than the scale Λ_0 over which the turbulence structure or mean strain rate S varies (e.g. $\Lambda_0^0 = (\overline{|u|^2}/\nabla|u|^2$ or $S/\nabla S)$ (Durbin 1999). The degree of this non-locality is defined by $-\tilde{L}_x = L_x/\Lambda_0$.

(iv) In large eddy simulations (LES) of the flow field it is usual to filter out the eddies with scales smaller than ℓ_f, a measure of this approximation being $\alpha_f = (\ell_f/L_x)$.

(v) In some models of the full flow field and stochastic simulations, small-scale dynamics are represented by random processes (Mason & Thomson 1992) with a time scale τ_s, a measure of this approximation being $\alpha_s = (\tau_s/T_1)$.

Note that these measures (i–v) are based on integral-scale quantities of turbulence because it is assumed that the models and simulations are being used primarily to calculate the mean flow and energy-containing eddies. To some extent these quantities also determine the smallest-scale motions even though the computations may not describe them exactly (e.g. on the Kolmogorov microscale l_k and time scale τ_k).

(vi) Most calculations are numerical and involve making approximations over spatial scales and temporal discretization scales Δx, Δt, a measure of the errors involved being $\Delta x^* = \Delta x/\Lambda_0$. Current research indicates how, although many earlier model calculations were greatly affected by numerical approximations, it is now possible to reduce Δx^* and Δt^* to sufficiently small values, thanks to greater computer power now available, and that most approximations or errors are caused by the model assumptions rather than their numerical approximations (Hills & Gould 1999). Studies of the errors of the filtering and discretization approximations have shown that for large-eddy and stochastic simulations it is generally necessary that the discretization scales are smaller than the filtering and stochastic time scales (i.e. $\Delta x^* \leq \frac{1}{2}\alpha_f$; $\Delta t^* \geq \frac{1}{2}\alpha_s$) (e.g. Mason & Callen 1986; Guerts 1999).

There are three main areas of development in modelling the full flow field:

(a) answering some of the basic statistical questions about turbulence (such as those in Section 1) by reducing the number of assumptions made to calculate, at some defined level of accuracy, key statistical quantities;

(b) calculating low-order statistics for non-stationary, non-local turbulent flows near rigid boundaries and near interfaces with non-turbulent or other kinds of turbulent or fluctuating flows;

(c) approximate simulations, especially reducing the errors in the resolved scales and
improving the statistical or stochastic modelling of the subgrid-scale motions and its
conditional dependence on the larger resolved scales.

Since turbulence has different physical and functional forms (e.g. in the relations
between statistical properties) depending on the type of the flow (e.g. Lumley 1999),
we now consider models in the context of these types, classified in terms of the
parameters just introduced. General models and simulation methods need to be
assessed over a range of turbulent flows, whereas those that are designed to be
applicable in a narrow range should be assessed accordingly.

(a) *Short time and rapidly changing turbulence* ($\tilde{T}_L \gg 1$, $|\tilde{A}|\tilde{T}_L \sim 1$, $\tilde{L}_X \sim 1$). For a short
time $t > t_0$ after turbulence is generated within a domain or is advected into it (i.e. $\tilde{T}_L \gg 1$)
the nonlinear terms have only a small influence on the velocity field, while distortions
by linear effects, such as by gradients of the mean velocity, impact on boundaries or
body forces, may have a large effect. Note that, because of the linearity, the turbulence
is mostly sensitive to the initial conditions, though for some types of distortion (such
as mean shear) certain of the resulting turbulence statistics are quite insensitive to their
initial state. The linearity enables many features of the turbulence to be calculated analytically, including even the first-order corrections in the nonlinear terms. Research into
these rapidly changing flows is progressing in two main directions. First, increasingly
complex forms of linear distortion are being analysed, particularly where the scale of the
turbulence L_x is larger than the distance Λ_T over which mean strain is varying (Leonard
1999; Hunt & Durbin 1999) (i.e. $\tilde{L}_X \gg 1$), or where combinations of distortion are
being applied (e.g. mean shear and stable stratification) (Hanazaki 1999).
   Second, several complex nonlinear effects can be studied because, when $\tilde{T}_L \gg 1$, they
are weak enough to be calculated by expansion methods, starting with the non-stationary
inhomogeneous linear solutions. This enables the growth of the back-reaction of the
turbulence on the mean or larger scale flow to be estimated, including the growth of
instabilities and waves, the development of large-scale flows driven by Reynolds stresses
of distorted turbulent flow fields, and the effects of the distortion on the nonlinear vortex
stretching and advection (see Section 4). The effects of these terms may remain quite
small for a large enough time (or distance) such that $\tilde{T}_L \sim 1$, and the linear processes can
effectively determine the flow structure even in fully developed flows. This is why these
studies provide insights into fundamental mechanisms. They also have more practical
objectives such as to provide an exact limiting case as a comparator for general statistical
models or to calculate in detail the effects of distortion on particular flows with given
upstream conditions.

(b) *Statistical models for small-scale fully developed turbulence.* There continues to
be incremental progress in constructing 'theoretical-physics' models for calculating
moments of the small-scale velocity field, from the Navier–Stokes equation, by making
the fewest possible assumptions about the mechanisms and no explicit assumptions
about numerical coefficients (McComb 1990). Although the ultimate aim is to describe

the statistics completely, at present the systematic renormalization and perturbation methods are limited to computing spectra in the Kolmogorov inertial and viscous ranges.

Earlier calculations in a fixed Eulerian frame using the direct interaction approximations (see Kraichnan 1959) needed to be modified to apply in a Lagrangian frame in order to agree with the observed form of the inertial-range spectrum, in which $E(k) = \alpha_k \bar{\varepsilon} k^{-p}$, where $p \simeq 5/3$. However, in recent RNG calculations of McComb (1999), in which different initial assumptions are made, this modification is not necessary. Although the amplitude α_k agrees well with the measurements to within the 10% accuracy of the experiments, there is no prediction about the small correction of p dependent on the intermittency of the energy dissipation rate. Perhaps this is consistent with Kida's (1998) conclusion that these methods provide satisfactory approximations for second- (and presumably other low-) order moments because they only depend on a subset of all the possible nonlinear interactions between Fourier modes or 'eddies'. These are not the most locally intense, such as those which lead to large fluctuations in the dissipation rate and the small corrections to p.

There is still an active stream of turbulence research that is mainly based on the methods of statistical physics because of the difficulties of developing reductionist theories based on the equations of motion. One observational justification is that joint probability distributions for particular combinations of variables have similar forms in different types of turbulence (e.g. Chatwin 1999). The theoretical basis essentially follows from the arguments of Landau & Lifshitz (1959) and Batchelor (1953), quoted in Section 1.

(c) *Statistical models for the overall structure – two-point moments.* Increasingly, the solutions of engineering and environmental problems involving turbulent flows require calculations of the spectra (or two-point moments) of the velocity. Also needed is some knowledge of the sensitivity of calculations of one-point moments (such as Reynolds stresses) to the variations in the spectra, anisotropy and inhomogeneity of the turbulence (a possible source of error emphasized in the account of one-point methods by Launder & Spalding 1972). The most extensively developed model for spectra is the eddy damped quasi-normal Markovian (EDQNM) coupled differential equation model for the second- and third-order moments. The basic theory (e.g. Lesieur 1990) includes some ideas from statistical physics, for example the relaxation time scale for relating second and third moments. For the case of isotropic turbulence, an alternative approach is to use renormalization methods, which avoids the assumption of this time scale (McComb 1990). EDQNM has been extended to uniformly distorted turbulent flows, such as large-scale compression, shear or stable stratification, etc. The method is consistent with linear theory for large strain rates and has been extensively verified (Cambon & Scott 1999). In our notation the conditions for this locally homogeneous theory are $\tilde{T}_L \sim 1$, $\tilde{A}\tilde{T}_L \sim 1$, $\tilde{L}_X \ll 1$. Bertoglio (1999) and Cambon (1999) described how the method can be extended with the aid of further approximations to weakly inhomogeneous turbulence (i.e. $|\tilde{A}|\tilde{T}_L \lesssim 1$) (when it is still possible to define three-dimensional spectra). It was shown how in some cases of strong deformation rate, such as during compression in an engine, one-point moments derived by integrating these spectra give more accurate calculations than those based on simpler equations for one-point models. This is because, except by adjustment of their empirical coefficients, such one-point models allow for similar variations in the turbulence structure (Launder & Spalding 1972).

Where the turbulence is highly non-local or non-stationary, such as near rigid or flexible interfaces (i.e. $|\tilde{A}|\tilde{T}_L \gg 1$, $\tilde{L}_X \gtrsim 1$) the above methods do not apply. However, linear methods can be used in some of these flows even though, formally, they are only valid when $\tilde{T}_L \lesssim 1$. In some cases the nonlinear effects are suppressed and the turbulence is changing slowly with time. In other cases, a correction to the linear model can be estimated using a relaxation time that varies with the eddy scale (Mann 1994). Banerjee (1999) reviewed recent work showing how these methods largely predicted the turbulence structure near density interfaces, verified by experiments and direct numerical simulation.

(d) *One-point closures for flows in engineering and the environment.* In the majority of practical turbulent flow problems the main objective is still to calculate the mean velocity $U(x, t)$, temperature $\theta(x, t)$, or some other mean scalar such as mass fraction, and some approximate measures of the amplitude and scale of the turbulence (e.g. $u_0 = \sqrt{\overline{u_1^2}}$, and L_x). From the mean momentum equation it follows that the gradients of the Reynolds stresses $\overline{u_i u_j}$ should be calculated as accurately as the required gradients of U, while other measures of turbulence may be derived less accurately (e.g. L_x). While $\overline{u_i u_j}(x, t)$ are moments at one point, L_x is formally a measure of two-point moments. But it is often assumed to be a quantity defined at a point (assuming local homogeneity, i.e. $\alpha_{NL} \lesssim 1$) or estimated from the local value of mean dissipation $\varepsilon (\simeq u_0^3/L_x)$.

Practical models devised to predict these quantities have tended to be Eulerian and have not changed fundamentally over the past 25 years (Lumley 1999; Pope 2000). Research into their rationale and limitations has led on to incremental modifications, usually designed for different types of turbulent flows. The numerous workshops for testing the models against experiments and systematic comparisons of all the elements of the models with direct numerical simulations have all contributed to these advances. Following Prandtl's (1925) physical arguments based on his flow visualization studies of eddies in a shear flow, in the simplest statistical models, the Reynolds stresses $(-u_i u_j)$ are assumed to be proportional to the gradients of mean velocity $\partial U_k/\partial x_j$ and to the 'bremsweg' (later 'mischungsweg') length scale ℓ_m. This depends on the type of flow and the location within it. Such relations, which are assumed to be independent of initial and boundary conditions of turbulence, and any non-local and non-stationary effects (i.e. $\tilde{T}_L \ll 1$, $|\tilde{A}|\tilde{T}_L \ll 1$, $\tilde{L}_X \ll 1$), are still widely applied for calculating nearly unidirectional flow because of their computational efficiency and convenience, for example over turbomachinery blades and in certain environmental flows. In many such cases the largest errors are caused by the 'wall-layer' approximations (e.g. Eq. (2.5)) in the boundary conditions for U near smooth or rough rigid surfaces, which can only be derived by local analyses for the viscous processes in the former case or for the flow through the roughness elements in the latter case. In unidirectional shear flows the Reynolds stress $-\overline{u_1 u_3}$ parallel to the mean shear $\partial U_1/\partial x_3$ is sometimes used to estimate the stresses in other directions and the total turbulence kinetic energy $K = \frac{1}{2}\overline{u_i u_i}$ or the components of the Reynolds stress $\overline{u_i u_j}$. The factors considered are the gradients of the mean velocity, the fluctuating pressure gradients, the eddy-induced transport of turbulence across the shear flow, and the mean rate of dissipation $\bar{\varepsilon}$ by viscous processes. The last three terms have to be approximated in terms of the same second-order moments and the mean flow or by the use of additional equations. For general-purpose calculations of industrial and some environmental flows with a mean velocity that is large relative to the turbulent fluctuations (so that $u_0/U \ll 1$) but which may vary in direction, only two coupled differential equations for K and $\bar{\varepsilon}$ are used, and often the Reynolds stresses are estimated by the local eddy viscosity approximation

$$\left(\overline{u_k u_k}\frac{1}{3}\delta_{ij} - \overline{u_i u_j} \right) = C_\mu (K^2/\bar{\varepsilon})(\partial U_i/\partial x_j + \partial U_j/\partial x_i), \qquad (2.9)$$

where C_μ is a coefficient determined by comparison with experiments.

For turbulent flows that are far from equilibrium, such as turbulence in a strongly diverging flow approaching a stagnation point (i.e. $|\tilde{A}|\tilde{T}_L \geq 1$), the turbulence structure changes rapidly, and the form of its anisotropy differs significantly from that in shear flows. Then the local relation (2.9) is incorrect and the k–ε equations give quite misleading information about the turbulence. However, in using the full Reynolds stress transport equations (RSTE) (Hanjalić 1999; Craft 1999), $-\overline{u_i u_j}$ develops as a result of the history of the mean strain $\partial U_i/\partial x_j$. Some industries are making use of models based on these RSTEs (Laurence 1999), while others consider that the possibility of extra accuracy does not compensate for the extra complexity. In flows that are highly inhomogeneous (i.e. $\tilde{L}_X \geq 1$) the higher relative contribution of the eddy transport, such as occurs in natural convection with low mean velocity (so that $u_0/U_0 \leq 1$), can be estimated more accurately by calculating explicitly transport equations for the third moments that are otherwise approximated in the RSTE in terms of second moments (André *et al.* 1976; Ilyushin 1999; Launder 1999).

The k–ε and RSTE model equations have been formulated quite generally so that all the tensors satisfy invariance properties and the dimensionless coefficients are the same in different types of flow and different ranges of *Re*. It is implicitly assumed that the effects of non-locality and non-equilibrium are small (i.e. $\tilde{L}_X < 1$, $|\tilde{A}|\tilde{T}_L < 1$), and therefore that (see Launder & Spalding 1972) any changes in the turbulence structure (e.g. spectra, anisotropy, etc.) only have a small effect on the second moments. However, in most recent developments, these assumptions about the universal and localized nature of second-order turbulence dynamics are being relaxed in various ways.

(i) The variations in the eddy structure of turbulent flow resulting from initial anisotropy of the moments $\overline{u_i u_j}$ (Townsend 1976), from their spectra or from various types of straining (see Cambon 1999), can be large enough to diminish the tendency of the pressure-strain term $\overline{p\,\partial u_i/\partial x_j}$ to reduce the anisotropy of the turbulence. This is now well understood in terms of the eddy dynamics (see Section 3), and can be succinctly described in an anisotropy diagram of the second and third variants II_b, III_b of the anisotropy tensor $b_{ij} = ((\overline{u_i u_j}/\overline{u_k u_k}) - \frac{1}{3}\delta_{ij})$ (Lumley 1978) or of the anisotropy of the spectra (cf. Kassinos, Reynolds & Rogers 2001).

This sensitivity is now allowed for in some models so that effectively the isotropic tendencies of the pressure-strain terms are assumed to be negligible if $III_b < 0$ (Launder 1999).

(ii) Another way in which variations in the turbulence structure are being modelled within the same general methodology is (following Hanjalić, Launder & Schiestel 1980; Schiestel 1987) to split the energy (K) equation into two parts, one for calculating the large scales which are dependent on the production of energy by the mean shear and buoyancy and the transfer of energy to the small scales, and the other calculating the smaller scales. The differences in the development times of these two parts of the spectrum approximately model the effects of changes in the form of the energy spectrum on the large-scale turbulence and the mean flow (Laurence 1999; Hanjalić 1999).

(iii) Where there are local gradients of the mean velocity $\partial U_i/\partial x_j$ the second 'fast' component of the pressure-strain terms is usually approximated as KG, where G is a tensor proportional to the local value of $(\partial U_i/\partial x_j)$ (which naturally follows from the Poisson equation for the fluctuating pressure). However, where these mean gradients vary rapidly over the length scale of the turbulence (i.e. $\alpha_{NL} \geq 1$), the assumptions of the modelling of these terms are invalidated. This causes errors, for example in the ratios of Reynolds stress $\overline{u_1^2}/\overline{u_3^2}$ etc. in high Reynolds number turbulence near a 'wall' and thence in calculations of heat transfer. One way of modelling this non-local effect is to allow quadratic and cubic products of Reynolds stress to appear in the linear rapid part of the pressure-strain term. This effectively accounts for the fact that mean strain distorts the shape of the two-point velocity correlation surfaces (which modify the pressure fluctuations). In this way one can formally arrange to satisfy the two-component limit to which turbulence reduces at a wall (Lumley 1978; Shih, Lumley & Janicka 1987; Launder & Li 1994; Craft & Launder 1996; Craft 1999). To extend such calculations into the buffer layer, one approach is to introduce higher derivatives of the mean velocity gradients ($\partial^2 U_1/\partial x_3^2$ etc.) into the approximation for the pressure strain (Launder & Li 1994; Craft 1999). Another approach is to introduce an auxiliary inhomogeneous differential equation for the variable $G(x_3)$ with the right-hand side being proportional to $\partial U_1/\partial x_3$. This robust numerical approximation for the pressure Poisson equation is being applied in several engineering applications (Durbin 1999; Laurence 1999). It is consistent with the normal velocity eddies being blocked by the wall, which becomes an increasingly significant process at high Re (see Sections 2 and 3). Similar non-local effects of blocking in stably stratified inversion layers (Banerjee 1999) have been handled by the two-component limit approach noted above (Craft & Launder 1996; Launder 1999). It has alternatively been modelled in some atmosphere and ocean calculations by expressing the length scale (needed for momentum or scalar flux calculations such as (2.9)) as integral expressions which have to be evaluated implicitly (Bougeault & Lacarrère 1989).

(iv) The relation (2.9) between Reynolds stress and mean velocity gradients, used in conjunction with the k–ε pair of equations, is not only local, but also linear in these gradients. Calculations using linearized theory for rapidly changing turbulence (Townsend 1976) show that this approximation is also in doubt if the gradients change significantly on a time scale of the order of that of turbulence T_L (i.e. $|\tilde{A}|\tilde{T}_L \geq 1$), which commonly occurs, for example, in aeronautical boundary layer flows. The errors are such that 'nonlinear eddy viscosity' formulations for (2.9) are being introduced in which $-\overline{u_i u_j}$ is expressed as an expansion, up to third order, in $\|\partial U_i/\partial x_j\|$ (Gatski 1999). However, even these steps are insufficient to represent correctly the sensitivity to streamline curvature. One proposal is that (2.9) be augmented by cubic-level products of mean strain and vorticity in ways that cannot simply be interpreted as modifying C_μ (Craft, Launder & Suga 1997).

(e) *Approximate simulations and derived statistics.* There have also been interesting developments in the application of approximate calculations of individual realizations of turbulent flows. These methods are used mainly to derive statistics from their ensembles. Most of the research is focused on approximating the small scales, especially those close to boundaries, but because all simulations are limited by the size of the domain and the period of the computation, there continues to be some uncertainty about the approximations involved in simulating the largest scales of turbulence comparable to those of the space-time domain and in estimating the effects on these scales of the errors at the smaller scales.

For example, what is the effect on the statistical properties of boundary layer simulations if the domain is not large enough to model the eddy structures, which can be 18 times larger than the boundary layer thickness (see Section 3), which have not so far been observed in large-eddy simulations? However in some 'low-order' simulations such structures are postulated in the basic assumptions of the simulation (Holmes *et al.* 1996) and consequently are represented. If numerical simulations are used to calculate the space-time development of individual realizations of turbulent flow, then specific initial data are required. But they are incomplete in most applications such as environmental forecasting or for control of turbulent flows. Kreiss (1999) posed the bold hypothesis that only the large-scale velocity field (with scale of order of L_x) need be specified (but at frequent time intervals much less than $T_L = L_x/u_0$) in order for the velocity field at small scales to adjust dynamically to the same form whatever its initial form, given the same large-scale field. He largely verified this conjecture in numerical simulations of two- and three-dimensional homogeneous turbulence forced at the large scale. The moderate Reynolds number, $Re \sim 200$, of the simulation may explain why no obvious small-scale instabilities were observed. This result is consistent with other studies, which show that large-scale eddy structures can dominate the evolution of the

flow field and provide a rational basis for flow control (Holmes *et al.* 1996), at least
in this range of *Re*. It may also be in accordance with the greatly reduced estimates
for error growth in high-*Re* flow systems. When eddy structures are free to move, any
'error' in their position eventually grows algebraically, and not exponentially; this leads
to lower growth in the errors of simulation of environmental flows than was originally
estimated by Lorenz (1963). Indeed Kreiss went so far as to argue that a much smaller
amount of data is required for environmental real time prediction than is usually being
supplied! (But see Hunt 1999.)

Large-eddy simulations, which are reviewed by Sandham (2003), have for more
than 30 years been producing striking new insights into turbulence, particularly the
structure of the energy-containing eddies. Active research is still needed to calculate
reliably the errors in the resolved motions on the scale of the filter l_F. Some new corrections have been proposed based on the assumption of similarity of eddy motion over a
range of length scales (e.g. Geurts 1999). Although such filtering has some effect on the
larger scales (so that the simulation cannot be a prediction for a particular realization
far into the future), many tests show that because of the downscale cascade in three-
dimensional turbulence, the structure and statistics of the large-scale eddy motion is
insensitive to these small-scale modelling assumptions (Jiménez 1999). (The upscale
cascade in two-dimensional turbulence makes the structures sensitive to smaller-scale
processes, e.g. Dritschel 1993.) The interactions between large and small eddy motions
are more critical near rigid boundaries within a surface layer of thickness h_s; shear
layers and coherent eddy structures form which extend over distances parallel to the wall
that greatly exceed h_s and often exceed L_x. These affect the level of the turbulent fluctuations near the wall and the large-scale motions in the interior of the flows, especially in
thermal convection. It is now realized that the earlier approach for modifying LES near
a wall by damping the velocity components is quite incorrect at very high *Re* ($\geq 10^4$)
because the wall amplifies the small-scale turbulence. A replacement by a local,
quasi-steady, boundary-layer velocity profile with associated small-scale turbulence
(Thomson 1999; Moin 1999) at least provides a reasonable first-order correction. But
this does not completely model the structure of the surface-layer eddies, which is significant for calculating the transition between the turbulence in the boundary layer at
the trailing edge of an aerofoil and in the wake. Currently, various attempts are being
made to model the layer with local boundary layer equations so that the solutions match
with the resolved motions above the surface (Moin 1999).

Because LES methods use grids that are much smaller than the integral scale (at
least 1/10) they require enormous computation time. Yet many studies show that for
unsteady, non-local, turbulent flows, where $\tilde{L}_X \geq 1$ and $|\tilde{A}|\tilde{T}_L \geq 1$, such as thermal con-
vection with a low mean velocity or unsteady wake flows, the k–ε or RSTE equations are
too inaccurate to be used as statistical models (since they require $\tilde{L}_X \ll 1$, $|\tilde{A}|\tilde{T}_L \ll 1$).
However, these can be used as unsteady equations to calculate a very large-eddy sim-
ulation (VLES) approximation to the realization of the unsteady velocities of the large
eddies. The 'turbulence' in the statistical equations is assumed to represent the small-
scale turbulence in the VLES unsteady calculations. Thus the method involves an
assumption about the independence of large- and small-scale motions. This assumption

is similar to that made in LES about the resolved and 'subgrid' scale eddies. But in the case of VLES, because more of the dynamics at the resolved scale are 'modelled' (e.g. eddy transport), the discretization scales and (implicit) separation scale are both much greater than for LES. This speed-up in computations more than compensates for any loss in accuracy for some flows (Hanjalić 1999), though others at the meeting disagreed.

As with other statistical models, appropriate approximations are made for specific flows to simplify the calculations sufficiently for particular features to be explored in more depth, for example by analytical solutions or using dynamical systems theory to study how the solutions evolve in time (Holmes *et al.* 1996). Research using these analytical solutions is showing how, below each large eddy, the surface layers of the interior flow have a complex internal structure at very high *Re* that would need an unrealistically small grid size and large computational time to be completely resolved (see Plate *et al.* 1998).

The ways in which small scales affect calculations of the large-eddy structure are quite sensitive to the assumptions about the surface boundary conditions and the small-scale statistical models; for example it depends on whether mean streamwise vortical eddies or other secondary flows exist, driven by normal stresses of large-scale turbulence. It remains to relate these calculations (e.g. Townsend 1976) to LES and other direct simulations of the large-scale eddy structure.

All the previous approximate simulations were based on filtering, averaging and approximating the Navier–Stokes equation. However, by analogy with statistical models that resemble the stochastic phenomena of turbulence, there are also methods based on phenomenological arguments for simulating the random or mean velocity fields without solving or approximating the Navier–Stokes equations. These may provide the full field or partial elements of it, such as the random fluctuations smaller than the resolved scales of large-eddy simulations. (For a review of Monte Carlo or kinematic simulations see Elliott & Majda (1996).) A novel approach for calculating the mean velocity in shear flows as *Re* varies from 10^3 to 10^6 by heuristically adapting the Navier–Stokes equations to reflect the motion of large eddies conveying momentum across the shear flow has been proposed by Holm (1999). Its solution agrees to within 5% with measured profiles of the mean velocity over this range of *Re*.

6 Concluding remarks

Can we conclude that current research is making progress towards answering the main questions, about turbulence? The evidence of this review suggests that the answer is yes, but that we are still some way off finding complete answers. Some particular advances are summarized in the abstract.

The main question, on which hang many others, is to what degree fluid turbulence is a universal phenomenon. On the one hand, research continues to provide more evidence (Section 3) and better dynamical explanations (Section 4), though still no complete theory, for the occurrence of similar qualitative features at the smaller scales in all types of three-dimensional turbulent flows; namely random, intermittent motions on wide ranges of length and time scales, where the very smallest scales are determined in part by molecular processes, a net cascade of energy to these scales, and a mean rate of dissipation $\hat{\varepsilon}$, that, when normalized on large-scale motions, is approximately independent of the Reynolds number. On the other hand, there are many non-universal aspects; not only are there qualitative differences in the structure of the large-scale eddy motions, but there is also increasing evidence (Section 3) that in quantitative terms (e.g. statistical two-point, Eulerian measures) the above small-scale phenomena depend to some degree on the particular types of large-scale flow. The dependence of the eddy's structure and statistics on the flow type is also found (both theoretically and experimentally) in the underlying dynamics that determines the eddy motions. This is reflected in the sensitivity to the type of flow of the space-time development of the moments, as for example in Reynolds stress transport equations (Section 5).

A consequence of this increasing evidence and understanding of non-universality is a change in the direction of research. There is now more emphasis on studying various types of turbulence within distinct parameter ranges, leading to a variety of statistical models and approximate simulations that typically depend on the characteristic forms of the large-eddy structures in each type.

At the same time, some models that have been formulated on the assumption of great generality are being adapted to particular types of flow, for example by allowing for significant non-local and non-equilibrium effects (Section 5). Some industrial users of statistical models have called for their systematic classification in terms of their assumptions and ranges of validity, so as to provide initial guidance to those applying such models to any particular type of flow. It was suggested that research along these lines may be at least as useful as conducting further comparisons among very general models applied to test case flows, which inevitably are quite idealized in terms of the geometry of the flow and the types of initial and boundary conditions.

Nevertheless, the search for some kind of universality is the goal of much experimental research in high Reynolds number turbulence; it tends to be restricted to studies of whether statistics of the relative velocity Δu (when suitably normalized and corrected for large-scale motions), are approximately the same in all types of fully developed turbulence at very high Re ($\geq 10^4$). Consequently much theoretical research is still directed towards establishing a dynamically based, universal statistical theory for motions in the inertial and viscous microscale ranges. Accurate Lagrangian and multi-point Eulerian measurements, which will require the new facilities now being proposed (e.g. Nieuwstadt 1999), could provide the critical tests for any such theory, especially if (see Section 4) it is based on an analysis of the velocity field around moving fluid elements.

Our review shows that although, regrettably, there are rather few innovative experimental research projects, there is certainly a diversity of theoretical and computational methods in turbulence. Most of them contribute some insight into the varied manifestations of turbulence. It is noticeable that some authors explicitly estimate the range of validity of their methods and results, but many others still leave this task to the reader or, worse, the person applying the method in practice. Another reason for the variety of methods, and for the focus of research on statistical models and approximate simulations, is that (Section 5) the scientific and practical applications vary greatly. Not only do the nonlinear interactions between turbulence and other processes vary, but also the level of complexity that is appropriate changes depending on the availability of computing capacity (for the turbulence part of the total calculation) and of sufficiently detailed or accurate input data.

Currently this diversity of methods and assumptions is not sufficiently understood or valued by those engaged in turbulence research and its applications. Some interesting combinations of different and hitherto competing approaches are now being tried, such as integrating the statistical and dynamical analyses of eddies by using Reynolds stress transport models in large-eddy simulations. At the same time, new hypotheses and general questions still have to be rigorously and competitively examined, such as those that emerged this year about the transition of very high Reynolds number eddy structure and the mathematical properties of singularities. Clearly this field of research is flourishing!

Acknowledgements

The Isaac Newton Institute Programme on Turbulence was generously supported by the following organizations: the Engineering and Physical Sciences Research Council (EPSRC), Royal Academy of Engineering, Rolls Royce, BAE SYSTEMS, DERA, Met Office, British Gas Technology and British Energy. Some workshops were sponsored by ERCOFTAC. Associated Companies were BNFL Magnox Generation, Schlumberger, Ove Arup, the major commercial code vendors, FLUENT Ltd., AEA Technology Ltd., CHAM Ltd. and Computational Dynamics Ltd. We are grateful for individual support as follows: J. C. R. H. to Trinity College, and Delft University of Technology; J. C. V. to the Royal Society. We are particularly grateful to Teresa Cronin without whom this paper would not have seen the light of day because she typed and retyped endless versions of it and carried out most of the reference search seemingly tirelessly.

References

Adrian, R. H. & Moser, R. 2000 Report of a workshop on the development of large eddy simulation. www.uiuc.edu/People/Adrian.html.

André, J. C., De Moor, G., Lacarrère, P. & Du Vachat, R. 1976 Turbulence approximations for inhomogeneous flows. *J. Atmos. Sci.* **33**, 476–481.

Angilella, J. R. & Vassilicos, J. C. 1999 Time-dependent geometry and energy distribution in a spiral vortex layer. *Phys. Rev.* E. **59**, 5427–5439.

Arneodo, A., Manneville, S., Muzy, J. F. & Roux, S. G. 1999 Revealing a lognormal cascading process in turbulent velocity statistics with wavelet analysis. *Phil. Trans. R. Soc. Lond.* A. **357**, 2415–2438.

Atkin, C., Monkewitz, P. & Sandham, N. 1999 Breakdown to turbulence and its control – introduction. Talk given at the Isaac Newton Institute.

Baht, G. S. & Narasimha, R. 1996 A volumetrically heated jet: large-eddy structure and entrainment characteristics. *J. Fluid Mech.* **325**, 303–330.

Banerjee, S. 1999 Closure strategies for modelling turbulent and transitional flows: turbulence near density interfaces – 2. Talk given at the Isaac Newton Institute.

Barenblatt, G. I. & Chorin, A. J. 1998 New perspectives in turbulence: scaling laws, asymptotics, and intermittency. *SIAM Rev.* **40**, 265–291.

1953 *The Theory of Homogeneous Turbulence*. Cambridge University Press.

Bayly, B. J. 1986 Three dimensional instabilities of elliptical flow. *Phys. Rev. Lett.* **57**, 2160–2163.

Belcher, S. E. & Vassilicos, J. C. 1997 Breaking waves and the equilibrium range of wind-wave spectra. *J. Fluid Mech.* **342**, 377–401.

Bertoglio, J. P. 1999 Models based on 2-point closures for inhomogeneous turbulence. Talk given at the Isaac Newton Institute.

Betchov, R. 1956 An inequality concerning the production of vorticity in isotropic turbulence. *J. Fluid Mech.* **1**, 497–504.

Bisset, D. K., Hunt, J. C. R., Cai, X. & Rogers, M. M. 1998 Interfaces at the outer boundaries of turbulent motions. *Annual Research Briefs*, pp. 125–135, Center for Turbulence Research, Stanford University.

Bonnet, J.-P. 1999 Compressibility effects in free shear flows. Talk given at the Isaac Newton Institute.

Bonnet, J. P. & Glauser, M. N. (eds.) 1993 *Eddy Structure Identification in Free Turbulent Shear Flows*. Kluwer.

Borée, J., Marc, D., Bazile, R. & Lecordier, B. 1999 On the behaviour of a large scale tumbling vortex flow submitted to a compression. *European Series in Applied and Industrial Mathematics*; Vol. 7, www.emath.fr/Maths/Proc. ESAIM Proceedings.

Bougeault, P. & Lacarrère, P. 1989 Parameterisation of orography-induced turbulence in a meso-beta scale model. *Mon. Wea. Rev.* **177**, 1872–1890.

Brasseur, J. 1999 Physical-scale-space dynamics of intermittent Burgers shocklets. Talk given at the Isaac Newton Institute.

Brethouwer, G., Hunt, J. C. R. & Nieuwstadt, F. 2003 Microstructure and Lagrangian statistics of the scalar field with a mean gradient in isotropic turbulence. *J. Fluid Mech.* **474**, 193–225.

Brown, G. L. & Thomas, A. S. W. 1977 Large structure in a turbulent boundary layer. *Phys. Fluids Suppl.* **20**, S243.

Cambon, C. 1999 Turbulence structure and vortex dynamics: stability of vortex structures in a rotating frame. Talk given at the Isaac Newton Institute.

Cambon, C. & Scott, J. F. 1999 Linear and nonlinear models of anisotropic turbulence. *Ann. Rev. Fluid Mech.* **31**, 1–53.

Carpenter, P. W. 1999 The use of wall compliance to control transition and turbulence. Talk given at the Isaac Newton Institute.

Castaing, B., Gunaratne, G., Heslot, F., Kadanoff, L., Libchaber, A., Thomas, S., Wu, X., Zaleski, S. & Zanetti, G. 1989 Scaling of hard thermal turbulence in Rayleigh–Bénard convection. *J. Fluid Mech.* **204**, 1–30.

Cazalbou, J. B., Spalart, P. R. & Bradshaw, P. 1994 On the behaviour of two-equation models at the edge of a turbulent region. *Phys. Fluids* **6**, 1797–1804.

Chatwin, P. C. 1999 Perspective in the understanding of turbulent systems: can turbulence researchers learn from probability theory? Talk given at the Isaac Newton Institute.

Chertkov, M., Pumir, A. & Shraiman, B. I. 2000 Statistical geometry and Lagrangian dynamics in turbulence. In *Intermittency in Turbulent Flows* (ed. J. C. Vassilicos). Cambridge University Press.

Ciliberto, S., Leveque, E. & Ruiz Chavaria, G. 2000 Nonhomogeneous scalings in boundary layer turbulence. In *Intermittency in Turbulent Flows* (ed. J. C. Vassilicos). Cambridge University Press.

Constantin, P. 2000 Navier–Stokes equations and fluid turbulence. *Proc. Intl. Congr. of Industrial & Applied Maths* (ed. J. Ball & J. C. R. Hunt). Clarendon.

Couder, Y. 1999 The interaction between stretching and rotation: from the laminar to the turbulent situations. Talk given at the Isaac Newton Institute.

Craft, T. J. 1999 Closure strategies for modelling turbulent and transitional flows; closure modelling near two-component limit: recent developments in second moment closure for buoyancy affected flows. Talk given at the Isaac Newton Institute.

Craft, T. J. and Launder, B. E. 1996 A Reynolds stress closure designed for complex geometries. *Int. J. Heat Fluid Flow* **17**, 245–254.

Craft, T. J., Ince, N. & Launder, B. E. 1996 Recent developments in second-moment closure for buoyancy-affected flows. *Dyn. Atmos. Oceans* **23**, 99–114.

Craft, T. J., Launder, B. E. & Suga, K. 1997 The prediction of turbulent transitional phenomena with a non-linear eddy viscosity model. *Int. J. Heat Fluid Flow* **18**, 15–28.

Davenport, A. G. 1961 The spectrum of horizontal gustiness near the ground in high winds. *Q. J. Roy. Metl. Soc.* **LXXXVII**, 194–211.

Davidson, P. A. 2004 *Turbulence.* Oxford University Press.

Derbyshire, S., Thomson, D. J. & Woods, N. 1999 Turbulent systems: problems and opportunities: atmospheric/environmental flows. Talk given at the Isaac Newton Institute.

Devenport, W. 1999 Characteristic eddy decompositions in real and model turbulent wakes. Talk given at the Isaac Newton Institute.

Doering, C. R. & Constantin, P. 1998 Bounds for heat transport in a porous layer. *J. Fluid Mech.* **376**, 263–296.

Doering, C. R. & Gibbon, J. D. 2000 Scale separation and regularity of the Navier–Stokes equations. In *Intermittency in Turbulent Flows* (ed. J. C. Vassilicos). Cambridge University Press.

Douady, S., Couder, Y. & Brachet, M. E. 1991 Direct observation of the intermittency of intense vorticity filaments in turbulence. *Phys. Rev. Lett.* **67**, 983–986.

Dritschel, D. G. 1993 Vortex properties of two-dimensional turbulence. *Phys. Fluids* **5**, 984–997.

Durbin, P. 1999 Mathematics of closure bifurcation of equilibria in second-moment closure. Talk given at the Isaac Newton Institute.

Elliott, F. & Majda, A. J. 1996 Pair dispersion over an inertial range spanning many decades. *Phys. Fluids* **8**, 1052–1060.

Eyink, G. 1999 Intermittency in turbulent flows and other dynamical systems: intermittency in turbulence: what is it and does it matter? Talk given at the Isaac Newton Institute.

Ferré, J. A., Mumford, J. C., Savill, A. M. & Giralt, F. 1990 Three-dimensional large-eddy motions and fine-scale activity in a plane turbulent wake. *J. Fluid Mech.* **210**, 371–414.

Ffowcs Williams, J. E., Roseblat, S. & Stuart, J. T. 1969 Transitions from laminar to turbulent flow. *J. Fluid Mech.* **39**, 547–559.

Flohr, P. & Vassilicos, J. C. 1997 Accelerated scalar dissipation in a vortex. *J. Fluid Mech.* **348**, 295–317.

Frisch, U. 1995 *Turbulence; the Legacy of A. N. Kolmogorov.* Cambridge University Press.

Fuehrer, P. L. & Friehe, C. A. 1999 A physically-based turbulent velocity time series decomposition. *Boundary-layer Met.* **90**, 241–295.

Fung, J. C. H., Hunt, J. C. R., Malik, N. A. & Perkins, R. J. 1992 Kinematic simulation of homogeneous turbulent flows generated by unsteady random Fourier modes. *J. Fluid Mech.* **236**, 281–318.

Gartshore, I. S. 1966 An experimental examination of the large-eddy equilibrium hypothesis. *J. Fluid Mech.* **24**, 89–98.

Gatski, T. B. 1999 Closure strategies for modelling turbulent and transitional flows: linear/non-linear eddy viscosity models. Talk given at the Isaac Newton Institute.

Gawedzki. 1999 Turbulent advection and breakdown of the Lagrangian flow. Talk given at the Isaac Newton Institute.

George, W. K. 1999 Perspective in the understanding of turbulent systems: similarity and turbulent solutions to the Navier–Stokes equations; perspectives in the understanding of turbulent systems: some new ideas for wall-bounded flows. Talks given at the Isaac Newton Institute.

George, W. K. & Castillo, L. 1997 Zero-pressure-gradient turbulent boundary layer. *App. Mech. Rev.* **50** (12), 689–729.

Geurts, B. J. 1999 Direct and large-eddy simulation: balancing errors in LES. Talk given at the Isaac Newton Institute.

Geurts, B. J. & Leonard, A. 1999 Is LES ready for complex flows? *Rep.* NI99009-TRB, Isaac Newton Institute, Cambridge, UK (also in *Closure Strategies for Transitional and Turbulent Flows* (ed. B. E. Launder & N. D. Sandham), Cambridge University Press, 2001).

Gibbon, J. D. 1999 Another look at Lundgren's transformation and stretched vortex solutions of 3D Navier–Stokes and Euler; turbulence structure and vortex dynamics vorticity alignment in the three-dimensional Euler and Navier–Stokes equations. Talks given at the Isaac Newton Institute.

Gibbon, J. D., Fokas, A. & Doering, C. R. 1999 Dynamically stretched vortices as solutions of the Navier–Stokes equations. *Physica* D **132**, 497.

Gledzer, Y. B., Dolzhansky, F. W., Obukhov, A. M. & Pomonanev, V. M. 1975 An experimental and theoretical study of the stability of motion in an elliptical cylinder. *Izv. Atm. Oc. Phys.* **11**, 617–622.

Gyr, A., Kinzelbach, W. & Tsinober, A. (eds.) 1999 *Fundamental Problematic Issues in Turbulence.* Birkhauser.

Hanazaki, H. 1999 Turbulence structure and vortex dynamics: linear processes in unsteady stably stratified turbulence with mean shear. Talk given at the Isaac Newton Institute.

Hanjalić, K. 1999 Closure strategies for modelling turbulent and transitions flows: modelling double diffusion phenomena; VLES approaches; mathematics of closure expanding the frontiers of single-point RANS application: some recent experience at

TU Delft (transition, multi-scales and VLES). Talks given at the Isaac Newton Institute.

Hanjalić, K., Launder, B. E. & Schiestel, R. 1980 Multiple time-scale concepts in turbulent transport modelling. In *Turbulent Shear Flows 2* (ed. K. Hanjalić and B. E. Launder), pp. 36–49. Springer.

Hatakeyama, N. & Kambe, T. 1992 Statistical laws of random strains and vortices in turbulence. *Phys. Lett. Rev.* **79**, 1257–1260.

Healey, J. 1999 Breakdown to turbulence and its control using asymptotic methods to make practical nonparallel stability calculations. Talk given at the Isaac Newton Institute.

Herring, J. 1999 Problems and issues in 2-D turbulence. Talk given at the Isaac Newton Institute.

Hewitt, G. F. 1999 Computational modelling of multidimensional phenomena in two-phase flow. Talk given at the Isaac Newton Institute.

Hills, D. & Gould, A. 1999 The INI programme on turbulence: the British Aerospace view. Talk given at the Isaac Newton Institute.

Hodson, H. 1999 Breakdown to turbulence and its control: late stage transition on the low pressure turbine. Talk given at the Isaac Newton Institute.

Högström, U. 1990 Analysis of turbulence structure in the surface layer with a modified similarity formulation for near neutral conditions. *J. Atmos. Sci.* **47**, 1949–1972.

Holm, D. 1999 The Navier–Stokes alpha model for fluid turbulence: subgrid scale modelling of fluctuation effects in the equations of motion. Talk given at the Isaac Newton Institute.

Holmes, P., Lumley, J. L. & Berkooz, G. 1996 *Turbulence, Coherent Structures, Dynamical Systems and Symmetry*. Cambridge University Press.

Hopfinger, E. J., Browand, F. K. & Gagne, Y. 1982 Turbulence and waves in a rotating tank. *J. Fluid Mech.* **125**, 505–534.

Hoxey, R. P. & Richards, P. J. 1992 Spectral characteristics of the atmospheric boundary layer near the ground. *First UK Wind Engineering Conference, Engineering Department, University of Cambridge, UK, September.*

Huerre, P. & Monkewitz, P. A. 1990 Local and global instabilities in spatially developing flows. *Ann. Rev. Fluid Mech.* **22**, 473–537.

Hunt, J. C. R. 1995 Practical and fundamental developments in the computational modelling of fluid flows. T. Hawksley memorial lecture at the Inst. Mech. Eng., 13th December 1994. *J. Mech. Eng. Sci.* C **209**, 297–314.

1999a Environmental forecasting and turbulence prediction. *Physica* D **133**, 270–295.

1999b Vortex dynamics and statistics of eddies in turbulence. Talk given at the Isaac Newton Institute.

Hunt, J. C. R. & Durbin, P. 1999 Perturbed vortical layers and shear sheltering. *Fluid Dyn. Res.* **24**, 375–404.

Hunt, J. C. R. & Morrison, J. F. 2000 Eddy structure in turbulent boundary layers. *Eur. J. Mech. B/Fluids* **19**, 673–694.

Hunt, J. C. R. & Savill, A. M. 2005 Guidelines and criteria for the use of turbulence models in complex flows. Chapter 8 *Prediction of Turbulent Flows*, pp. 291–343. Cambridge University Press.

Hunt, J. C. R. & Vassilicos, J. C. 2000 *Turbulence Structure and Vortex Dynamics*. Cambridge University Press.

Hunt, J. C. R., Kaimal, J. C. & Gaynor, J. E. 1988 Eddy structure in the convective boundary layer – new measurements and new concepts. *Q. J. R. Met. Soc.* **114**, 821–858.

Hunt, J. C. R., Moin, P., Lee, M., Moser, R. D., Spalart, P., Mansour, N. N., Kaimal, J. C. & Gaynor, E. 1989 Cross correlation and length scales in turbulent flows near surfaces. *Advances in Turbulence 2* (ed. H. Feidler & H. Fernholtz), pp. 128–134. Springer.

Hunt, J. C. R., Sandham, N. D., Vassilicos, J. C., Launder, B. E., Monkewitz, P. A. & Hewitt, G. F. 2001 Development in turbulence research: a review based on the 1999 Programme of the Isaac Newton Institute, Cambridge, *J. Fluid Mech.* **436**, 353–391.

Hussain, F. 1986 Coherent structures and turbulence. *J. Fluid Mech.* **173**, 303–356.

Ilyushin, B. 1999 Closure strategies for modelling turbulent and transitional flows; use of higher moments to construct pdf's in stratified flows. Talk given at the Isaac Newton Institute.

Jiménez, J. 1999 Limits of performance of eddy viscosity subgrid models. Talk given at the Isaac Newton Institute.

Jiménez, J., Wray, A. A., Saffman, P. G. & Rogallo, R. S. 1993 The structure of intense vorticity in isotropic turbulence. *J. Fluid Mech.* **255**, 65–90.

Kaneda, Y. 1993 Lagrangian and Eulerian time correlations in turbulence. *Phys. Fluids* A **5**, 2835–2845.

1999 Spectral closure approach – (non)-universality of the equilibrium range of turbulence in 2D/3D. Talk given at the Isaac Newton Institute.

Kassinos, S. C., Reynolds, W. C. & Rogers, M. M. 2001 One-point turbulence structure tensors. *J. Fluid Mech.* **428**, 213–248.

Kellog, R. M. & Corrsin, S. 1980 Evolution of a spectrally local disturbance in grid-generated, nearly isotropic turbulence. *J. Fluid Mech.* **96**, 641–669.

Kerr, R. M. 1999 Turbulence structure and vortex dynamics: applications of vortex dynamics to MHD current dynamics; direct and large eddy simulation: dynamic backscatter tests of decaying isotropic turbulence. Talks given at the Isaac Newton Institute.

Kerr, O. S. & Dold, J. W. 1994 Periodic steady vortices in a stagnation-point flow. *J. Fluid Mech.* **276**, 307–325.

Kerswell, R. R. 1999 A variational principle for the Navier–Stokes equations. *Phys. Rev.* E **59**, 5482–5494.

Kevlahan, N. K. R. & Hunt, J. C. R. 1997 Nonlinear interactions in turbulence with strong irrotational straining. *J. Fluid Mech.* **337**, 333–364.

Kholmyansky, M. & Tsinober, A. 2000 On the origins of intermittency in real turbulent flows. Talk given at the Isaac Newton Institute.

Kida, S. 1998 *Turbulence* (in Japanese).

Kida, S. & Tanaka, M. 1994 Dynamics of vortical structures in a homogeneous shear flow. *J. Fluid Mech.* **274**, 43–68.

Kida, S., Miura, H. & Adachi, T. 2000 Flow structure visualization by low-pressure vortex. In *Intermittency in Turbulent Flows* (ed. J. C. Vassilicos). Cambridge University Press.

Kim, K. C. & Adrian, R. J. 1999 Very large-scale motion in the outer layer. *Phys. Fluids* **11**, 417–422.

Kit, E. L. G., Strang, E. J. & Fernando, H. J. S. 1997 Measurement of turbulence near shear-free density interfaces. *J. Fluid Mech.* **334**, 219–234.

Kiya, M., Ohyama, M. & Hunt, J. C. R. 1986 Vortex pairs and rings interacting with shear layer vortices. *J. Fluid Mech.* **172**, 1–15.

Kraichnan, R. H. 1959 The structure of isotropic turbulence at very high Reynolds numbers. *J. Fluid Mech.* **5**, 497–543.

Kreiss, H.-O. 1999 A numerical investigation of the interaction between the large and small scales of the two-dimensional incompressible Navier–Stokes equations. Talk given at the Isaac Newton Institute.

Landau, L. E. & Lifshitz, L. M. 1959 *Fluid Mechanics.* Pergamon.

Launder, B. E. 1999 Application of TCL modelling to stably stratified flows. Talk given at the Isaac Newton Institute.

Launder, B. E. & Li, S. P. 1994 On the elimination of wall topography parameters from second-moment closure. *Phys. Fluids* **6**, 999–1006.

Launder, B. E. & Sandham, N. D. 2001 *Closure Strategies for Transition and Turbulent Flows.* Cambridge University Press.

Launder, B. E. & Spalding, B. 1972 *Mathematical Models of Turbulence.* Academic Press.

Laurence, D. 1999 Closure strategies for modelling turbulent and transitional flows: closure modelling for industrial flows; LES modelling of industrial flows. Talks given at the Isaac Newton Institute.

Le Dizes, S. 1999 Perspectives in the understanding of turbulent systems: stability of vortex filaments in strain fields. Talk given at the Isaac Newton Institute.

Leonard, A. 1999 Turbulence structure and vortex dynamics: evolution of localized packets of vorticity in turbulence; direct and large-eddy simulation: a tensor-diffusivity sub-grid model for large-eddy simulation. Talks given at the Isaac Newton Institute.

Leray, J. 1933 Etude de diverses équations intégrales non-linéaires et de quelques problèmes que pose l'hydrodynamique. *J. Math. Pure Appl.* **12**, 1–82.

Lesieur, M. 1990 *Turbulence in Fluids*, 2nd edn. Kluwer.
   1999 *Turbulence et Determinisme.* Press Université de Grenoble.

Leszchiner, M. 1999 Turbulence modelling. Talk given at the Isaac Newton Institute.

Leweke, T. & Williamson, C. H. K. 1998 Cooperative elliptic instability of a vortex pair. *J. Fluid Mech.* **360**, 85–119.

Lifschitz, A. & Hameiri, E. 1991 Local stability conditions in fluid dynamics. *Phys. Fluids* A **3**, 2644–2651.

Lingwood, R. 1999 Breakdown to turbulence and its control: swept-wing boundary layers: convectively or absolutely unstable? Talk given at the Isaac Newton Institute.

Loitsyanskii, L. G. 1939 Some basic laws for isotropic turbulent flow. *Trudy Tsentr. Aero-Giedrodin Inst.* **440**, 3–23.

Lorenz, E. N. 1963 Deterministic nonperiodic flow. *J. Atmos. Sci.* **20**, 130–141.

Lucchini, P. 1999 Transition prediction methods. Receptivity to low-frequency disturbances. Talk given at the Isaac Newton Institute.

Lumley, J. L. 1978 Computational modelling of turbulent flows. *Adv. Appl. Mech. Rev.* **18**, 124–176.
   1999 Prediction methods in turbulence. Talk given at the Isaac Newton Institute.

Lundgren, T. S. 1982 Strained spiral vortex model for turbulent fine structure. *Phys. Fluids* **25**, 2193–2203.
   1999 Perspectives in the understanding of turbulent systems: review of the spiral vortex turbulence model. Talk given at the Isaac Newton Institute.

Lundgren, T. S. & Mansour, N. N. 1997 Transition to turbulence in an elliptical vortex. *J. Fluid Mech.* **307**, 43–62.

Malik, N. A. & Vassilicos, J. C. 1996 Eulerian and Lagrangian scaling properties of randomly advected vortex tubes. *J. Fluid Mech.* **326**, 417–436.

Malkus, W. V. R. & Waleffe, F. 1991 Transition from order to disorder in elliptical flow: a direct path to shear flow turbulence. In *Advances in Turbulence* **3** (ed. A. V. Johansson & P. H. Alfredsson). Springer.

Mann, J. 1994 The spatial structure of neutral atmospheric surface-layer turbulence. *J. Fluid Mech.* **273**, 141–168.

Marušić, I. & Perry, A. E. 1995 A wall-wake model for the turbulence structure of boundary layers. Part 2. Further experimental support. *J. Fluid Mech.* **298**, 389–407.

Mason, P. J. & Callen, N. S. 1986 On the magnitude of the subgrid-scale eddy coefficient in large-eddy simulations of turbulent channel flow. *J. Fluid Mech.* **162**, 439–462.

Mason, P. J. & Thomson, D. J. 1992 Stochastic backscatter in large-eddy simulations of boundary layers. *J. Fluid Mech.* **242**, 51–78.

McComb, W. D. 1990 *Theory of Fluid Turbulence.* Clarendon.

1999 Can microscopic RG theory be applied to the large-eddy simulation of macroscopic fluid turbulence?; testing the 'hypothesis of local chaos' by calculating a renormalized dissipation rate for isotropic turbulence; renormalized perturbation theory: has the turbulence problem been solved? Talks given at the Isaac Newton Institute.

McGrath, J. L., Fernando, H. J. S. & Hunt, J. C. R. 1997 Turbulence, waves and mixing at shear-free density interfaces. Part 2. Laboratory experiments. *J. Fluid Mech.* **347**, 235–261.

Melander, M. V. & Hussain, F. 1993 Coupling between a coherent structure and fine-scale turbulence. *Phys. Rev. E.* **48**, 2669–2689.

Methven, J. & Hoskins, B. J. 1997 Spirals in potential vorticity. Part I: measures of structure. *J. Atmos. Sci.* **55**, 2053–2066.

Miyazaki, T. & Hunt, J. C. R. 2000 Turbulence structure around a columnar vortex; rapid distortion theory and vortex wave excitation. *J. Fluid Mech.* **402**, 349–378.

Moffatt, H. K. 1984 Simple topological aspects of turbulent vorticity dynamics. In *Turbulence and Chaotic Phenomena in Fluids* (ed. T. Tatsumi). Elsevier.

2004 The Burgers vortex: variations on a turbulent theme. Talk given at the Isaac Newton Institute.

Moin, P. 1999 Some future challenges in large eddy simulation. Talk given at the Isaac Newton Institute.

Moin, P. & Mahesh, K. 1999 Direct numerical simulation: a tool in turbulence research. *Ann. Rev. Fluid Mech.* **30**, 539–578.

Monin, A. S. & Yaglom, A. M. 1975 *Statistical Fluid Mechanics.* MIT Press.

Moser, R. D., Rogers, M. M. & Ewing, D. W. 1998 Self-similarity of time-evolving plane wakes. *J. Fluid Mech.* **367**, 255–289.

Mydlarski, L. & Warhaft, Z. 1998 Passive scalar statistics in high Péclet-number grid turbulence. *J. Fluid Mech.* **358**, 135–175.

Nazarenko, S., Kevlahan, N. K.-R. & Dubrulle, B. 1999 WKB theory for rapid distortion of inhomogeneous turbulence. *J. Fluid Mech.* **390**, 325–348.

Nicolleau, F. & Vassilicos, J. C. 1999 Wavelets for the study of intermittency and its topology. *Phil. Trans. R. Soc. Lond.* A **357**, 2439–2457.

Nie, Q. & Tanveer, S. 1999 A note on third-order structure functions in turbulence. *Proc. R. Soc. Lond.* A **455**, 1615–1635.

Nieuwstadt, F. 1999 Future strategies towards understanding and prediction of turbulent systems: a 200 bar wind tunnel for turbulence measurements at very high Reynolds numbers. Talk given at the Isaac Newton Institute.

Ohkitani, K. 1998 Stretching of vorticity and passive vectors in isotropic turbulence. *J. Phys. Soc. Japan* **67**, 44–47.

1999 Perspective in the understanding of turbulent systems: characterization of nonlocality in turbulence. Informal discussions on triad interactions in turbulence; triad interactions in forced turbulence. Talks given at the Isaac Newton Institute.

Ohkitani, K. & Gibbon, J. D. 2000 Numerical study of singularity formation in a class of Euler and Navier–Stokes flows. *Phys. Fluids* **12**, 3181–3194.

Ooi, A., Martin, J., Soria, J. & Chong, M. S. 1999 A study of the evolution and characteristics of the invariants of the velocity-gradient tensor in isotropic turbulence. *J. Fluid Mech.* **381**, 141–174.

Passot, T., Politano, H., Sulem, P. L., Angillela, J. R. & Meneguzzi, M. 1995 Instability of strained vortex layers and vortex tube formation in homogeneous turbulence. *J. Fluid Mech.* **282**, 313–338.

Perot, B. & Moin, P. 1995a Shear-free turbulent boundary layers. Part 1. Physical insights into near wall turbulence. *J. Fluid Mech.* **295**, 199–227.

1995b Shear-free turbulent boundary layers. Part 2. New concepts for Reynolds stress modelling of inhomogeneous flows. *J. Fluid Mech.* **295**, 229–245.

Perry, A. E. 1999 Breakdown to turbulence and its control: turbulent spots; mathematics of closure: closure for the streamwise evolution of turbulent boundary layers using classical similarity laws and structure based modelling. Talks given at the Isaac Newton Institute.

Plate, E. J., Federovich, E. E., Viega, D. X. & Wyngaard, J. C. (eds.) 1998 *Buoyant Convection in Geophysical Flows.* Kluwer.

Pope S. B. 2000 *Turbulent Flows.* Cambridge University Press.

Pradeep, D. J. & Hussain, F. 2002 Core dynamics of a structured vertex: instability and transition, *J. Fluid Mech.* **447**, 247–285.

Prandtl, L. 1925 Bericht über Untersuchung zur Ausgebildeten Turbulenz. *Z. Angew. Math. Mech.* **5**, 136–139.

Praskovsky, A. A., Gledzer, E. B., Karyakin, M. Yu. & Zhou, Y. 1993 The sweeping decorrelation hypothesis and energy-inertial scale interaction in high Reynolds number flows. *J. Fluid Mech.* **248**, 493–511.

Procaccia, I. 1999 Perspectives in the understanding of turbulent systems: the calculation of the anomalous scaling exponents in turbulence from first principles. Talk given at the Isaac Newton Institute.

Ptasinski, P. K., Boersma, B. J., Nieuwstadt, F. T. M., Hulsen, M. A., Van den Brule, B. H. A. A. & Hunt, J. C. R 2003 Turbulent Channel flows near maximum drug reduction; stimulation, experiments and mechanisms, *J. Fluid Mech.* **490**, 251–291.

Pullin, D. I. & Saffman, P. G. 1998 Vortex dynamics in turbulence. *Ann. Rev. Fluid Mech.* **30**, 31–51.

Pumir, A. 1999 Geometry of Lagrangian dispersion in turbulence. Talk given at the Isaac Newton Institute.

Pumir, A. & Siggia, E. D. 1990 Collapsing solutions in the 3D Euler equations. *Phys. Fluids* A **2**, 220–241.

Queiros-Conde, D. & Vassilicos, J. C. 2000 Turbulent wakes of 3-D fractal grids. In *Intermittency in Turbulent Flows* (ed. J. C. Vassilicos). Cambridge University Press.

Reeks, M. 1999 Closure approximations for dispersed particle flows. Talk given at the Isaac Newton Institute.

Richardson, L. F. 1926 Atmospheric diffusion shown on a distance-neighbour graph. *Proc. R. Soc. Lond.* A. **110**, 709–737.

Rodi, W. 1999 Closure strategies for modelling turbulent and transitional flows: large-eddy simulation of bluff body flows. Talk given at the Isaac Newton Institute.

Rotta, J. 1951 Statistische Theorie nichthomogeneous Turbulenz. *Z. Phys.* **129**, 547.

Saddoughi, S. G. & Veeravalli, S. V. 1994 Local isotropy in turbulent boundary layers at high Reynolds number. *J. Fluid Mech.* **268**, 333–372.

Saffman, P. G. 1967 The large scale structure of homogeneous turbulence. *J. Fluid Mech.* **27**, 581–589.

Sandham, N. D. 2005 Turbulence simulation. Chapter 6 *Prediction of Turbulent Flows*, pp. 207–235. Cambridge University Press.

   1999 Perspectives in the understanding of turbulent systems: exploiting direct numerical simulation of turbulence. Closure strategies for modelling turbulent and transitional flows: DNS studies of compressible flows. Talks given at the Isaac Newton Institute.

Savill, A. M. 1999 Breakdown to turbulence and its control, and Closure strategies for modelling turbulent and transitional flows: Talks given at the Isaac Newton Institute.

Schiestel, R. 1987 Multiple time-scale modeling of turbulent flows in one-point closures. *Phys. Fluids* **30**, 722–731.

Schoppa, W. & Hussain, F. 2002 Coherent structure generation in near wall turbulence. *J. Fluid Mech.* **453**, 57–108.

Schwarz, K. W. 1990 Evidence for organised small scale structure in fully developed turbulence. *Phys. Rev. Lett.* **64**, 415–418.

Shih, T.-H., Lumley, J. L. & Janicka, J. 1987 Second-order modelling of a variable-density mixing layer. *J. Fluid Mech.* **180**, 93–116.

Silverman, B. W. & Vassilicos, J. C. (eds.) 1999 Proceedings of the Royal Society Discussion Meeting on Wavelets: the key to intermittent information? *Phil. Trans. R. Soc. Lond.* A **357**, 2393–2625.

Smith, F. 1999 On nonlinear spots in favourable or adverse pressure gradients. Talk given at the Isaac Newton Institute.

Sreenivasan, K. R. 1999 On structures and statistics in small scale turbulence; polymers' effects on turbulence. Talk given at the Isaac Newton Institute.

Sulem, P. L., She, Z. S., Scholl, H. & Frisch, U. 1989 Generation of large-scale structures in three-dimensional flow lacking parity-invariance. *J. Fluid Mech.* **205**, 341–358.

Tabeling, P. 1999 Perspectives in the understanding of turbulent systems: cascades in 2-D turbulence; new scales in three dimensional turbulence. Talks given at the Isaac Newton Institute.

Taylor, G. I. 1938 Production and dissipation of vorticity in a turbulent fluid. *Proc. Roy. Soc. Lond.* A **164**, 15–32.

Tennekes, H. & Lumley, J. L. 1971 *First Course in Turbulence.* MIT Press.

Thomson, D. J. 1999 Understanding and predicting many environmental flows requires consideration of turbulence. Talk given at the Isaac Newton Institute.

Townsend, A. A. 1976 *Structure of Turbulent Shear Flow.* Cambridge University Press.

Trefethen, L. N., Trefethen, A. E., Reddy, S. C. & Driscoll, T. A. 1993 Hydrodynamic stability without eigenvalues. *Science* **261**, 578–584.

Tsinober, A. 2000 Vortex stretching versus production of strain/dissipation. In *Vortex Dynamics and Turbulence Structure* (ed. J. C. R. Hunt & J. C. Vassilicos). Cambridge University Press.

Turner, J. S. 1986 Turbulent entrainment: the development of the entrainment assumption, and its application to geophysical flows. *J. Fluid Mech.* **173**, 431–471.

Van Atta, C. 1991 Local isotropy of the smallest scales of turbulent scalar and velocity fields. *Proc. R. Soc. Lond.* A **434**, 139–147.

Vassilicos, J. C. 1999 Turbulence structure and vortex dynamics. Dynamics of near-singular vortex; informal discussion on triad interactions in turbulence: interscale energy transfer in a compact vortex; intermittency in turbulent flows and other dynamical systems: wind tunnel turbulence and intermittency behind a 3-D fractal grid; future strategies towards understanding and prediction of turbulent systems: intermittency. Talks given at the Isaac Newton Institute.

   2000a Near-singular flow structure: dissipation and deduction. In *Vortex Dynamics and Turbulence Structure* (ed. J. C. R. Hunt & J. C. Vassilicos). Cambridge University Press.

2000b (ed.) *Intermittency in Turbulent Flows.* Cambridge University Press.

Verzicco, R. & Jiménez, J. 1999 On the survival of strong vortex filaments in 'model' turbulence. *J. Fluid Mech.* **394**, 261–279.

Voke, P. 1999 LES of bypass and separated flow transition. Talk given at the Isaac Newton Institute.

Voke, P., Sandham, N. D. & Kleiser, L. (eds.) 1999 *Direct and Large-Eddy Simulation III, Proc. Isaac Newton Institute Symposium/ERCOFTAC Workshop, Cambridge, UK, 12–14 May 1999.* ERCOFTAC Series, volume 7, Kluwer.

von Kármán, Th. 1930 Mechanische Ähnlichkeit und Turbulenz. *Nachr. Ges. Wiss. Göttingen, Math. Phys.* K1, 58–76.

Voth, G. A., Satyanarayan, K. & Bodenschatz, E. 1998 Lagrangian measurements at large Reynolds numbers. *Phys. Fluids* **10**, 2268–2280.

2000 The issue of local isotropy of velocity and scalar turbulent fields. In *Turbulence Structure and Vortex Dynamics* (ed. J. C. R. Hunt & J. C. Vassilicos). Cambridge University Press.

Water, W. Van Der 1999 Angle-dependent structure functions for fully developed turbulence. Talk given at the Isaac Newton Institute.

Wu, X., Jacobs, R. G., Hunt, J. C. R. & Durbin, P. A. 1999 Simulation of boundary layer transition induced by periodically passing wakes. *J. Fluid Mech.* **398**, 109–153.

Wundrow, D. W. & Goldstein, M. E. 2001 Effect on a laminar boundary layer of small-amplitude streamwise vorticity in the upstream flow. *J. Fluid Mech.* **426**, 229–262.

Zagarola, M. V. & Smits, A. J. 1998 Mean-flow scaling of turbulent pipe flow. *J. Fluid Mech.* **373**, 33–79.

3

RANS modelling of turbulent flows affected by buoyancy or stratification

B. E. Launder

University of Manchester

1 Introduction

The aim of this chapter is to provide, in plain English, a guide to the capabilities and shortcomings of turbulence models for reproducing satisfactorily engineering flows where buoyant or stratification effects are important. While these two descriptors are often used interchangeably in the literature, in the present chapter *buoyant* is used to denote a situation where the effect of gravity is to cause a force field whose primary effect is on the **mean** flow, while *stratified* implies that the principal effects on the flow arise from gravitational action on the *turbulent fluctuating velocities*. The distinction is neither pedantic nor unimportant; for, a stratified flow will ordinarily require a more rigorous modelling of gravitational effects than a buoyant flow. Put another way, gravitational effects on horizontal flows are more troublesome than on vertical flows. The account may hopefully also be useful where the flows of interest are affected by other types of force field, perhaps particularly flows affected by Coriolis forces or swirl.

The chapter gives especial attention to two-equation models of turbulence as this is currently the main level of commercial CFD. Linear two-equation eddy-viscosity models are considered first, beginning with the situation where buoyant/stratification effects are absent. This is important to enable the reader to assess whether, for the flows of interest, a linear eddy viscosity model would be suitable even in a uniform density situation. Thereafter, the treatment of buoyant flows with linear two-equation models is considered. While the applicability of this level of modelling is not wide there are certain types of flow where satisfactory agreement can be obtained. The question of the treatment of a wall boundary condition in buoyancy affected flows is also examined.

Prediction of Turbulent Flows, eds. G. F. Hewitt and J. C. Vassilicos.
Published by Cambridge University Press. © Cambridge University Press 2005.

Section 3 goes on to consider the more elaborate path to modelling known as second-moment closure and to the simplifications of those schemes to two-equation forms that are more widely applicable than conventional eddy-viscosity models. Section 4 considers more elaborate treatments whose current use is largely confined to universities but which have been found to give markedly improved predictions over alternative schemes in several buoyant and complex shear flows. A more complete (and more mathematical) treatment of such modelling approaches is provided in another book arising from the INI Conference, Launder & Sandham (2002) (see chapters by Craft & Launder, 2002 and Ilyushin, 2002). Finally, Section 5 provides a brief overview of the main conclusions and recommendations.

2 The k–ε eddy-viscosity model

2.1 Origin and form

The k–ε eddy-viscosity model, as we now understand it, originated from research at Imperial College nearly 30 years ago in the group led by D. B. Spalding (Jones & Launder, 1972, 1973). Although not originally presented in these terms, it is perhaps helpful to arrive at it by considering the budget equation for turbulent kinetic energy, k, which in words may be written:

Rate of increase of $=$	**Rate of creation of** $+$	**Rate of gain of** $-$	ε, **the rate of**
k following a mean	**k by mean shear,**	**k by diffusion**	**viscous**
streamline	**buoyancy and**		**dissipation of k.**
	other mechanisms		

If the only contribution to turbulence energy generation is by mean shear and we have what is known as a **simple shear flow**, $U(y)$, the equation becomes:

$$\frac{Dk}{Dt} = -\overline{uv}\frac{\partial U}{\partial y} + (\text{diffusion}) - \varepsilon. \tag{3.1}$$

Further, if transport effects (convection and diffusion) can be neglected, the flow is referred to as in **local equilibrium** and so we may write:

$$-\overline{uv}\frac{\partial U}{\partial y} = \varepsilon. \tag{3.2}$$

Multiplication of each side of Eq. (3.2) by $(-\overline{uv}/\varepsilon)$ together with the definition:

$$c_\mu \equiv (-\overline{uv})^2/k^2 \tag{3.3}$$

enables Eq. (3.2) to be re-written:

$$c_\mu \frac{k^2}{\varepsilon} \frac{\partial U}{\partial y} = -\overline{uv}. \qquad (3.4a)$$

This is the constitutive equation between stress and strain adopted in two-dimensional eddy-viscosity models to compute thin shear flows where U is the primary velocity component which varies rapidly only in the perpendicular direction, y.

Evidently, the group $c_\mu k^2/\varepsilon$ plays the role of an effective turbulent viscosity ν_t,

$$\nu_t \equiv c_\mu k^2/\varepsilon. \qquad (3.5)$$

For wider use the formula is generalized to:

$$\left(\overline{u_i u_j} - \frac{2}{3}\delta_{ij}k\right) = -\nu_t\left(\frac{\partial U_i}{\partial x_j} + \frac{\partial U_j}{\partial x_i}\right), \qquad (3.4b)$$

though it is underlined that while Eq. (3.4b) is a *mathematically* valid generalization of Eq. (3.4a), it is not unique. That is to say, other more elaborate connections between the anisotropic part of the Reynolds stresses, $(\overline{u_i u_j} - \frac{2}{3}\delta_{ij}k)$ and the mean strain field may be proposed which reduce to Eq. (3.4a) for the specific case of a two-dimensional flow in simple shear.

If an 'eddy-viscosity' transport process is adopted for the turbulence-driven flux of momentum (as is implied by (3.4)), it is consistent to adopt the same form for the diffusional transport of a mean scalar quantity such as temperature or chemical species:

$$-\overline{u_j\theta} = \frac{\nu_t}{\sigma_\theta} \frac{\partial \Theta}{\partial x_j}, \qquad (3.6)$$

where Θ and θ denote the mean and fluctuating value of the scalar in question. The turbulent Prandtl/Schmidt number, σ_θ, is habitually treated as a constant, though, in fact, it is manifestly not. In free shear flows its value is usually taken in the range 0.5–0.7 while for flows near walls a value of approximately 0.9 is usually adopted.

An analogous gradient-diffusion representation is adopted for the diffusional flux of turbulence energy (which has only a minor influence on the mean field, at least if the generation term is substantial). The kinetic energy equation, Eq. (3.1), may then be written:

$$\frac{Dk}{Dt} = \nu_t\left(\frac{\partial U}{\partial y}\right)^2 + \frac{\partial}{\partial y}\left(\frac{\nu_t}{\sigma_k}\frac{\partial k}{\partial y}\right) - \varepsilon, \qquad (3.7a)$$

or, in generalized tensor form:

$$\frac{Dk}{Dt} = \underbrace{\nu_t \frac{\partial U_i}{\partial x_j} \left(\frac{\partial U_i}{\partial x_j} + \frac{\partial U_j}{\partial x_j} \right)}_{P_k} + \frac{\partial}{\partial x_j} \left(\frac{\nu_t}{\sigma_k} \frac{\partial k}{\partial x_j} \right) - \varepsilon. \tag{3.7b}$$

By analogy, the energy dissipation rate, ε, is supposed to be described by an equivalent transport equation:

$$\frac{D\varepsilon}{Dt} = c_{\varepsilon 1} \frac{\varepsilon}{k} P_k + \frac{\partial}{\partial x_j} \left(\frac{\nu_t}{\sigma_\varepsilon} \frac{\partial \varepsilon}{\partial x_j} \right) - c_{\varepsilon 2} \frac{\varepsilon^2}{k}, \tag{3.8}$$

where the coefficients $c_{\varepsilon 1}$, $c_{\varepsilon 2}$ and σ_ε are supposedly constants chosen empirically to give agreement with certain representative one- and two-dimensional flows.

The empirical coefficients are usually assigned the following values, provided the turbulent Reynolds number, $k^2/\nu\varepsilon$, is high (which in practice means greater than about 100):

c_μ	$c_{\varepsilon 1}$	$c_{\varepsilon 2}$	σ_k	σ_ε
0.09	1.44	1.92	1.0	1.3

The above form of the k–ε model will already be familiar to most readers of this book. The re-iteration is intended to serve as a reminder of the many intrinsic assumptions made. These may be usefully repeated:

(i) the flow is in simple shear;
(ii) the state of turbulence is near local equilibrium ($P_k \approx \varepsilon$);
(iii) there is no substantial effect of buoyant or other force fields on the turbulence.

If these criteria are met the only important Reynolds stress, $\overline{uv}$, is well described by Eq. (3.4). However, it is often assumed that:

(iv) The eddy-viscosity concept, as presented in Eq. (3.4b) may be applied to *all* the Reynolds stress components $(\overline{u_i u_j})$ in *arbitrary* shear flows (not just a simple shear) with the turbulent viscosity being determined from the solution of Eqs. (3.7b) and (3.8).

Clearly, this last set of assumptions represents a pretty large pill to swallow!

2.2 Shortcomings of the model – high Reynolds number form

Unsurprisingly, a model based on so many assumptions proves to have a very limited width of applicability. The reason for its widespread use, despite the shortcomings, arises from three features:

(i) In its mathematical form it is free from ambiguities (like distance from the wall).

(ii) It genuinely does offer an improvement on earlier models like the mixing length hypothesis (Prandtl, 1925).

(iii) Potentially more general treatments like second-moment closure have often proved to be disappointing in practice when applied to problems with flow regions of complicated shape, such as conventionally arise in industrial flows.

Here the principal shortcomings of the k–ε model are noted for the case of unstratified flows. (The case of flows significantly affected by buoyancy is taken up in Section 2.4.) Some of the failures of predictions can be identified predominantly with a single underlying assumption; others have multiple contributory causes.

In wakes and in other 'weak' shear flows where P_k is appreciably less than ε, the standard k–ε model underestimates turbulent mixing rates. This defect is sometimes corrected by causing the coefficients c_μ to increase as $\overline{P_k/\varepsilon}$ falls (the overbar on P_k and ε implies an average over the wake cross-section, Rodi, 1972). The practice has little generality, however. We note that simply making c_μ a function of the local ratio (P_k/ε) (as would be desirable in order to retain an invariant model) produces unsatisfactory shapes of wake profiles. This is because the ratio falls to zero at a plane or axis of symmetry leading to a marked increase in c_μ which, in turn, produces a much too flat velocity profile.

Another weakness associated with free shear flows is the so-called 'round jet anomaly'. While the standard k–ε model does reasonably well in predicting the *plane* jet in stagnant surroundings, for the corresponding axisymmetric jet the computed rate of growth with distance downstream is some 30% larger than measured. Pope (1978) has suggested a correction to the dissipation equation that introduces a mean strain tensor that vanishes in plane flows. The extra process is based on plausible physical arguments. However, his scheme works satisfactorily only for the round jet in stagnant surroundings: for a round jet in a moving stream (as measured, for example, by Forstall & Shapiro, 1950), Huang (1986) has shown that the standard model gives more satisfactory results.

It is pertinent to note that the same type of shortcoming in predicting these two types of flow (the round jet and the far wake) is also present with the most common form of second-moment closure. This is partly because this latter scheme, like the k–ε model, uses a transport equation for ε in which the local mean velocity gradient plays too prominent a role (or, in other words, the coefficient $c_{\varepsilon 1}$ is too large). Only with the more recent and more elaborate *TCL second-moment closure*, to be discussed later, are both these weaknesses satisfactorily removed.

Another weakness with the standard form of ε equation occurs in flow near walls. It is found that as the flow encounters a region of rising pressure (an 'adverse' pressure gradient) the ε equation leads to too low levels of ε near the wall, thus to too high viscosities, cf. Eq. (3.4), and, consequently, too high skin friction. As a result, a flow that, in fact, separates may be predicted to remain attached – a weakness that is intolerable if one wishes to predict the behaviour of highly loaded aerofoils (where, by design, one seeks to obtain as much lift as possible *without* separation occurring). An early attempt at eliminating both that defect and the round jet anomaly was proposed by Hanjalić & Launder (1980). Essentially, they suggested adding a term proportional to:

$$k \frac{\partial U_i}{\partial x_j} \frac{\partial U_m}{\partial x_n} \varepsilon_{ijk} \varepsilon_{mnk} \qquad (3.9)$$

to the right-hand side of Eq. (3.8). The net effect was to enhance the role of normal-stresses in the dissipation equation. While marked improvements resulted for the indicated flows, it was later discovered that the correction produced the opposite sign of correction in flows with curved streamlines to what was needed. Thus the term clearly does *not* represent a general improvement to the dissipation source. It does seem worth underlining, however, that there is no reason why the action of mean strain in raising ε should be assumed proportional to P_k. What, physically, one is trying to mimic is the role of mean-flow straining in hastening the breakdown of turbulent eddies (or, in other words, in promoting the flow of kinetic energy across the spectrum). The term containing the mean strain must be a scalar, but there are many forms that meet this requirement, of which P_k and Eq. (3.9) are just two. Thus, the writer firmly believes that in the future more diverse mean-strain contributions to the creation of ε need to be explored. In this connection, the analysis of Oberlack (1995) of the tensorial length-scale equation would appear to offer very helpful insights.

Wilcox (1993), whose customer base is principally in the aerospace industry, has found that the problem of too high skin friction as the flow develops in positive pressure gradient can be removed if one adopts as one's dependent variable the quantity $\omega \equiv \varepsilon/k$ instead of ε, using a transport equation for ω of precisely the same form as Eq. (3.8). While, for external flow over aerofoils, this is a beneficial switch, the ω equation gives rise to problems where high external stream turbulence is present. For that reason, other approaches to removing this weakness of the ε equation have been adopted by the writer's group. The first of these is the so-called 'Yap correction' (Yap, 1987; Launder, 1988) in which a further source term S_Y is added to the right-hand side of the ε equation, Eq. (3.8).

$$S_Y = \max\left[0.83\frac{\varepsilon^2}{k}(\alpha - 1)\alpha^2, 0\right], \qquad (3.10a)$$

where $\alpha \equiv k^{3/2}/(2.55\, y\varepsilon)$ and y denotes distance from the wall. The quantity α is the ratio of the computed turbulent length scale ($k^{3/2}/\varepsilon$) to the length scale that is found to occur in local-equilibrium, near-wall turbulence, 2.55 y. If the local value of ε is so small that α is greater than unity, the term will be a source (since $(\alpha - 1)$ will be positive), thus tending to increase ε. Besides adverse pressure gradients, this additional source has been found to be beneficial in separated flows and buoyant flows where most of the generation occurs away from the wall (Ince & Launder, 1989). However, it has a severe weakness: it introduces the normal distance to the wall. Iacovides & Raisee (1997) have removed this problem by re-casting the correction as:

$$S_{\text{YIR}} = \max(0.83F(F + 1)^2 \varepsilon^2/k, 0), \qquad (3.10b)$$

where

$$F \equiv \left[\sqrt{(\partial l/\partial x_j)^2} - 2.55 \right] \text{ and } l \equiv k^{3/2}/\varepsilon.$$

As before, if the length scale does equal 2.55y the correction is zero. If, however, $\partial l/\partial x_j$ is greater than in equilibrium, the source term is again positive, driving the level of ε up and the length scale down. While the consequences of employing the correction need exploring in a wider range of flows than the two-dimensional sepa-rated flows considered so far, this offers a potentially attractive route for removing one of the weaknesses of the ε equation.

Other weaknesses in the k–ε model are associated with the very severe shortcom-ings of the eddy-viscosity approximation itself, Eq. (3.4). As has long been known (Bradshaw, 1973), any eddy-viscosity approximation grossly underestimates the magnitude of minor stresses – that is, the size of stresses where there is only a very weak associated strain. One consequence of this is that it greatly underestimates the ability of the turbulent stresses acting in a plane perpendicular to the primary strain to induce secondary motions. While these motions may be small (1% or less of the primary velocity), in complex duct flows such as an AGR[*] cooling channel, they can nevertheless transport appreciable amounts of heat. In other flows, such as the three-dimensional wall jet (that is, a jet discharged into a stagnant medium adjacent and parallel to a plane wall), the ratio of the lateral to wall-normal rate of spread computed by the k–ε model may be too small by a factor of five or more (!) simply because the computed Reynolds stresses normal to the jet direction are essentially isotropic (Craft & Launder, 1999, 2001).

Similar shortcomings also arise in flows with weak streamline curvature or in swirling flows where, depending on whether the angular velocity increases or decreases with radius of curvature, the computed mixing may be either excessive or far too little. These latter weaknesses have been corrected by different workers in *ad hoc* ways by introducing to the ε equation additional source terms. However, such corrections are applicable, at best, to only a narrow class of flows. Thus, the proposal of Launder *et al.* (1977) improved prediction of boundary layers on curved

---

[*] 'Advanced Gas-coded Reactor'

surfaces and spinning cones but it did poorly when applied to the development of a swirling jet. It is in any event illogical to suppose that one can, in any general sense, correct errors to a second-rank tensor (the Reynolds stress tensor) by tinkering with the determination of a scalar quantity (ε). Thus, if such an approach is used in CFD predictions, it should only be adopted for interpolation: that is, used only to predict a particular flow whose principal features have already been satisfactorily accounted for in computing other similar flows with the same modelling strategy. If one seeks a more general predictive scheme, the only credible route for improvement is to adopt a more widely applicable connection between stress and strain than is provided by Eq. (3.4). Such strategies are discussed in Sections 3 and 4.

2.3 Handling the near-wall sublayer

It has been implicitly assumed in what has so far been written that, when a wall is present, turbulence in the vicinity of that surface is close to local equilibrium, so that turbulence energy is being created and destroyed at nearly the same rate. These conditions lead to a 'universal' state of turbulence where the velocity and other dependent variables are functions of distance from the wall provided all quantities are made non-dimensional by the local friction velocity, $\sqrt{\tau_w/\rho}$, and the kinematic viscosity.

The existence of such a universal near-wall region would be highly fortunate because it would mean that one's numerical solver did not have to resolve the sublayer immediately adjacent to the wall where flow variables are changing so rapidly that a very fine grid is needed. Instead, one could match to conditions prevailing a little further from the wall where viscous effects are negligible and where mean velocity and turbulence gradients are much less steep. The 'law of the wall', which proposes the mean velocity to increase with the logarithm of distance from the wall, is the best known of these relationships or 'wall functions':

$$U/(\tau_w/\rho)^{1/2} = (1/\kappa) \ln [E y (\tau_w/\rho)^{1/2}/\nu],$$

where κ is the so-called Von Karman constant and E is also usually taken as a constant determining the thickness of the viscous sublayer.

Unfortunately, CFD is used to resolve *difficult* flow problems and, in these circumstances, the near-wall flow is usually far from the idealized equilibrium state where the above simple wall function can give a reliable boundary condition. There have been several refined versions of wall functions proposed where attempts have been made to develop forms applicable over a wider range of flow conditions. The best known and the most widely used form is that due to Spalding (see Launder & Spalding, 1974) where essentially the wall friction velocity $\sqrt{\tau_w/\rho}$ is replaced by the square root of the near-wall turbulence energy, $k^{1/2}$, i.e.:

$$\rho U k^{1/2}/\tau_w = (1/\kappa^*)\ln(E^* y k^{1/2}/\nu)$$

or, in the notation we shall adopt:

$$U^* = (1/\kappa^*)\ln(E^* y^*),$$

where κ^* and E^* are constants. Two major advances over the conventional version arise from doing that: firstly, that no singularity arises at points of separation or reattachment where the wall shear stress is zero and, secondly, that the wall heat-transfer rate need not go to zero where the wall shear stress vanishes; indeed, it may often attain its maximum value there! Originally, the value of the turbulence energy was that computed at the node next to the wall. Chieng & Launder (1980) developed a variation on this approach in which (along with several other refinements) the kinetic energy was always evaluated at a particular position, the edge of the viscous sublayer, irrespective of the distance of the near-wall node from the wall. (The value of k at the sublayer edge is found by extrapolating linearly the levels of turbulence energy at the two nodes closest to the wall.) While this ensures that k is nominally evaluated at a single position irrespective of the size of the near-wall control volume, the extrapolation can cause convergence problems so the practice is rarely adopted.

More extensive improvements to wall-function strategy, including the case where buoyancy exerts an appreciable net force on fluid moving along a vertical (or near-vertical) surface, have been recently developed by Gerasimov (2004) (see also Craft *et al.*, 2002a,b) especially for the case of mixed and natural convection flows (though the scheme has also been shown to bring improvements under purely forced convection situations). The approach adopted is an analytical one, based on an assumed universal eddy viscosity that increases linearly with distance *from the outer edge of the viscous sublayer* (the turbulent viscosity being zero within the sublayer itself). Like earlier publications from the group, it recognized the need, effectively, to permit the thickness of the viscous sublayer to vary according to the near-wall variation of shear stress. However, to avoid the instabilities arising with Chieng & Launder's approach, instead of explicitly altering the viscous sublayer thickness, the mean kinetic energy dissipation rate in the sublayer was itself made to depend on the rate of variation of molecular shear stress across this viscous sublayer, a device that had a very similar overall effect, but without the numerical instabilities. By choosing such a simple assumed turbulent viscosity profile, analytical forms for the near-wall velocity and temperature profiles are obtained. For further details, the reader is referred to the cited papers. Examples of adopting the approach are provided in Sections 2.4 and 4.3.

The preceding approach to wall functions has greatly improved the capability for obtaining inexpensive engineering CFD results of sufficient accuracy in a range of circumstances, including buoyancy-driven flow. However, it must be emphasized that it deals with by no means all the types of wall flows encountered. The simple profile assumed for the turbulent viscosity means that many non-equilibrium phenomena cannot be captured. In the writer's group, therefore, a more general route to wall functions has, in parallel, been developed (Gant, 2002; Craft *et al.*, 2001, 2002a, 2004). However, an explanation of the strategy adopted is best deferred until after the alternative approach to handling the near-wall sublayer, that of *low-Reynolds-number modelling*, has been introduced. The 'Reynolds number' that is here referred to is the local *turbulent Reynolds number, $k^2/\varepsilon v$.*

The form of low-Reynolds-number model predominantly used at Manchester with linear eddy-viscosity models is that due to Launder & Sharma (1974). This choice is made because several independent studies, Patel *et al.* (1985), Betts & D'Affa Alla (1986) and Savill (1991) have concluded it was the most successful of the then available models within the context of linear EVMs. Historically, the model introduces only the relatively small changes to the original scheme of Jones & Launder (1972) needed to make the resultant model entirely compatible with that optimized for free shear flows (Launder *et al.*, 1973). The mathematical form is:

$$v_t = c_\mu f_\mu k^2/\tilde{\varepsilon}, \tag{3.11}$$

$$\frac{Dk}{Dt} = v_t \frac{\partial U_i}{\partial x_j} \left(\frac{\partial U_j}{\partial x_i} + \frac{\partial U_i}{\partial x_j} \right) - \varepsilon + \frac{\partial}{\partial x_k} \left[\left(\frac{v_t}{\sigma_k} + v \right) \frac{\partial k}{\partial x_k} \right], \tag{3.12}$$

$$\frac{D\tilde{\varepsilon}}{Dt} = c_{\varepsilon 1} v_t \frac{\partial U_i}{\partial x_j} \left(\frac{\partial U_i}{\partial x_j} + \frac{\partial U_j}{\partial x_i} \right) \frac{\tilde{\varepsilon}}{k} - c_{\varepsilon 2} f_\varepsilon \frac{\tilde{\varepsilon}^2}{k}$$

$$+ 2.0 v v_t \left(\frac{\partial^2 U_i}{\partial x_j \partial x_k} \right)^2 + \frac{\partial}{\partial x_k} \left[\left(\frac{v_t}{\sigma_\varepsilon} + v \right) \frac{\partial \tilde{\varepsilon}}{\partial x_k} \right], \tag{3.13}$$

where $\varepsilon \equiv \tilde{\varepsilon} + 2v \left(\frac{\partial k^{1/2}}{\partial x_j} \right)^2$; $f_\mu = \exp(-3.4/(1+0.02 R_t^2)); f_\varepsilon = (1 - 0.3 \exp - R_t^2).$

With this form the wall boundary conditions are homogeneous, i.e. $k = \tilde{\varepsilon} = 0$ since the dissipation rate at the wall is equal to $2v(\partial k^{1/2}/\partial x_j)^2$, Jones & Launder (1972).

The scheme is more successful than any wall-function approach described so far, especially where the non-equilibrium sublayer behaviour is associated with the preferential damping of the turbulent fluctuations normal to the wall or where the mean velocity vector undergoes skewing across the viscosity-affected sublayer (as may arise in rotating machinery or even in flow around bends). However, in separated flows, where, as noted above, the basic version of the ε equation leads to excessive near-wall length scales, integrating the equation to the wall makes the computed shear stress and, more importantly, the heat-transfer coefficient, much more sensitive to defects in the model of the outer region. Certainly, the scheme should be used only in conjunction with some type of length-scale limiter, whether that be the original 'Yap correction', Eq. (3.10a) or the corresponding differential form, Eq. (3.10b), due to Iacovides & Raisee (1997).

The principal argument against using a low-Reynolds-number two-equation model across the sublayer is that it requires a very considerable number of grids across the near-wall sublayer. With the two-equation model discussed above, one should, as a rough guide, aim to adopt around 25 nodes over the viscous and buffer region (out to y^* equal to 50, say). As a result, for the computation of a flow involving just a single wall (and thus requiring fine grids in just one direction), computational times are longer than with a wall-function treatment such as that described above by Gerasimov and co-workers by a factor that usually lies between 3 and 15. In three-dimensional flows this factor is certain to be significantly higher. Thus, we return again to the question of how to simplify the handling of this very expensive sublayer zone. One obvious option is to use a simpler near-wall turbulence model since a coarser mesh can then be employed. For example, Choi *et al.* (1989) used the mixing-length hypothesis in the near-wall zone (with just eight to ten nodes across the sublayer) to obtain reasonably satisfactory prediction of flow around an unseparated U-bend – considerably better than what had formerly been achieved with simple 'log-law' wall functions, Fig. 3.1. It is unlikely that the refined wall-function approach due to Gerasimov would have significantly improved on the simple log-law results as the main problem is that the flow in the sublayer undergoes *strong skewing*, a feature that neither the basic wall function nor the Gerasimov approach can capture (unless, in the latter case, the wall function is applied only over the buffer layer rather than the fully-turbulent regime where wall functions can most economically operate). Provided that the ability to represent this skewing is retained, however, the quality of the computation depends mainly on the model used in the outer region. For example, the use of an algebraic second-moment closure such as will be examined in Section 3.2.2 produces in Fig. 3.1(c) a marked improvement on the linear EVM. (It is noted, however, that for a tighter U-bend where separation results, Iacovides *et al.*, 1996, the use of the MLH across the sublayer leads to serious errors – though in this case even a full two-equation linear EVM across the sublayer is not entirely successful.)

Finally, in this summary of cheap approaches to handling the sublayer, we return to the *numerical* wall-function approach developed by Gant and other members of the Manchester team (Gant, 2002; Craft *et al.*, 2001, 2002). Essentially, the strategy entails using two quite separate grids and solvers: a relatively coarse primary grid

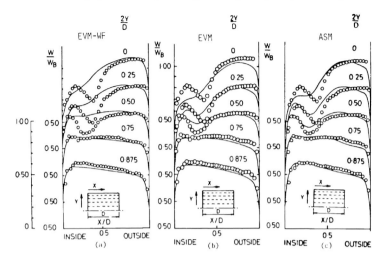

Fig. 3.1 Computation of flow around unseparated, square-sectioned U-bend. Velocity profiles at 90° around bend.
(a) k–ε EVM computations from Johnson (1984) using wall functions;
(b) k–ε EVM computations, Choi *et al.* (1989); mixing-length model in sublayer;
(c) k–ε ASM computations, Choi *et al.* (1989); mixing-length model in sublayer.

covering the whole flow domain (which by no means resolves the details of the turbulence interactions across the sublayer) and a very fine grid that just covers the near-wall control volume of the primary mesh. On this latter 'wall-function' grid, slightly simplified versions of the low-Reynolds-number equations are solved in which diffusion processes parallel to the wall are dropped and the pressure gradient parallel to the wall is taken as the same as that existing on the primary grid along its wall-adjacent cells. By this simplification the velocity component normal to the wall may be found from continuity rather than from the momentum equation for the wall-normal direction. The solution on the fine 'wall-function' grid employs the true boundary conditions of the relevant low-*Re* turbulence model at the wall and matches dependent variables to those of the main solver at the outer edge of the near-wall cell. In turn, the fine-grid 'wall-function' solution is used to compute the source terms needed for the wall-adjacent cells of the primary grid: terms like the wall shear stress and the relevant source terms for the turbulence energy and dissipation equations. The approach is not limited to linear eddy-viscosity models; indeed, Gant (2002) shows applications to the impinging jet using a non-linear eddy-viscosity model. Generally the predicted behaviour with numerical wall functions is very close to that of the corresponding complete low-Reynolds-number model. The only area where there is an important difference is in the matter of computing time. Generally, depending on the test flow, the numerical wall function requires computing times that are only between one-eighth and one-thirteenth of the corresponding complete low-*Re* solution.

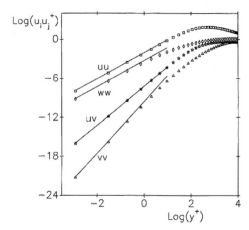

Fig. 3.2 Near-wall variation of the Reynolds stresses. Symbols: DNS data (Kim *et al.*, 1987). Solid lines are of slope 2 (for $\overline{u^2}$ and $\overline{w^2}$), 3 (for $\overline{uv}$) and 4 (for $\overline{v^2}$).

Before leaving the subject of 'low-Reynolds-number' effects in linear eddy-viscosity models, it is appropriate to note two further elaborations that have been proposed. The first was introduced by Cotton & Ismael (1995) who took the coefficient c_μ as a function of two parameters (rather than just one): the turbulent Reynolds number and the strain rate S defined as:

$$S \equiv \frac{k}{\varepsilon} \sqrt{\left(\frac{\partial U_i}{\partial x_j} + \frac{\partial U_j}{\partial x_i} \right)^2}.$$

Now, the parameter S is a group not containing the molecular viscosity, so it may become significant in situations where a wall is not present. In fact, testing both by Cotton & Ismael (1995) and Craft *et al.* (1997) (the latter in connection with a non-linear eddy-viscosity treatment) confirms that the observed data of rapidly strained turbulence does provide firm support for the two-parameter characterization of c_μ (that is, $c_\mu(R_t, S)$). We return to this subject when the topic of non-linear eddy-viscosity models is addressed in Section 3.4.

The second approach is the so-called k–ε–$\overline{v^2}$ model of Durbin (1993, 1995). Launder (1986, 1988) had earlier demonstrated that the 'viscous damping' in the sublayer of the coefficient c_μ in the formula for turbulent viscosity was not strictly due to viscous effects but, rather, because the viscosity formula *should* have been $\nu_t = c'_\mu k \overline{v^2}/\varepsilon$. If that form is used, c'_μ is essentially constant because $\overline{v^2}$ falls to zero as the wall is approached as y^4 rather than y^2 (which is the exponent with which k varies), Fig. 3.2. The novelty of Durbin's approach is that he proposes an elliptic relaxation equation from which the level of, effectively, $\overline{v^2}/k$ may be obtained. The companion book in the INI series (Launder & Sandham, 2002) contains a chapter by Gatski & Rumsey (2002) giving further details of the approach. It may also be

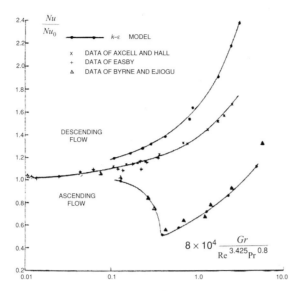

Fig. 3.3 Normalized Nusselt number variation versus buoyancy parameter for flow through a vertical pipe. From Cotton & Jackson (1987). (Nu_0 is Nu for zero buoyancy; Gr, Re, and Nu are conventional Grashof, bulk Reynolds and Nusselt numbers based on pipe diameter.)

noted that Durbin's elliptic-relaxation strategy has been successfully applied (on a more rigorous and transparent basis) to the higher-order approach to closure offered by stress-transport modelling (i.e. second-moment closure).

2.4 Successes and shortcomings in buoyancy affected flows

2.4.1 Vertical flows near heated walls

The ability of the k–ε model to predict buoyancy-modified flows depends very much on whether the primary action of the buoyant force is to modify the mean momentum or the turbulent velocity fluctuations. In the latter case, what we have termed *stratified flow*, the conventional k–ε model by no means captures the observed effects. Where the mean motion is adjacent to a wall and is predominantly vertical, however, the addition of a buoyant force that aids or impedes that motion may be well captured. A case in point is that of mixed convection in a vertical pipe, explored extensively by Cotton & Jackson (1987) employing the Launder–Sharma (1974) form of the low-Reynolds-number k–ε model. For upflow, strong heating of the pipe wall causes the near-wall layers to receive a significant buoyant upthrust which results in a velocity maximum developing close to the wall. This, in turn, leads to a thickening of the near-wall sublayer due to the very rapid decrease of shear stress across the sublayer. The consequence of this is that the heat-transfer coefficient is appreciably reduced, Fig. 3.3. It is seen from the same figure that agreement was less satisfactory for downflow. A later paper (Jackson *et al.*, 1989) reported, however, that agreement was improved by including the length-scale limiter,

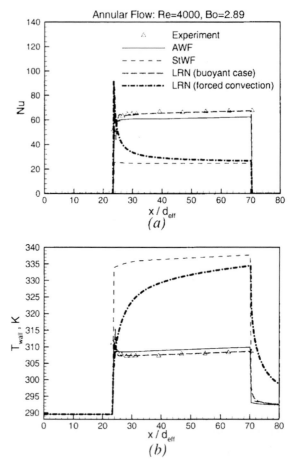

Fig. 3.4 Prediction of *Nu* in buoyancy-modified downward flow in an annulus with heated core, from Gerasimov (2004). AWF = analytical wall function; StWF = standard log wall function.

Eq. (3.10a). The writer's group has also completed an extensive series of computations of buoyantly modified vertical flows to explore the capabilities of the analytical wall functions (AWF) developed by Gerasimov (2004) in his Ph.D. research. Figure 3.4 shows an example from this work relating to downward flow in an annulus heated at the inner radius with the outer wall insulated. We note that there is excellent agreement both with the experimental data and with the results obtained with the low-Reynolds-number form of the k–ε model. The usual logarithmic-law wall function (StWF) gives a very poor prediction, however. (A further application of these wall functions, used in conjunction with a stress-transport model for the main flow region, is presented in Section 4.)

The linear k–ε EVM has also been successfully applied to purely natural convection on a vertical plate and in tall rectangular cavities, Ince & Launder (1989,

1995). These papers, however, adapted the treatment of generation in the turbulent kinetic energy equation from that conventionally adopted in order to account more accurately for the influences of buoyancy on the turbulence. This is the subject now considered.

The presence of density fluctuations (ρ') leads to a fluctuating force ($\rho' g_i$) acting in direction x_i. This leads to the buoyant creation of turbulent kinetic energy, G_k, at a rate

$$G_k = \frac{g_i}{\rho} \overline{\rho' u_i}.$$
(3.14)

If the density variations are due only to temperature changes, it is convenient to replace density fluctuations by temperature fluctuations by way of the dimensionless expansion coefficient, α:

$$\alpha \equiv -\frac{\partial \rho}{\partial \Theta}\bigg|_p \frac{\Theta}{\rho}.$$

Thus:

$$G_k = -\alpha \frac{g_i}{\Theta} \overline{\theta u_i}.$$
(3.15)

This term is added to that due to shear generation in Eq. (3.1) or (3.7b). If the turbulent kinematic heat flux is approximated by the usual gradient transport model, Eq. (3.6), the buoyant generation rate reduces to:

$$G_k = \frac{\alpha g_i}{\Theta} \frac{v_t}{\sigma_\theta} \frac{\partial \Theta}{\partial x_i}.$$
(3.16)

Let us note that the repeated 'i' suffix implies summation in all three Cartesian directions, so G_k amounts to:

$$G_k = \frac{\alpha v_t}{\Theta \sigma_\theta}\left[g_1 \frac{\partial \Theta}{\partial x_1} + g_2 \frac{\partial \Theta}{\partial x_2} + g_3 \frac{\partial \Theta}{\partial x_3}\right].$$
(3.17)

Now, let us suppose we take x_1 as vertically up, with x_2 and x_3 being horizontal. The gravitational vector g_i acts *vertically* downwards, whereas, for the flows considered so far, the mean flow is directed upwards and rapid temperature variations occur only in the *horizontal* direction. So, it is clear from Eq. (3.17) that in the direction where the gravitational vector acts there is negligible temperature gradient, while in the horizontal plane where rapid temperature variations arise there is no gravitational component. Thus from Eq. (3.17) the gravitational term makes a negligible contribution to the generation of turbulent kinetic energy, at least if a simple gradient-diffusion model is adopted for the heat flux.

It is true that in many vertical shear flows direct buoyant contributions to k-generation are indeed small compared with that due to shear; however, they are not as small as the simple isotropic gradient-diffusion model leads one to conclude. The problem is that Eq. (3.6) only gives a fair indication of the turbulent heat flux in directions where the temperature gradient is large (as such, it suffers from similar weaknesses to the simple eddy-viscosity concept, Eq. (3.4)). In fact, measurements have conclusively shown that, in a pipe flow or boundary layer on a flat plate, the turbulent heat flux in the direction of flow is substantially *greater* than that normal to the wall even though the temperature gradient in the flow direction is small. The reason that fact has not wrecked thermal boundary-layer computations based on Eq. (3.6) is that, in the streamwise direction, the net turbulent diffusion of heat is negligible compared with convection by the *mean* motion (and, in the boundary-layer approximation, for computational convenience, the streamwise diffusion is accordingly dropped).

In the present consideration, however, we are interested in the role of the heat flux in G_k. In a naturally driven boundary layer or plume, this contribution can be significant even though (in solving the thermal energy equation) the net vertical heat transport by turbulence is unimportant (because, as noted above, it is swamped by mean flow convection).

The task is to adopt a better approximation for $\overline{u_i\theta}$ than is provided by Eq. (3.6). A simple advance, widely used by the writer's group, is to employ what has become known as the generalized gradient-diffusion hypothesis (GGDH), Daly & Harlow (1970):

$$\overline{u_i\theta} = -c_\theta \frac{\overline{u_ju_i}}{\varepsilon} k \frac{\partial\Theta}{\partial x_j},$$ (3.18)

where the coefficient c_θ is approximately 0.30. The use of this approximation in the vertically directed flow (with again x_1 vertical) leads to:

$$\overline{u_1\theta} = -c_\theta \frac{k}{\varepsilon} \left[\overline{u_1^2} \frac{\partial\Theta}{\partial x_1} + \overline{u_1u_2} \frac{\partial\Theta}{\partial x_2} + \overline{u_1u_3} \frac{\partial\Theta}{\partial x_3} \right].$$ (3.19)

So, while the first term on the right of Eq. (3.19) may be small (because Θ varies only slowly with x_1), one or both of the other components containing horizontal temperature gradients may be substantial.

On incorporating the GGDH into Eq. (3.15), the buoyant term may be written:

$$G_k = \frac{c_\theta\alpha}{\Theta} \frac{k}{\varepsilon} g_i \overline{u_ju_i} \frac{\partial\Theta}{\partial x_j}.$$ (3.20)

As noted, it is vital to account for the generation by horizontal temperature gradients which involve the primary shear stress or shear stresses. These, within the context of the present treatment, may be represented by the linear eddy-viscosity formula, Eq. (3.4b).

Thus:

$$G_k = \frac{c_\theta \alpha}{\Theta} \frac{k}{\varepsilon} g_i \frac{\partial \Theta}{\partial x_j} \left[\frac{2}{3} \delta_{ij} k - \nu_t \left(\frac{\partial U_i}{\partial x_j} + \frac{\partial U_j}{\partial x_i} \right) \right]. \qquad (3.21)$$

This was the formulation used in the Ince–Launder study of flow in closed, tall rectangular cavities. It provides a simple approximation for G_k which, while still generally underestimating the importance of buoyant generation, is distinctly better than Eq. (3.16). To put matters in context, however, the inclusion of length-scale damping is an order of magnitude more important in this flow for, without Eq. (3.10a), levels of *Nu* were 35% too high!

2.4.2 *Other buoyancy-modified flows*

The rather surprising success of the k–ε model in buoyancy-driven flow near vertical heated or cooled walls does not carry over to the various other types of buoyantly modified flows arising in engineering practice. However, most research workers have been aware of the inadequacies of an eddy-viscosity model for mimicking transport in such flows and so their calculations have started from a, physically, better founded approach, usually based on second-moment closure. One case where several computational results are available is the stably-stratified horizontal jet or mixing layer. Craft *et al.* (1996) have shown that the usual k–ε model, in which the total production of turbulent kinetic energy (that is $P_k + G_k$) acts in the ε equation with a coefficient $c_{\varepsilon 1}$, does not reproduce the very great reduction in mixing rate in a stably-stratified mixing layer that occurs in practice. Thus, for the case of mixing between a fresh-water and a saline stream (Uittenbogaard, 1988), by the downstream end of the 40 m tank the computed vertical salinity concentration (shown by the dashed line) had become nearly uniform whereas a very sharp interface between fresh and salt water still remained in the experiment, Fig. 3.5.

Viollet (1980) has looked at a similar flow where density variations arose from a vertical temperature gradient. His predictions (not shown) superficially suggest a quite different capability of the model, for agreement between measurement and computation is satisfactory. That result was achieved, however, by neglecting the contribution of G_k in the dissipation equation (while still including it in the k equation). Now, since the stratification is stable, G_k will be negative (in other words the vertical fluctuations have to do work against the stable density stratification). So, by omitting this term, ε will be increased; hence the level of k and, more emphatically, the turbulent viscosity ($c_\mu k^2/\varepsilon$) will be reduced, greatly diminishing mixing. This brings computations more into line with experiments for the horizontal shear layer. However, McGuirk and Papadimitriou (1985) have examined a two-dimensional, stably-stratified free-surface jet employing the same device (of dropping the buoyant source from the ε equation) but they still found a substantially too rapid spreading rate and dilution of the light fluid, Fig. 3.6.

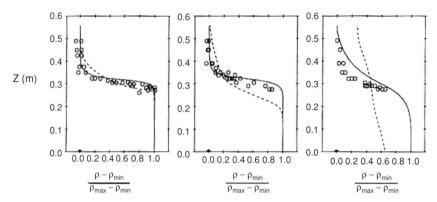

Fig. 3.5 Development of horizontal, stably-stratified mixing layer showing normalised density profiles at 5 m, 10 m and 20 m downstream of origin. Experiments
(symbols) Uittenbogard (1988); Computations (lines) Craft *et al.* (1996): continuous lines – TCL closure; broken lines k–ε EVM.

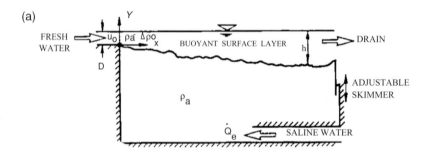

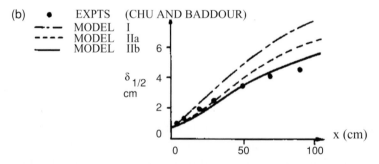

Fig. 3.6 Mixing of fresh water surface layer with saline fluid from McGuirk
& Papadimitriou (1985). (a) Flow configuration. (b) Variation of jet half-width
with distance. Note: Model I is a k–ε model with the buoyant term in ε equation
discarded; Model II is a basic second-moment closure without and with 'wall
reflection' terms applied at the free surface (see Section 3).

The above comparison has shown that even for stably-stratified, horizontal flows merely dropping G_k from the ε equation does not provide a generally satisfactory route for computing stratified flows with the k–ε model. Moreover, different practices are also required depending on whether the flow is stable or unstable (Viollet, 1987), vertical or horizontal (Hossain & Rodi, 1982). As remarked in Section 2.2 in relation to correcting the model to account for the effects of streamline curvature, it is not possible to remedy, in any general sense, a defect that principally arises from the treatment of the Reynolds-stress *tensor* (in the present case associated with the gravitational force affecting vertical fluctuations much more than horizontal ones) by making adjustments to the *scalar* quantity, ε. This becomes especially evident when attention is shifted to recirculating flows or, at least to situations where the thin-shear-flow approximation is not valid, such as in the tunnel-fire scenarios examined by Cordier (1999) and Woodburn (1995).

While it *is* possible to improve eddy-viscosity models so that they give better agreement with buoyantly modified second moments than the methods noted above, the rationale for this approach can be best explained in terms of the second-moment transport equations, i.e. for $\overline{u_i u_j}$, $\overline{u_i \theta}$ and $\overline{\theta^2}$. Thus, Section 3 presents the exact second-moment equations and discusses their approximation before reporting the application of simplified two-equation truncations of these models.

3 Second-moment closure and related models

3.1 Introduction

The strong effect of the earth's gravitational field on turbulence structure has been one of the main motivations for the development of second-moment closure; for the primary roles of the gravitational vector appear explicitly in the second-moment transport equations. The exact second-moment transport equations are readily obtainable by multiplying the first-moment equations (the Navier–Stokes equations, the thermal energy equation, etc.) by the fluctuating velocity or temperature. Thus:

$$\frac{\mathrm{D}\overline{u_i u_j}}{\mathrm{D}t} = \overline{u_i \frac{\mathrm{D}\tilde{U}_j}{\mathrm{D}t}} + \overline{u_j \frac{\mathrm{D}\tilde{U}_i}{\mathrm{D}t}}, \tag{3.22}$$

where $\tilde{U}_i \equiv U_i + u_i$ is the instantaneous velocity in direction x_i. The above equality applies because $\overline{u_i \frac{\mathrm{D}U_j}{\mathrm{D}t}} = 0$. Likewise:

$$\frac{\mathrm{D}\overline{u_i \theta}}{\mathrm{D}t} = \overline{u_i \frac{\mathrm{D}\tilde{\Theta}}{\mathrm{D}t}} + \overline{\theta \frac{\mathrm{D}\tilde{U}_i}{\mathrm{D}t}}, \tag{3.23}$$

where $\tilde{\Theta} \equiv \Theta + \theta$ is the instantaneous value of temperature, mass fraction or whatever scalar quantity is in question.

By forming the second-moment equations in the way indicated above by Eqs. (3.22)–(3.23) and (regarding the buoyant term in the Navier–Stokes equation) by expressing density fluctuations in terms of fluctuations in the scalar Θ:

$$\rho' = -\alpha\theta\rho/\Theta, \qquad (3.24)$$

where $\alpha \equiv \left.\dfrac{\partial\rho}{\partial\Theta}\right|_p \dfrac{\Theta}{\rho}$ is the dimensionless volumetric expansion coefficient,[+] the transport equations for $\overline{u_i u_j}$ and $\overline{u_i\theta}$ may be expressed in the form given below:

$$\underbrace{\frac{D\overline{u_i u_j}}{Dt}}_{C_{ij}} = - \left(\overline{u_i u_k}\frac{\partial U_j}{\partial x_k} + \overline{u_j u_k}\frac{\partial U_i}{\partial x_k}\right) \qquad\qquad P_{ij}$$

$$- \frac{\alpha}{\Theta}(\overline{u_j\theta}\,g_i + \overline{u_i\theta}\,g_j) \qquad\qquad G_{ij}$$

$$+ \frac{\overline{p}}{\rho}\left(\frac{\partial u_i}{\partial x_j} + \frac{\partial u_j}{\partial x_i}\right) \qquad\qquad \phi_{ij}$$

$$- \frac{\partial}{\partial x_k}\left(\overline{u_i u_j u_k} + \frac{\overline{pu_j}}{\rho}\delta_{ik} + \frac{\overline{pu_i}}{\rho}\delta_{jk} - \nu\frac{\partial\overline{u_i u_j}}{\partial x_k}\right) \quad d_{ij}$$

$$- 2\nu\overline{\frac{\partial u_i}{\partial x_k}\frac{\partial u_j}{\partial x_k}}, \qquad\qquad \varepsilon_{ij} \quad (3.25)$$

$$\frac{D\overline{u_i\theta}}{Dt} = - \overline{u_i u_k}\frac{\partial\Theta}{\partial x_k} - \overline{\theta u_k}\frac{\partial U_i}{\partial x_k} \qquad\qquad P_{i\theta} \equiv P_{i\theta 1} + P_{i\theta 2}$$

$$- \frac{\alpha\overline{\theta^2}g_i}{\Theta} \qquad\qquad G_{i\theta}$$

$$+ \frac{\overline{p}}{\rho}\frac{\partial\theta}{\partial x_i} \qquad\qquad \phi_{i\theta}$$

$$- \frac{\partial}{\partial x_k}\left(\overline{u_k u_i\theta} + \frac{\overline{p\theta}}{\rho}\delta_{ik}\right) \qquad\qquad d_{i\theta}$$

$$- (\lambda + \nu)\overline{\frac{\partial u_i}{\partial x_k}\frac{\partial\theta}{\partial x_k}}. \qquad\qquad \varepsilon_{i\theta} \qquad (3.26^\circ)$$

[+] If Θ denotes temperature and the fluid is an ideal gas, α is unity.
[º] This omits for clarity molecular transport terms that are of no significance in fully turbulent regions.

Contracting Eq. (3.25) produces an equation for $\overline{u_i^2}$, i.e. twice the turbulent kinetic energy. It is noted that in Eq. (3.26) the gravitational vector multiplies $\overline{\theta^2}$. Thus a complete second-moment closure requires that that quantity too be computed from its own transport equation, an exact form of which is readily obtained as:

$$\frac{\mathrm{D}\overline{\theta^2}}{\mathrm{D}t} = \underbrace{-2\overline{\theta u_k}\frac{\partial\Theta}{\partial x_k}}_{P_{\theta\theta}} - \underbrace{2\lambda\overline{\frac{\partial\theta}{\partial x_k}\frac{\partial\theta}{\partial x_k}}}_{\varepsilon_{\theta\theta}} - \underbrace{\frac{\partial}{\partial x_k}\left(\overline{u_k\theta^2} - \lambda\frac{\overline{\partial\theta^2}}{\partial x_k}\right)}_{d_{\theta\theta}}. \tag{3.27}$$

In Eqs. (3.25)–(3.26) the direct generation terms* due to shear, mean scalar gradients and buoyancy (P_{ij}, $P_{i\theta 1}$, $P_{i\theta 2}$, $P_{\theta\theta}$, G_{ij}, $G_{i\theta}$) do not require approximation with a complete second-moment closure. This is, to a large extent, the attraction of this level of modelling. Qualitatively correct responses of turbulence to, say, streamline curvature or a buoyant source can be inferred merely by examining the role of the different generative agencies in the Cartesian component equations. Figure 3.7 shows the very different interconnections that the buoyant terms exert, depending on whether the shear is horizontal or vertical. In both cases direction x_1 is the direction of primary velocity and x_2 the principal direction of property gradients, whether mean velocity or scalar. Thus, in the absence of buoyancy, the normal stress acting down the mean velocity gradient ($\overline{u_2^2}$) acts directly in the shear-stress equation (through the generation term $-\overline{u_2^2}\partial U_1/\partial x_2$). The shear-stress $\overline{u_1 u_2}$ in turn acts on the streamwise normal stress, $\overline{u_1^2}$, through the generation term $-\overline{u_1 u_2}\partial U_1/\partial x_2$. Finally, some of the energy in streamwise fluctuations is diverted to $\overline{u_2^2}$ by the process ϕ_{ij} which we consider in Section 3.2.1. Different components of the stress tensor act on the scalar flux as indicated by the solid lines in Fig. 3.7. The broken lines in the figure show the further couplings arising from the buoyant generation. It is noted that for a horizontal flow the buoyant term directly opposes (or assists) the same interconnections arising from mean velocity or temperature gradients. For a vertically directed flow, however, the interconnections are more diverse. Noting the very different buoyant couplings in the two cases helps one appreciate why a simple modification of the ε equation within an eddy-viscosity approach can only fit a narrow range of buoyant flows.

Before the second-moment equations can be used within a CFD solver, models must be adopted for the unknown processes. In Section 3.2 the so-called Basic Model is presented with the various simplified versions – some re-packaged in eddy-viscosity format. For completeness, other non-linear eddy-viscosity models are presented in Section 3.3.

---

* While referred to as 'generation terms' there is no requirement that the terms should be positive.

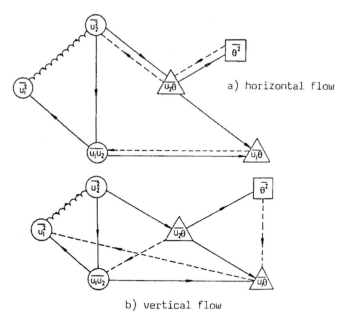

a) horizontal flow

b) vertical flow

Fig. 3.7 Contrasting buoyant couplings in horizontal and vertical thin shear flows.
– – – – Buoyant coupling
_ _ _ _ _ Coupling through mean velocity/scalar gradient
∩∩∩∩ Pressure-strain coupling
From Launder (1989b).

3.2 The 'Basic Model' and some simpler derivatives

3.2.1 The Basic Model in free shear flows

What is today known as the Basic Model evolved from contributions over a span of
forty years by Rotta (1951), Rodi (1972), Launder *et al.* (1975), Launder (1975),
Gibson & Launder (1978) and Craft & Launder (1992). A complete tensorial statement of the model appears as Appendix 1 though the individual approximations
are discussed below.

The most important term to model is the pressure-strain process, ϕ_{ij}. By examining the exact Poisson equation for the fluctuating pressure, it may be concluded
that any reasonably general model of the process should consist of three parts:

- a *turbulence* or 'slow' part, ϕ_{ij1}
- a *mean-strain* or 'rapid' part, ϕ_{ij2}
- a *buoyant* part, ϕ_{ij3}.

If other types of force field act on turbulence (e.g. magnetic), they too will exert a
corresponding influence on the pressure-strain process, while the choice of a rotating
coordinate frame from which to examine the flow means that a straightforward

'rotation adaptation' of ϕ_{ij2} is required (Bertoglio, 1982). In the discussion that follows, we assume, however, that for free shear flows:

$$\frac{p}{\rho}\left(\frac{\partial u_i}{\partial x_j} + \frac{\partial u_j}{\partial x_i}\right) \equiv \phi_{ij} = \phi_{ij1} + \phi_{ij2} + \phi_{ij3}, \qquad (3.28)$$

where the component parts have the meanings indicated above.

The turbulence part of ϕ_{ij} is usually modelled adopting the original proposal of Rotta (1951):

$$\phi_{ij1} = -c_1 \frac{\varepsilon}{k}\left(\overline{u_i u_j} - \frac{2}{3}\delta_{ij}k\right). \qquad (3.29)$$

Sometimes c_1 is made a function of dimensionless stress invariants in order that the coefficient should vanish when the turbulent fluctuations in one direction vanish (what is known as the *two-component limit*). This strategy and the rationale behind it are explained in Section 4. For the moment we note simply that the model proposed in Eq. (3.29) always tends to drive the Reynolds-stress tensor towards isotropy (the isotropic state being $2/3\,\delta_{ij}k$).

The corresponding form of ϕ_{ij2} was first proposed by Naot *et al.* (1970) but as a *replacement for*, rather than as a supplement to, ϕ_{ij1}:

$$\phi_{ij2} = -c_2\left(P_{ij} - \frac{2}{3}\delta_{ij}P_k\right). \qquad (3.30)$$

Nowadays, what is known as the Basic Model adopts both (3.29) *and* (3.30) though Fig. 3.8 indicates there has been a wide range of values adopted for c_1 and c_2. We shall return shortly to the origin of the diagonal line in the figure.

The dissipation rate, ε_{ij}, is usually assumed to be isotropic because it takes place in the smallest scales of turbulent motion:

$$\varepsilon_{ij} = \frac{2}{3}\delta_{ij}\varepsilon. \qquad (3.31)$$

Some workers have instead preferred the choice:

$$\varepsilon_{ij} = \frac{\overline{u_i u_j}}{k}\varepsilon. \qquad (3.32)$$

This form does not respect the deeply rooted idea that, as turbulent eddies get broken down into smaller and smaller scales, they become more isotropic. However, from a practical point of view, this violation is academic since (3.32) can clearly be re-written:

$$\varepsilon_{ij} = \frac{\varepsilon}{k}\left(\overline{u_i u_j} - \frac{2}{3}\delta_{ij}k\right) + \frac{2}{3}\delta_{ij}\varepsilon,$$

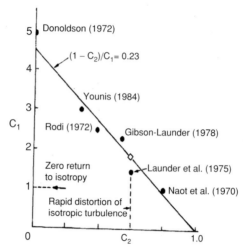

Fig. 3.8 Map of proposals for coefficients c_1– c_2 in the Basic Model. Adapted from Launder (1985).

and the first term is evidently of the same form as the approximation for ϕ_{ij1}. In other words, Eq. (3.31) or (3.32) gives identical results provided, in the latter case, the magnitude of the coefficient c_1 is smaller by 1.0 than if (3.31) is used.

For the case of a local-equilibrium shear flow in the absence of buoyant effects, the approximations made so far imply:

$$0 = P_{ij} - c_1 \frac{\varepsilon}{k} \left(\overline{u_i u_j} - \frac{2}{3} \delta_{ij} k \right) - c_2 \left(P_{ij} - \frac{2}{3} \delta_{ij} P_k \right) - \frac{2}{3} \delta_{ij} \varepsilon$$

or, on re-grouping:

$$a_{ij} \equiv \frac{\left(\overline{u_i u_j} - \frac{2}{3} \delta_{ij} k \right)}{k} = \frac{(1 - c_2)}{c_1} \frac{\left(P_{ij} - \frac{2}{3} \delta_{ij} P_k \right)}{P_k}. \tag{3.33}$$

In words, Eq. (3.33) states that the component anisotropy of the Reynolds-stress tensor is proportional to the corresponding component anisotropy of the stress-generation tensor. One of the most interesting features of this result is that the two empirical coefficients from the models of ϕ_{ij1} and ϕ_{ij2} coalesce into a single unknown coefficient $(1 - c_2)/c_1$. Experimental data suggest that in a simple shear flow that coefficient needs to be approximately 0.23 to produce the correct level of shear stress. Now, returning to Fig. 3.8, the diagonal line appearing on that graph (which is quite a good mean fit to the 'experimental data' of the pairings of c_1 and c_2 in various models) corresponds to:

$$c_1 = (1 - c_2)/0.23.$$

In other words, in local equilibrium, any of the different pairings shown in Fig. 3.8 would lead to very nearly the correct result since the 'data points' lie close to the line. Thus, one needs to shift attention to non-equilibrium situations to determine the values of the coefficients more precisely. Within the framework of such a simple model, the optimum choices for c_2 are in the range 0.5–0.6, with c_1 in the range 1.7 to 2.0, with the pairing being chosen to accord broadly with the above-noted interrelationship, i.e.

$$c_1 \cong (1 - c_2)/0.23.$$

Following Launder (1975), buoyant effects are, by analogy with Eq. (3.30), approximated as:

$$\phi_{ij3} = -c_3 \left(G_{ij} - \frac{2}{3} \delta_{ij} G_k \right), \tag{3.34}$$

a form which Schumann (1976) has shown is applicable also when the force field arises from a magnetic field. The usual practice nowadays is to set c_3 equal to c_2.

Finally, diffusive transport is approximated by the 'GGDH' model of Daly & Harlow (1970) introduced in Section 2 (viz. Eq. (3.18)):

$$d_{ij} = \frac{\partial}{\partial x_k} \left(c_s \frac{k}{\varepsilon} \overline{u_k u_m} \frac{\partial \overline{u_i u_j}}{\partial x_m} \right). \tag{3.35}$$

The energy dissipation rate, ε, which appears as an unknown in many of the above terms is, with this Basic Model, found from solution of the same transport equation as in Section 2 except that the more accurate representation of P_k and G_k obtainable with second-moment closure is used instead of the eddy-viscosity model.

Precisely analogous approximations are made for the unknown terms in the turbulent scalar flux equation (Launder, 1976). The corresponding modelled forms are:

$$\frac{\overline{p}}{\rho} \frac{\partial \theta}{\partial x_i} \equiv \phi_{i\theta} = \phi_{i\theta1} + \phi_{i\theta2} + \phi_{i\theta3}, \tag{3.36}$$

where:

$$\phi_{i\theta1} = -c_{1\theta} \frac{\varepsilon}{k} \overline{u_i \theta}, \tag{3.37}$$

$$\phi_{i\theta2} = -c_{2\theta} P_{i\theta2}, \tag{3.38}$$

$$\phi_{i\theta3} = -c_{3\theta} G_{i\theta}, \tag{3.39}$$

and where the recommended values for the empirical coefficients are:

$$c_{1\theta} = 3.0; \quad c_{2\theta} = 0.5; \quad c_{3\theta} = 0.5.$$

(Notice in Eq. (3.38) that it is only the heat flux generated by mean velocity gradients that is included in $\phi_{i\theta2}$ since only mean velocity gradients (not mean scalar gradients) contribute to pressure fluctuations.)

Closure is completed by the assumption of isotropy in the dissipative motions:

$$\varepsilon_{i\theta} = 0$$

and the usual GGDH approximation of diffusive transport

$$d_{i\theta} = \frac{\partial}{\partial x_k} \left(c_\theta \frac{k}{\varepsilon} \overline{u_k u_m} \frac{\partial \overline{u_i \theta}}{\partial x_m} \right). \tag{3.40}$$

Finally, in the transport equation for the mean square scalar fluctuations:

$$\varepsilon_{\theta\theta} = \frac{\varepsilon}{k} \frac{\overline{\theta^2}}{R}, \tag{3.41}$$

$$d_{\theta\theta} = \frac{\partial}{\partial x_k} \left(c_\theta \frac{k}{\varepsilon} \overline{u_k u_m} \frac{\partial}{\partial x_m} \overline{\theta^2} \right). \tag{3.42}$$

The quantity R is the turbulent time-scale ratio for the dynamic and scalar field. It is usually assumed to be constant (but see Section 3.3) though the different choices made (ranging from about 0.5 to 1.0) indicate that even for free flows, this is far from a universal prescription.

While there has been some use of the complete closure as presented above for buoyant flows (e.g. Haroutunian & Launder, 1988; Craft *et al.*, 1996) the usual approach has been to reduce the closed forms of Eqs. (3.25)–(3.27) to algebraic forms by omitting the transport terms (convection and diffusion) or by approximating them by way of Rodi's (1976) proposal:

$$(C_{ij} - d_{ij}) = \frac{\overline{u_i u_j}}{k}(C_k - d_k)$$

$$= \frac{\overline{u_i u_j}}{k}(P_k - \varepsilon). \tag{3.43}$$

Some of the more widely cited simplifications are presented below together with test cases to which they have been applied.

3.2.2 Simplified second-moment closures for buoyant flows

Purely local-equilibrium studies for buoyant flows have been reported by Mellor (1973) and Launder (1975) but the first attempt to provide a simplification of the second-moment equations in situations where transport and buoyant effects were appreciable was made by Gibson & Launder (1976). Equation (3.43) was used to eliminate transport terms from the Reynolds stress equation while, for the scalar flux, it was assumed that:

$$(C_{i\theta} - d_{i\theta}) = \frac{\overline{u_i\theta}}{2\overline{\theta^2}}(P_\theta - \varepsilon_\theta) + \frac{\overline{u_i\theta}}{2k}(P_k - \varepsilon). \tag{3.44}$$

Finally, using Eq. (3.41), the local-equilibrium assumption (in this case $\varepsilon_{\theta\theta} = P_{\theta\theta}$) was used to reduce Eq. (3.27) to:

$$\overline{\theta^2} = -2R\frac{k}{\varepsilon}\overline{u_k\theta}\frac{\partial\Theta}{\partial x_k}. \tag{3.45}$$

For the case of two-dimensional shear flows, the resultant algebraic truncations for the erstwhile transport equations can be re-packaged into a buoyancy-dependent eddy-viscosity and turbulent Prandtl number (see Equations (31) and (32) from the Gibson–Launder 1976 paper). For general applications, however, additional terms would need to be included and this would preclude the representation of the diffusion fluxes or Reynolds stress through isotropic viscosity or diffusion coefficients. Within its limitations (two-dimensional flow), the model led to satisfactory prediction of several horizontal shear flows. Figure 3.9 shows, for example, the computed rapid damping of entrainment rate into a stably-stratified surface jet as the mean Richardson number increases (that is, as the stable stratification becomes progressively stronger). Reasonably close agreement is obtained with the experimental data of Ellison & Turner (1959).

A more extensive development of essentially the same model has been presented by Hossain & Rodi (1982) which was applied to different vertical free-shear flows. Table 3.1 compares their results for the rate of growth of half-width of the plane jet and plume with the recommended experimental values (Chen & Rodi, 1980).

It is evident that the buoyant plume is far more sensitive than the plane isothermal jet to whether an ASM or linear eddy-viscosity approach is used. In fact Hossain & Rodi (1982) report that most of the differences from using the 'pure' k–ε approximation for this flow are removed by the inclusion of G_k in the source terms of the k and ε equations (the difference between the M1 and M2 results for the plane plume) rather than by adopting 'ASM' relations for the stresses and fluxes. However, exchanges with Professor Rodi have clarified that, in the expression for the vertical heat flux that appears in G_k, an ASM type of approximation was nevertheless introduced (apparently similar to the form adopted by Ince & Launder, 1989, discussed above, viz. Eq. (3.21)). This is important, for otherwise, if the usual eddy-diffusivity approach had been used (flux proportional to vertical temperature gradient), the buoyant term would have been too small to exert a significant effect on the spreading rate.

Table 3.1 *Measured and computed rates of spread for the*
asymptotic plane jet and plume

	$dy_{1/2}/dx$			
Flow	Exp't	M1	M2	M3
Plane isothermal jet	0.110	0.116	0.116	0.116
Plane plume	0.120	0.087	0.115	0.127

M1 $\equiv$ k–ε EVM *without* buoyant terms in k and ε equations.
M2 $\equiv$ k–ε EVM including buoyant terms in k and ε equations.
M3 $\equiv$ buoyant terms included in algebraic second-moment (ASM) closure.

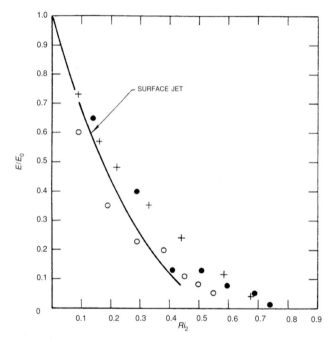

Fig. 3.9 Entrainment rate in a plane surface jet. Line: computations; symbols: different experimental runs of Ellison & Turner (1959) (from Gibson & Launder, 1976). E/E_0 denotes entrainment rate relative to unstratified flow; Ri_2 denotes an average gradient Richardson number across the jet.

Finally, mention is made of very similar, more recent work by Hanjalić and co-workers (Hanjalić & Vasić, 1993; Hanjalić & Musemic, 1997). Given the range of flows considered by them, it seems appropriate to give their modelling proposals in more detail. The closure is essentially at the same level as the M2 model of Hossain & Rodi (1982). That is, while a full ASM approach is applied to the heat fluxes, a purely isotropic viscosity is assumed for the turbulent stresses:

$$-\overline{u_i u_j} + \frac{2}{3}\delta_{ij}k = \nu_t \left(\frac{\partial U_i}{\partial x_j} + \frac{\partial U_j}{\partial x_i} \right). \tag{3.46}$$

The underlying rationale must presumably have been that, when buoyancy affects turbulence, the most important effects enter through the buoyant term in the turbulence energy equation. The writer does not share entirely that perception but, in modelling at this level, it is probably as well not to refine (3.46) by including buoyant terms in addition to the strain rate, as the frailties of that equation must have been partially compensated by the assumed role of buoyancy elsewhere in the model. At least, the use of Eq. (3.46) makes it straightforward to employ a more widely valid stress-deformation hypothesis for use in complex strain fields without disturbing the well-calibrated buoyant effects (see Section 3.4). The turbulent heat fluxes are then obtained from:

$$\overline{u_i\theta} = -\phi_\theta \frac{k}{\varepsilon} \left(\overline{u_i u_j}\frac{\partial \Theta}{\partial x_j} + (1 - c_{2\theta})\left(\overline{\theta u_j}\frac{\partial U_i}{\partial x_j} + \beta \overline{g_i \theta^2} \right) + \varepsilon_{i\theta} \right), \tag{3.47}$$

where the coefficient ϕ_θ denotes:

$$\phi_\theta \equiv \left[c_{1\theta} + \frac{1}{2}\left(\frac{P_k + G_k}{\varepsilon} - 1 \right) + \frac{1}{4R}\left(\frac{P_{\theta\theta}}{\varepsilon_{\theta\theta}} - 1 \right) \right]^{-1}. \tag{3.48}$$

The dissipation rate of $\overline{u_i\theta}$ is neglected while $\varepsilon_{\theta\theta}$ is obtained from Eq. (3.41) with the time-scale ratio R taken as constant:

$$R = 0.80 \tag{3.49}$$

(though the alternative strategy proposed by Haroutunian & Launder, 1988, Eq. (3.71) below, was also tested and noted as giving 'some improvements'). Finally, the transport equation for $\overline{\theta^2}$ was simplified to the following:

$$\overline{\theta^2} = -c_\theta^\theta \frac{k}{\varepsilon}\overline{\theta u_j}\frac{\partial \Theta}{\partial x_j}, \tag{3.50}$$

where

$$c_\theta^\theta = 2\left(\frac{P_k + G_k}{\varepsilon} - 1 + R^{-1} \right)^{-1}. \tag{3.51}$$

Equations (3.50) and (3.51) may be recognized as a slightly more elaborate form of Eq. (3.45) which reduce to the latter when there is a balance between generation and dissipation rates. The empirical coefficients implicitly contained in the k–ε model are unchanged from the standard values given in Section 2, while the values of the other coefficients used in the model are:

$$c_{1\theta} = 3.0; \quad c_{2\theta} = 0.4; \quad c_{\varepsilon 3} = 0.8.$$

Note that this last choice is well below the value of 1.44 chosen by Gibson & Launder (1976) and Hossain & Rodi (1982). However, Hanjalić & Vasić (1993) did note that for vertical flows near walls 'the results seem to be less sensitive to the choice of $c_{\varepsilon 3}$ [and so] in line with the practice of other authors, we adopted for vertical walls: $c_{\varepsilon 3} = c_{\varepsilon 1} = 1.44$'. Whether the authors are *really* saying that to get good agreement for vertical natural convection, they had to use 1.44 rather than 0.8 is not known. It is evidently not satisfactory to employ different model coefficients depending on the orientation of the shear flow.

3.3 Remarks on wall-reflection effects

In fact, the Basic Model presented so far does not function satisfactorily near walls or, indeed, on either side of liquid–gas interfaces. Of course, in the immediate vicinity of the wall there is the viscous sublayer and, like the simple k–ε EVM considered in Section 2, adaptations are needed to stress-transport closures to make these schemes sensitive to the local turbulence Reynolds number. However, the weakness of the Basic Model, as presented, extends much more extensively than just the viscous sublayer. For an unstratified simple shear flow directed parallel to the wall, as one approaches the wall, while still in a region where viscous effects are negligible, one finds from experiments or direct numerical simulations that the velocity fluctuations normal to the wall (x_2) are progressively reduced compared with those parallel to the wall (but at right angles to the flow direction) x_3. Now, in the x_2–x_3 plane perpendicular to the flow's motion, there is no direct creation of Reynolds stress; so the presence of *any* fluctuating energy in the x_2–x_3 plane is a consequence of transfer by way of ϕ_{ij} (viz. Eqs. (3.28)–(3.30)). The nature of the approximations made for these processes in the Basic Model means that energy is transferred equally into $\overline{u_2^2}$ and $\overline{u_3^2}$. In a free flow this is usually not too far from the truth; but, as noted above, it becomes strikingly incorrect close to a wall. The pressure reflections from the rigid wall essentially impede the transfer of energy into directions normal to the wall.

In order to mimic this wall-reflection effect a correction must be applied to the Basic Model. The first such proposal was by Shir (1973) who, having included only the process ϕ_{ij1} in his pressure-strain model (Eq. (3.29)), devised a wall correction of broadly similar form. He added two ingredients:

• the unit vector normal to the wall, n_k, which, in a Cartesian system with x_2 normal to the wall, has component values (0, 1, 0);

- the ratio of the local turbulent length scale, l ($\equiv k^{3/2}/\varepsilon$) to the normal distance from the wall, x_n.

Later, Gibson & Launder (1978), working with the more complete form of ϕ_{ij}, added corresponding wall-damping ingredients involving the mean shear and buoyant parts, ϕ_{ij2} and ϕ_{ij3}.

Thus, the form usually adopted in software implementations of the Basic Model is:

$$\phi_{ij} = \phi_{ij1} + \phi_{ij2} + \phi_{ij3} + \phi_{ij1}^w + \phi_{ij2}^w + \phi_{ij3}^w, \quad (3.52)$$

where

$$\phi_{ij1}^w = c_1'(\varepsilon/k)\left(\overline{u_k u_m}\, n_k n_m \delta_{ij} - \frac{3}{2}\overline{u_k u_i} n_k n_j - \frac{3}{2}\overline{u_k u_j} n_k n_i\right) f\left(\frac{l}{x_n}\right), \quad (3.53)$$

$$\phi_{ij2}^w = c_2'\left(\phi_{km2} n_k n_m \delta_{ij} - \frac{3}{2}\phi_{ik2} n_k n_j - \frac{3}{2}\phi_{jk2} n_k n_i\right) f\left(\frac{l}{x_n}\right), \quad (3.54)$$

$$\phi_{ij3}^w = c_3'\left(\phi_{km3} n_k n_m \delta_{ij} - \frac{3}{2}\phi_{ik3} n_k n_j - \frac{3}{2}\phi_{ij3} n_k n_i\right). \quad (3.55)$$

The function f takes the value unity in the vicinity of the wall (where l increases linearly with x_n) and diminishes to zero as l/x_n becomes very small. Sometimes a linear relation, $(l/2.5x_n)$, or a quadratic form, $(l/2.5x_n)^2$, has been adopted. With either choice the coefficients are taken as:

$$c_1' = 0.5; \quad c_2' = 0.3; \quad c_3' = 0.$$

The absence of any wall correction to the buoyant part ($c_3' = 0$) is surprising but is surely due to the absence of a sufficient body of accurate near-wall data for stratified flows. An application of this closure to a stratified surface jet has appeared in Fig. 3.6, where McGuirk & Papadimitriou (1985) show that even at a free surface it is important to include 'wall-reflection' effects.

More recently, Craft & Launder (1992) discovered that the proposed redistribution ϕ_{ij2}^w did not work at all well for flow impinging on a wall. For a stagnation flow, on the axis of symmetry the generation rate of $\overline{u_2^2}$ (the stress perpendicular to the wall) is $-2\overline{u_2^2}\partial U_2/\partial x_2$. This is evidently positive since $\partial U_2/\partial x_2$ is negative. The action of the wall-reflection model of Eq. (3.54) is to reduce the ability of the pressure-strain term to redistribute energy. As a result, more of the originally produced energy remains in $\overline{u_2^2}$. So, instead of dampening velocity fluctuations normal to the wall, it *enhances* them – the opposite of what was required to conform with experiment. A modified version of ϕ^w_{ij2} was therefore proposed:

$$\phi^w_{ij2} = -0.08\frac{\partial U_l}{\partial x_m}\overline{u_l u_m}(\delta_{ij} - 3n_i n_j)f$$

$$- 0.1ka_{lm}\left(\frac{\partial U_k}{\partial x_m}n_l n_k \delta_{ij} - \frac{3}{2}\frac{\partial U_i}{\partial x_m}n_l n_j - \frac{3}{2}\frac{\partial U_j}{\partial x_m}n_l n_j\right)f$$

$$+ 0.4k\frac{\partial U_l}{\partial x_m}n_l n_m\left(n_i n_j - \frac{1}{3}\delta_{ij}\right)f, \tag{3.56}$$

where

$$f = l/2.5x_n \quad \text{and} \quad a_{ij} \equiv \left(\overline{u_i u_j} - \frac{2}{3}\delta_{ij}k\right)\bigg/ k.$$

Another inconsistency in the Basic Model is that most authors have chosen not to add wall-reflection terms to the scalar-flux model; that is Eqs. (3.37)–(3.39) are used unmodified. In fact, in cases where one is principally concerned with heat transfer rates from a wall and buoyant effects are weak, the writer's group has generally preferred to adopt the still simpler algebraic GGDH formulation:

$$-\overline{u_i\theta} = c_\theta\frac{k}{\varepsilon}\overline{u_i u_k}\frac{\partial\Theta}{\partial x_k}, \tag{3.57a}$$

which, close to a heated or cooled wall whose normal points in the x_2 direction, reduces to:

$$-\overline{u_2\theta} = c_\theta\frac{k}{\varepsilon}\overline{u_2^2}\frac{\partial\Theta}{\partial x_k}. \tag{3.57b}$$

Thus, one adopts a second-moment closure to determine the Reynolds stresses as accurately as possible (and, in particular, $\overline{u_2^2}$), but then resorts to a gradient-transport representation for $\overline{u_i\theta}$.

A weakness in applying the above wall-reflection approach is that the corrections have been developed empirically to give reasonable agreement for the case of flow parallel to what is (so far as a turbulent eddy is concerned) an infinite flat surface. But, that idealized situation is not one that pertains in most of the flows to which CFD is applied. It must be partly for that reason that, while Gibson & Launder (1978) produced an algebraic model from their closure (by neglecting entirely the transport terms) which included their wall-reflection proposals, subsequent applications of algebraic second-moment closures (ASMs) to buoyancy-modified turbulence have discarded any such wall-reflection elements. What modellers have generally done, e.g. Hanjalić & Vasić (1993) is permit themselves a modest reoptimization of coefficients to suit the flows considered. As we have noted, Hanjalić & Vasić (1993) actually employed a linear eddy-viscosity model embedding all the buoyant influences in the algebraic forms for the heat fluxes and in the transport equations for k and ε.

In fact, a linear eddy-viscosity model is known from Section 2 to be ill-suited for representing complex strain fields. Thus, if one decides to retain the Hanjalić–Vasić strategy for dealing with buoyant effects, one should at least employ a more general characterization of the effects of strain on the stress field by adopting a non-linear eddy viscosity, an approach that is briefly summarized in the next subsection.

3.4 A non-linear connection between stress and deformation

The great advantage of adopting an eddy-viscosity model is that it is both relatively stable and simple to code. But, as has been noted in Section 2, it falls well short of providing the accuracy, over a broad range of flows, that is needed. The question that arises is whether, by treating the linear-strain rate as just the leading term in a series providing an *explicit* relation between the turbulent stresses and the mean velocity gradient, a more generally applicable formula could be achieved. The original formulation by Pope (1975) included terms up to the fourth power in mean velocity gradient, but the four or five initial proposals that were actually used in computations all adopted only terms up to quadratic level, an extension which introduced three additional unknown coefficients. However, there is a large variation in the numerical values of the coefficients adopted by different workers: even the signs of the coefficients differed. The reason appears to be the following: each group of workers had had their minds on one particular weakness of the linear stress-strain relation that they wished to remove by the addition of the quadratic terms and they chose the coefficients of the non-linear terms to achieve that goal. The writer's conclusion from seeing the disparity in values was that to make a widely useful extension of the linear EVM, one would have to include higher order vorticity and strain products. For that reason our work, Craft *et al.* (1993, 1996, 1997), Suga (1995), has been based on the following cubic formulation:

$$
a_{ij} \equiv \left(\overline{u_i u_j} - \frac{2}{3}\delta_{ij}k\right)\Big/ k = -\frac{\nu_t}{k}S_{ij} + \frac{\nu_t}{\varepsilon}\left(c_1\left(S_{ik}S_{kj} - \frac{\delta_{ij}}{3}S_{mk}S_{km}\right)\right.
$$
$$
+ c_2(\Omega_{ik}S_{kj} + \Omega_{jk}S_{ki})\Big) + c_3\left(\Omega_{ik}\Omega_{jk} - \frac{\delta_{ij}}{3}\Omega_{mk}\Omega_{mk}\right)
$$
$$
+ \frac{\nu_t k}{\varepsilon^2}\left(c_4(S_{ki}\Omega_{mj} + S_{kj}\Omega_{mi})S_{km} + c_5\left(\Omega_{il}\Omega_{lm}S_{mj} + S_{il}\Omega_{lm}\Omega_{mj}\right.\right.
$$
$$
\left.\left. - \frac{2}{3}\delta_{ij}S_{lm}\Omega_{mn}\Omega_{nl}\right) + c_6 S_{ij}S_{kl}S_{kl} + c_7 S_{ij}\Omega_{kl}\Omega_{kl}\right),
$$

where

$$
S_{ij} \equiv \frac{k}{\varepsilon}\left(\frac{\partial U_i}{\partial x_j} + \frac{\partial U_j}{\partial x_i}\right); \quad \Omega_{ij} \equiv \frac{k}{\varepsilon}\left(\frac{\partial U_i}{\partial x_j} - \frac{\partial U_j}{\partial x_i}\right) - \varepsilon_{ijk}\Omega_k; \qquad (3.58)
$$

and Ω_k is the rotation rate of the coordinate axes.

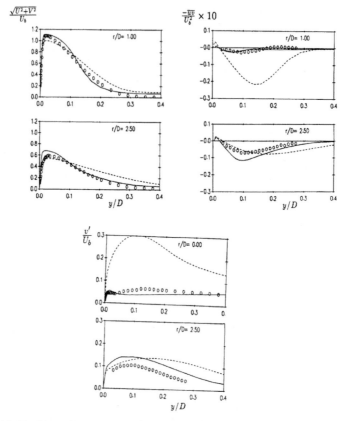

Fig. 3.10 Profiles of mean velocity, shear stress and rms velocity fluctuations normal to wall near the stagnation point of an impinging axisymmetric turbulent jet. (Fully developed pipe flow at discharge from jet, two diameters above impingement surface.) Symbols: experiments (Cooper *et al.*, 1993); broken line: linear EVM (Launder & Sharma, 1974); solid line: NLEVM (Craft *et al.*, 1996b).

Subsequently, Horiuti (1994, personal communication), who has transformed an algebraic version of the Basic stress-transport models into an explicit form, confirmed that cubic elements naturally arise. These cubic elements are essential to give the model the appropriate sensitivity to curvature. Robinson (2001) and Iacovides & Raisee (1999), in numerical studies covering a wide range of separated flows (a class not examined by Suga), have recommended the following slightly amended values for the coefficients:

c_μ	c_1	c_2	c_3	c_4	c_5	c_6	c_7
$1.2/(1 + 3.5 [\max(S, \Omega)]^{1.3})$	-0.1	0.1	0.26	$-10c_\mu^2$	0	$-5c_\mu^2$	$5c_\mu^2$

where $S \equiv \sqrt{S_{ij} S_{ij}}$ and $\Omega \equiv \sqrt{\Omega_{ij} \Omega_{ij}}$.

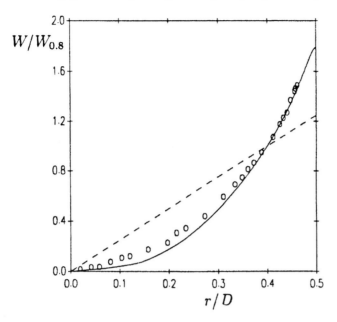

Fig. 3.11 Tangential velocity profile in a pipe rotating about its own axis ($Re =$ 45 000). Broken line: any linear EVM; solid line: Craft *et al.* (1997b); symbols: experiments (Cheah *et al.*, 1993).

The quoted form of the model is designed for free flows or near-wall flow employed with wall-functions. Craft *et al.* (1997) give the more elaborate form where integration across the viscous sublayer is required.

Figures 3.10 and 3.11 provide an impression of the success achievable with this scheme in relation to the linear k–ε model. For the impinging jet, Fig. 3.10, use of the NLEVM avoids the spurious rise in turbulence energy that the linear k–ε model generates due to the flow's deceleration as it approaches the stagnation point. In flow through a pipe rotating about its axis, Fig. 3.11, the cubic EVM predicts the strongly non-linear variation in swirl velocity observed by experiment while any linear EVM gives purely solid-body rotation.

Thus, if one were planning to adopt the Hanjalić–Vasić model, it would seem that the use of Eq. (3.58) in place of Eq. (3.46) would provide a better basis for computing other than thin shear flows. Nevertheless, as a final word, one needs to remind oneself that the direct role of the vertical scalar flux multiplied by the gravitational acceleration clearly has an important impact on the vertical velocity fluctuations (see Fig. 3.7). It is difficult to imagine that a non-linear eddy-viscosity model (with buoyant effects entering only through the k and ε equations) provides a generally satisfactory way of simulating these effects. Section 4, therefore, presents more elaborate routes for treating buoyant flows that have been found to provide superior agreement for a number of difficult test cases.

4 More elaborate routes to closure

4.1 Introduction

The Basic Model and the simplifications of it discussed in the previous section evidently have many weaknesses. For this reason, to handle the buoyancy-modified flows studied at Manchester over the last decade, new, more general models have been developed. It is acknowledged at the outset that our goal has been to develop models of *very wide* applicability. This is not the core need of the industrial user nor the research student, both of whom generally require accurate CFD calculations of *very particular* configurations and flow conditions. Nevertheless, as a guide to directions that may be taken to improve the reliability of the modelling, it is hoped that the present section is seen as helpful.

The rationale stimulating UMIST's advanced modelling work may be summarized as follows.

- It makes little sense to try and 'package' a simplified second-moment model so that it reduces to a buoyancy-sensitive eddy-viscosity closure because a different model is needed for each different class of flow (vertical/horizontal; free/near wall). Such 'zonal models' have little utility for the general circumstances in which CFD software must be applied.
- Using a truncated ASM form is also unattractive because the resultant algebraic equations can become excessively 'stiff', particularly under stratified conditions. Although, in principle, an algebraic equation should be faster to solve than a differential equation, this may not in fact be the case due to the greater difficulties of securing convergence.
- While, in principle, full second-moment closure is a more attractive route, the Basic Model clearly has too many underlying weaknesses (in free shear flows) and inconveniences (in confined flows or other situations where rigid surfaces of complex shape are present) to serve as the basis of a general modelling tool.

The conclusion drawn from the above three levels of shortcomings was that one should try and develop a second-moment closure that did not suffer from the weaknesses of the Basic Model – in other words, a form that could be applied both near walls[+] and in free flows without modification and which achieved a much wider range of validity in buoyant free flows. As it happened, in the late 1980s, research was already in progress in our group to develop a more satisfactory model for near-wall (non-buoyant) flows and swirling free jets. The approach being followed was to apply the so-called *two-component limit* (TCL) principle in modelling the unknown terms in the turbulent transport equations, an approach that had first been advocated by Lumley (1978). This principle seemed admirably suited to account

---

[+] Excluding, however, the very thin viscous sublayer adjacent to the wall itself.

for effects of gravitational forces on turbulence since, in an increasingly stable stratification, vertical velocity fluctuations are progressively damped relative to those in the horizontal plane, so that the turbulent fluctuations do indeed approach the two-component limit (that is, where one component of velocity fluctuations is vanishingly small compared with the other two). The underlying modelling idea is that, in this limit, models of the processes should themselves take on the correct limiting values. If one ensures *that*, there is no risk of any of the Reynolds normal stresses taking on negative values* or, equivalently, of the shear stress violating the Schwartz inequality. An outline of the TCL model is given in the next section, while applications to buoyancy-modified and other complex shear flows are presented in Section 4.3.

In situations where turbulence is decaying with downstream distance due to the persistence of strongly stable conditions, it may be the case that the diffusion terms in the second-moment equations become especially important in predicting the development of the flow. Our experience is that in these circumstances, it is necessary to adopt a substantially more rigorous modelling of second-moment diffusion than is offered by the GGDH formula (e.g. Eq. (3.18)). To illustrate the approach, a partial third-moment closure is applied to a stably-stratified mixing layer. Finally, Section 4 closes with a view of other ideas in closure modelling that have seen some success for particular applications but which to date have not been widely applied.

4.2 TCL closure of the second-moment equations

While Schumann (1977) and, in more detail, Lumley (1978) had set out the principles of applying the two-component limit, it was only some seven years later that Lumley's group (Shih & Lumley, 1985) first advanced suggestions for how to apply these in model formulation. The main attention in that work was directed at the mean-strain parts of the pressure-strain and pressure-scalar-gradient terms (ϕ_{ij2} and $\phi_{i\theta 2}$) in the stress-and-scalar-flux-transport equations.

In what follows, only an outline presentation is given; the constraints applied in determining the numerous coefficients are those adopted by Fu *et al.* (1987) and by later UMIST work (a full account is provided in different chapters of Launder and Sandham, 2002). First we note that from the Poisson equation describing pressure fluctuations the mean-strain part of the pressure-strain correlation may be written (Chou, 1945):

$$\phi_{ij2} = \frac{1}{2\pi} \int \frac{\partial U_i'}{x_m'} \frac{\overline{\partial u_m'}}{x_j'} \left(\frac{\partial u_i}{\partial x_j} + \frac{\partial u_j}{\partial x_i} \right) \frac{\mathrm{d}\,Vol}{r}, \tag{3.59}$$

---

* Evidently, for $i = j$, $\overline{u_i u_j}$ cannot in reality be negative.

where quantities with a prime superscript are evaluated at x' while those without are evaluated at x (see inset sketch).

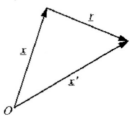

Next we assume that the flow can be regarded as homogeneous so that the mean velocity gradient at x' can be assumed to be that at x and thus taken outside the integral sign. It seems that in most regions of the flow this assumption is adequate.[*]

$$\phi_{ij2}^* = \frac{\partial U_l}{\partial x_m}(a_{lj}^{mi} + a_{li}^{mj}),$$
(3.60)

where

$$a_{lj}^{mi} \equiv \frac{1}{2\pi}\int \overline{\frac{\partial u_m'}{\partial x_l'}\left(\frac{\partial u_i}{\partial x_j} + \frac{\partial u_j}{\partial x_i}\right)}\frac{\mathrm{d}\,Vol}{r}.$$
(3.61)

Thus, the task of devising an approximation for the process ϕ_{ij2}^* has been replaced by that for approximating the fourth-rank tensor a_{lj}^{mi}. On the face of it, it does not seem that progress has been made. The advantage of focusing on Eq. (3.61), however, is that, for any modelled form of a_{lj}^{mi}, one may impose many constraints which the exact equation must satisfy: for example if i is set equal to j (that is, one contracts the equation), continuity demands that $a_{lj}^{mi} = 0$ in a flow where the turbulence is incompressible. There is also the *two-component limit* to apply. First, however, a general form of approximation is needed. Noting that the dimensions of Eq. (3.61) are those of Reynolds stress, and that the two-point velocity correlation appears in the numerator, naturally suggests one should look for an approximation in terms of the stress anisotropy tensor. Thus, we postulate:

$$a_{lj}^{mi} = k \begin{bmatrix} \alpha_1\delta_{mi}\delta_{lj} + \alpha_2(\delta_{ml}\delta_{ij} + \delta_{il}\delta_{nj}) \\ + \alpha_3 a_{mi}\delta_{lj} + \dots \\ + \alpha_6 a_{mi}\delta_{lj} + \dots \\ + \alpha_{13}a_{mi}a_{kl}a_{kj}\dots\alpha_{20}a_{pq}a_{qp}a_{mi}\delta_{lj} \\ + \text{higher order terms} \end{bmatrix}.$$
(3.62)

If we truncate the expansion with third-order terms, even after including symmetry constraints (i.e. the form must, as implied, be unchanged by interchanging m and i or *l* and *j*) there remain twenty unknown coefficients as indicated in (3.62). However, by imposing the following constraints that Eq. (3.61) must satisfy:

$$a_{li}^{mi} = 0 = a_{mj}^{mi}, \tag{3.63a}$$

$$a_{kk}^{mi} = 2\overline{u_m u_i}, \tag{3.63b}$$

and by enforcing the two-component limit, that is:

$$a_{lj}^{m\alpha} = 0 \quad \text{if} \quad u_\alpha = 0 \quad \text{(and thus } \overline{u_\alpha u_j} = 0), \tag{3.63c}$$

eighteen of the twenty unknowns may be determined. After extensive re-grouping of terms, the result may be presented as:

$$
\begin{aligned}
\phi_{ij2}^* = {} & -0.6\left(P_{ij} - \frac{1}{3}\delta_{ij}P_{kk}\right) + 0.3\varepsilon\, a_{ij}\left(\frac{P_{kk}}{\varepsilon}\right) \\
& -0.2\left[\frac{\overline{u_k u_j}\,\overline{u_l u_i}}{k}\left(\frac{\partial U_k}{\partial x_l} + \frac{\partial U_l}{\partial x_k}\right) - \frac{\overline{u_l u_k}}{k}\left(\overline{u_i u_k}\frac{\partial U_j}{\partial x_l} + \overline{u_j u_k}\frac{\partial U_i}{\partial x_l}\right)\right] \\
& - c_2[A_2(P_{ij} - D_{ij}) + 3a_{mi}a_{nj}(P_{mn} - D_{mn})] \\
& + c_2'\left\{\left(\frac{7}{15} - \frac{A_2}{4}\right)\left(P_{ij} - \frac{1}{3}\delta_{ij}P_{kk}\right)\right. \\
& + 0.1\varepsilon\left[a_{ij} - \frac{1}{2}\left(a_{ik}a_{kj} - \frac{1}{3}\delta_{ij}A_2\right)\right]\left(\frac{P_{kk}}{\varepsilon}\right) - 0.05a_{ij}a_{lk}P_{kl} \\
& + 0.1\left[\left(\frac{\overline{u_i u_m}}{k}P_{mj} + \frac{\overline{u_j u_m}}{k}P_{mi}\right) - \frac{2}{3}\delta_{ij}\frac{\overline{u_l u_m}}{k}P_{ml}\right] \\
& + 0.1\left[\frac{\overline{u_l u_i}\,\overline{u_k u_j}}{k^2} - \frac{1}{3}\delta_{ij}\frac{\overline{u_l u_m}\,\overline{u_k u_m}}{k^2}\right]\left[6D_{lk} + 13k\left(\frac{\partial U_l}{\partial x_k} + \frac{\partial U_k}{\partial x_l}\right)\right] \\
& + \left.0.2\frac{\overline{u_l u_i}\,\overline{u_k u_j}}{k^2}(D_{lk} - P_{lk})\right\}, \tag{3.64}
\end{aligned}
$$

where $D_{mn} \equiv -\left(\overline{u_m u_k}\dfrac{\partial U_k}{\partial x_n} + \overline{u_n u_k}\dfrac{\partial U_k}{\partial x_m}\right)$ and A_2 is the second invariant of the stress anisotropy defined in Eq. (3.66).

In this form we note that the only linear term is in fact the form adopted in the Basic Model (but note that now the coefficient '0.6' is not an empirical constant but a direct product of the analysis). The groups of terms multiplied by the two undetermined coefficients are of very unequal length. For that reason our early work with the TCL model put c_2' to zero (Launder, 1989a). However, it was later discovered that, by retaining both groups of terms, not only could some of the hard-to-predict features of free-shear flows be better resolved but, more importantly, the approximation could be applied also close to walls (though, as noted above, outside the viscous/buffer region where spatial variations in mean-velocity gradient are so

rapid that an 'inhomogeneous' correction is needed). The recommended values
(Launder & Li, 1994) are:

$$c_2 = 0.55; \quad c_2' = 0.6.$$

While analysis has proved effective for determining the form of ϕ_{ij2}^*, for the turbulence part of ϕ_{ij} a more 'brute force' approach is adopted. It is assumed that:

$$\phi_{ij1} = -c_1\varepsilon \left(a_{ij} + c_1' \left(a_{ik}a_{jk} - \frac{1}{3}A_2\delta_{ij} \right) \right), \tag{3.65}$$

which will be recognized as the form adopted by the Basic Model (Eq. (3.29)) with
an additional non-linear term. In order that this expression should conform with the
two-component limit, it is necessary that c_1 should fall to zero whenever this limit is
reached. A way of doing this was proposed in the original paper by Lumley (1978).
We can form two independent invariant measures of the anisotropy of Reynolds
stress from the tensor a_{ij} which may conveniently be taken as:

$$A_2 \equiv a_{ij}a_{ji} \quad \text{and} \quad A_3 \equiv a_{ij}a_{jk}a_{ki}. \tag{3.66}$$

What, for present purposes, is extremely helpful is that the invariant quantity A,
defined as:

$$A \equiv 1 - \frac{9}{8}(A_2 - A_3), \tag{3.67}$$

has the property that it always equals zero in two-component turbulence (and, of
course, unity in isotropic turbulence). Thus, by making c_1 a function of A, it is easy
to ensure compliance with the TCL. The values of these coefficients adopted in
most of our studies are:

$$c_1 = \left(3.75A_2^{1/2} + 1\right)A; \quad c_1' = 0.7. \tag{3.68}$$

A complete statement of the TCL model for ϕ_{ij} together with the corresponding
form for $\phi_{i\theta 1}$, $\phi_{i\theta 2}$ and $\phi_{i\theta 3}$ are given in Appendix 2, Table 3.5. Derived originally
by Craft (1991) (see also Craft & Launder, 1989), the TCL modelling of $\phi_{i\theta 2}$ leads
to the interesting result that all the coefficients are determined by the analysis,
their values being shown in the table. (The numerical values for $\phi_{i\theta 1}$, however, are
determined by computer optimization.)

The buoyant contributions to ϕ_{ij} and $\phi_{i\theta}$ are determined in a similar manner, the
forms being listed in Tables 3.4 and 3.5. Again, all the free coefficients are fixed
by kinematic constraints, including compliance with the two-component limit.

There is a further important novelty concerning the TCL model and that relates
to the dissipation equation. With the TCL model, while superficially the form of

the ε transport equation is conventional:

$$\frac{D\varepsilon}{Dt} = d_\varepsilon + c_{\varepsilon 1}\frac{\varepsilon}{k}(P_k + G_k) - c_{\varepsilon 2}\frac{\varepsilon^2}{k}, \tag{3.69}$$

the coefficients have been optimized to the values:

$$c_{\varepsilon 1} = 1.0; \quad c_{\varepsilon 2} = \frac{1.92}{1 + 0.7A_2^{1/2}A_{25}}, \tag{3.70}$$

where $A_{25} = \max{(A, 0.25)}$.

The reduction of $c_{\varepsilon 1}$ (from the more usual value of approximately 1.44) means that local mean velocity gradients have a diminished role in the equation compared with transport. The sink term has also been reduced except for the case of isotropic turbulence for then, since $A_2 = 0$, the conventional value of 1.92 is returned for $c_{\varepsilon 2}$. Among the various benefits that this choice brings over the usual form is that the same coefficient values can be used in both horizontal and vertical buoyant flows.

While solution of a transport equation for $\varepsilon_{\theta\theta}$ should, for consistency, be the path for finding this unknown quantity, in common with others active in this field, the writer recommends that at present this be found by prescribing the time-scale ratio, R. Unlike the previously proposed practice of assigning a constant value to R, however, following Haroutunian and Launder (1988), it is suggested that it be obtained from:

$$R = \frac{3}{2}(1 + A_{2\theta}), \quad \text{where } A_{2\theta} \equiv \frac{\overline{u_i\theta}\,\overline{u_i\theta}}{k\overline{\theta^2}}. \tag{3.71}$$

$A_{2\theta}$ is just the dimensionless correlation coefficient between velocity and scalar fluctuations.

4.3 *Some applications of the TCL Model*

We begin this comparison by considering the computations of Craft (1991) of a horizontal, stably-stratified flow that had been created by passing a uniform flow past a screen of differentially heated horizontal rods, Webster (1964), Young (1975). The computations shown in Fig. 3.12 have assumed local equilibrium (i.e. $(P_k + G_k) = \varepsilon$) which, judging from the scatter in the experimental data, was not always close to the truth. Nevertheless, since the data were simply plotted showing the dependence of various non-dimensional groups involving the Reynolds stresses and heat fluxes on the gradient Richardson number, Ri, there is no other supposition that could have been made. (Note that in this figure v denotes vertical velocity fluctuations and u streamwise fluctuations.) Evidently, as the gradient Richardson number increases, the shear-stress and horizontal-heat-flux correlation coefficients decay while the

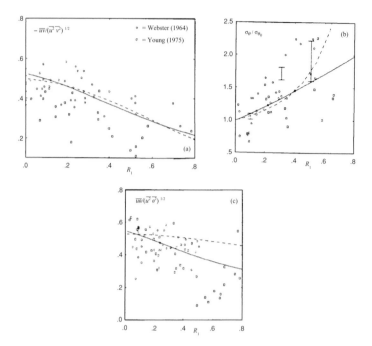

Fig. 3.12 Second-moment sensitivity to stable buoyant stratification in horizontal, nominally-equilibrium homogeneous shear flows, Craft *et al.* (1996).
(a) Shear-stress correlation coefficient.
(b) Normalized turbulent Prandtl number.
(c) Correlation coefficient of horizontal heat flux.
Symbols: experiments, o Webster (1964); □ Young (1975).
Computations: - - - - - Basic Model, _____ TCL Model.

turbulent Prandtl number increases. The TCL Model reproduces this behaviour at least as accurately as the Basic Model even though the two empirical coefficients in the buoyant terms of the latter model were optimized by reference to these data (Launder, 1975). In contrast, the TCL Model has no empirically adjustable coefficients in the buoyant parts of ϕ_{ij} and $\phi_{i\theta}$.

Attention now shifts to two-dimensional inhomogeneous flows, so transport equations have been solved for the continuity and momentum equations, for four Reynolds stresses, two scalar fluxes, for $\overline{\theta^2}$ and ε; that is, eleven dependent variables in all. First we examine the spreading rate of five self-preserving free shear flows covering a range of flow types (Cresswell *et al.*, 1989). Although these flows are usually regarded as 'parabolic' (i.e. that streamwise diffusion is neglected thus permitting a forward-marching solution to be applied) in the results presented in Table 3.2, an elliptic solver has been used for both the buoyant plumes and the jets. The differences between a parabolic and an elliptic solution for these cases are, in fact, quite significant especially for the axisymmetric flows (see el Baz *et al.*, 1993).

Table 3.2 *Spreading rates of five self-preserving shear flows*

Flow	Basic Model	Recommended exptl. values	TCL Model
Plane plume	0.078	0.120	0.118
Round plume	0.088	0.112	0.122
Plane jet	0.100	0.110	0.110
Round jet	0.105	0.093	0.101
Plane wake	0.078	0.098	0.100

For example, a parabolic solution leads to spreading rates for the round jet some 14% greater than that returned for an elliptic treatment.

It is evident from Table 3.2 that the TCL Model is far more successful at capturing the spreading behaviour of these five free flows than the Basic Model. Even though there is some discrepancy in the computed rate of spread of the two axisymmetric flows, the error is at least very similar in these two cases and small enough ($\approx 8\%$), for many applications, to be within the required computational accuracy. (The Basic Model, in contrast, has larger errors for each flow and, more seriously, while the computed spreading rate of the round plume is too low, that for the round jet is too high.) The same workers extended their study to the case of a hot round jet discharged vertically downwards into a cold-water environment moving upward at less than 2% of the mean jet discharge velocity. The experimental study was one that had been undertaken at the CEGB's Berkeley Nuclear Laboratories specifically to assess capabilities of alternative modelling schemes (Cresswell, 1988). Figure 3.13a provides a vector-velocity plot of the flow in the vicinity of discharge showing the reversal in the jet direction, while Figs. 3.13b & c compare the measured and computed shear-stress and mean velocity profiles at a position one diameter downstream of the jet discharge. It is clear that the TCL Model captures the mixing more accurately than the Basic Model. Moreover, an unexpected advantage from the use of the TCL scheme was that, because that model (even during iterations of an un-converged solution) could not generate negative normal stresses, convergence rates were much faster. Thus, despite the many extra arithmetic operations per iteration (associated with the complex forms of ϕ_{ij} and $\phi_{i\theta}$) total computing times for convergence were 30% less than for the Basic Model.

We return finally to the case of the horizontal, stably stratified mixing layer. When considering Fig. 3.5, attention in Section 2.4.2 was directed purely at the inadequacy of computations with the k–ε eddy-viscosity model (shown by a broken line). Clearly, the use of the TCL second-moment closure (solid line) gives much closer agreement with the experimental data presented here in normalized form. In

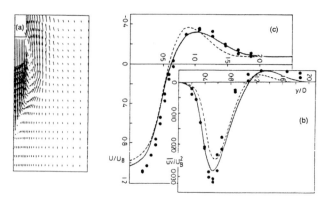

Fig. 3.13 Axisymmetric, downward directed buoyant jet, Cresswell *et al.* (1989).
(a) Computed velocity vectors.
(b) Turbulent shear stress one diameter below jet discharge.
(c) Mean velocity at same position.
Symbols: experiments, Cresswell (1988);
computations: - - - - - Basic Model, _____ TCL Model.

fact, at the most downstream position ($x = 40$ m), it is evident that the computed salinity profile is too broad even with the TCL model. The reason for this has been the subject of a detailed study by Kidger (1999) (see also Craft, Kidger and Launder, 1997; Craft and Launder, 2002), from which it was concluded that the problem principally lay with the rudimentary handling of diffusive transport of the scalar fluxes. This is a topic examined further in Section 4.4.

Before that, it may be of interest to draw attention to other (non-buoyant) applications of the TCL Model in three-dimensional and other complex strain fields. The first relates to the fully-developed flow through a duct of rectangular cross-section where a portion of the wall is roughened, Fig. 3.14. The experiment by Hinze (1973) was perhaps contrived to achieve particular effects but, nevertheless, ducts composed of partly rough and partly smooth surfaces are not uncommon in the nuclear industry – for example, in the cooling channels of AGRs. In this experiment only the axial velocity contours were measured, though the strong uplift of these contours near the centre of the lower wall is a clear indication of an appreciable secondary motion induced by the anisotropy of the Reynolds stresses acting in the cross-sectional plane of the duct. An isotropic viscosity model of turbulence entirely fails to predict any secondary motion because, if we take x_2 and x_3 in the plane of the cross section, the major contribution to the secondary motion arises from variations in the *differences* between $\overline{u_2^2}$ and $\overline{u_3^2}$ over the cross section. Now, an isotropic viscosity model gives equal values to these normal stresses, thus resulting in there being zero source of streamwise vorticity (which is another way of looking at the secondary flow). By solving transport equations for the Reynolds stresses, one

Table 3.3 *Asymptotic spreading rates for 3D wall jet*

	Normal	Lateral	
	$\dfrac{\mathrm{d}y_{1/2}}{\mathrm{d}z}$	$\dfrac{\mathrm{d}x_{1/2}}{\mathrm{d}z}$	$\dfrac{\dot{x}_{1/2}}{\dot{y}_{1/2}}$
Expt. Abrahamsson *et al.* (1997)	0.065	0.32	4.94
k–ε EVM	0.079	0.069	0.88
Basic DSM	0.053	0.814	15.3
TCL DSM	0.060	0.51	8.54

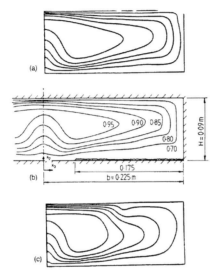

Fig. 3.14 Axial velocity contours in flow through a partially-roughened rectangular channel, Launder & Li (1994).
(a) Computation, TCL Model.
(b) Experiment, Hinze (1973).
(c) Computation with Basic Model.

escapes this problem, though one notes from Fig. 3.14 that while the Basic and TCL Models both predict that there should be an upwelling of the streamwise velocity contours above the smooth patch on the lower wall, only the TCL Model gives the correct shape of the contours due to the different secondary velocity profiles predicted by the two models.

An even stronger manifestation of the effects that an induced streamwise vorticity can exert occurs in the three-dimensional wall jet or free-surface jet: that is, the shear flow arising from the discharge of a round jet parallel and adjacent to a plane surface (which may be a wall or, say, a gas–liquid interface). Table 3.3

from Craft & Launder (1999) compares asymptotic, self-similar growth rates for the three-dimensional wall jet obtained with different levels of modelling with the experimental data of Abrahamsson *et al.* (1997). The observed lateral rate of spread parallel to the wall is some five times as great as that normal to it. The eddy-viscosity model in contrast predicts a spreading-rate ratio of less than unity. The solution of stress-transport equations has a very large effect, the Basic Model giving a rate of spread that is now too large by more than a factor of three! The TCL Model does better both in the computed rates of spread and in the shapes of the velocity profiles (not shown) though the quoted spreading rates are evidently still appreciably too large. The reason for this discrepancy, it transpired, was that, contrary to the experimenters' supposition, the measured wall jet had not reached its fully-developed state (an assumption which had reduced the computation from three-dimensional to effectively two-dimensional thus enabling a more precise numerical solution). Craft & Launder (1999, 2001) show that, when they subsequently computed the flow with a 'marching' three-dimensional solver, the TCL Model gave a computed rate of spread that accurately matched the measured behaviour at 50 nozzle diameters downstream (though to reach asymptotic conditions required a further 100 diameters of development).

Similar modelling challenges arise with the three-dimensional, isothermal free-surface jet, for which experiments were made by Rajaratnam & Humphreys (1984). Because in this case the experimental data were evidently not fully developed, a developing flow computation was adopted from the beginning (Craft *et al.*, 1999). One notes from Fig. 3.15 that the vertical and lateral growth rates are captured much more satisfactorily with the TCL Model, a superiority that carries over to the mean velocity profiles too, Fig. 3.16. The writer was, indeed, surprised to find that the Basic Model performed so poorly when applied to jets near plane surfaces; for the wall-reflection terms, Eqs. (3.52), (3.56), had been calibrated for just such surfaces. It helps underline the advantages of adopting a model that does not have to apply such wall-reflection corrections.

As a final example, the opposed wall-jet study of Craft *et al.* (2003) is shown in Fig. 3.17. The main problem is that of predicting correctly how far the downward-directed wall jet can penetrate against the slow upward-moving stream. Altogether three different turbulence models and three different wall treatments were employed. Because the figure is printed in monochrome both the 'blue' (downward vertical velocity) and the 'red' (upward vertical velocity) ends of the scale appear as black. Our attention is, however, purely on the distance that the downward-directed wall jet on the right side of each figure penetrates. The target data are the LES simulations of Addad *et al.* (2003). Partly the RANS results confirm what has already been reported in Section 3; namely, that the analytical wall-function approach of Gerasimov (2003) used in conjunction with the high-*Re* k–ε EVM (not shown)

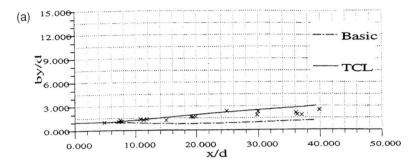

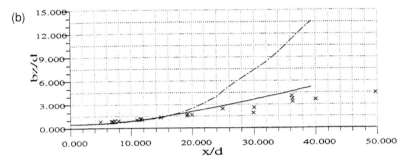

Fig. 3.15 Vertical and horizontal spreading rates in a free-surface, non-buoyant jet, Craft *et al.* (1999).
(a) Vertical spread.
(b) Horizontal spread.
Symbols: experiment, Rajaratnam & Humphreys (1984).
Computations: _ ___ _ ___ Basic Model,
        _____ TCL Model.

does well in mimicking the behaviour of the low-*Re k–ε* EVM. However, for this colliding flow neither scheme predicts a fast enough detachment and turning of the wall jet. If the *k–ε* EVM is replaced by the TCL second-moment closure, however, agreement with the LES data of Addad *et al.* (2003) is improved, particularly when used with the analytical wall function. Gerasimov (2004) has also considered cases where the wall-jet temperature is higher than the up-flowing stream, causing a buoyant up-thrust to be applied to the wall jet which reduces the penetration depth. Again, the TCL closure with analytical wall functions achieved the best agreement with the reference data.

4.4 Diffusive transport in stably-stratified flows

In considering the stably-stratified saline mixing layer in Section 4.3, it was asserted that the too wide dispersal of the salinity mixing layer 40 m downstream was a consequence of the rudimentary approximation of diffusion processes

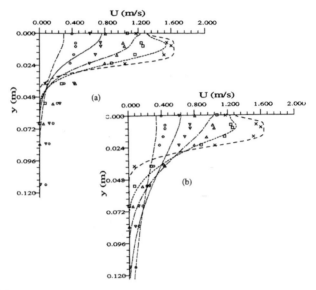

Fig. 3.16 Mean velocity-profile development in free-surface jet, Craft *et al.* (1999).
Symbols: experiment, Rajaratnam & Humphreys (1984).
Lines: computations
(a) Basic Model; (b) TCL Model.

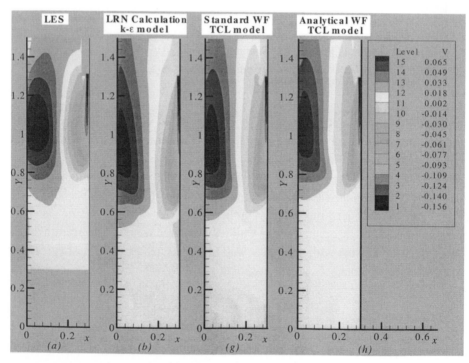

Fig. 3.17 Prediction of the downward-directed wall jet. (Note: wall jet descends
at right of each figure.) From Craft *et al.* (2003).

in the second-moment equations. Craft *et al.* (1996) developed a model for the diffusion of $\overline{\theta^2}$ from approximating and substantially simplifying the exact transport equation for $\overline{u_i \theta^2}$; the form proposed was:

$$\overline{u_i \theta^2} = -c_\theta' \frac{k}{\varepsilon} \left[\left(\overline{u_i u_k} - \frac{3}{4} \frac{\overline{\theta^2} \beta_i}{\varepsilon_\theta} \overline{\theta u_k} \right) \frac{\partial \overline{\theta^2}}{\partial x_k} + 2 \overline{\theta u_k} \frac{\partial \overline{u_i \theta}}{\partial x_k} \right]. \tag{3.72}$$

The crucial element in the above is the presence of the buoyant term within parentheses which, in a stably-stratified medium, reduces the diffusion of $\overline{\theta^2}$. While this form certainly led to marked improvement in the mean density profile, it has proved to be tiresomely unstable. Consequently, in his Ph.D. research, Kidger (1999) opted for solving transport equations for all the triple moments containing θ (i.e. salinity fluctuations). A discussion of the closure strategy employed is not attempted here but the relevant (edited) extract from Kidger's thesis is included as Appendix 3. Thus, in addition to transport equations for all second moments, further equations are solved for:

$$\begin{array}{ll} \overline{\theta^3} & \text{(which appears in the buoyant term of the } \overline{u_i \theta^2} \text{ equation),} \\ \overline{u_i \theta^2} & \text{(i.e. for } \overline{u_1 \theta^2} \text{ and for } \overline{u_2 \theta^2} \text{ in a two-dimensional flow)} \\ \text{and} \quad \overline{u_i u_j \theta} & \text{(i.e. in two dimensions, } \overline{u_1^2 \theta}, \overline{u_2^2 \theta} \text{ and } \overline{u_1 u_2 \theta}). \end{array}$$

Thus six extra transport equations were solved for the above variables while the six non-zero velocity triple moments $\overline{u_i u_j u_k}$ were obtained from the conventional GGDH approach. Kidger (1999) reported much greater difficulties in getting the second-moment version to converge compared with the third-moment closure; so, that may be noted as one advantage of the more elaborate scheme. A very stark difference was also found in the actual computed behaviour, Fig. 3.18. As with the earlier results of Craft *et al.* (1996), the second-moment results show a too great dispersion and, what is worse, the salinity levels do not remain within the upper and lower bounds of saline concentration at the start of the mixing region. The cause of this unphysical behaviour is the buoyant terms in the vertical density-flux equation leading to a counter-gradient transport of the denser (saltier) fluid.[+]

The same buoyant terms are also present, of course, in the second-moment equations when one adopts the third-moment closure. However, by determining

---

[+] The reader may note that the TCL2 results in Fig. 3.17 are different from those shown in Fig. 3.5. The latter computations were performed ten years ago and, understandably from that distance, memories have faded. It seems, however, that in Fig. 3.5 the local computed values of $\rho_{\min}$ and $\rho_{\max}$ were used (and thus the over- and undershoots did not become apparent) whereas, in Fig. 3.17, experimental values have been used for these quantities, plainly exposing the unphysical nature of the results.

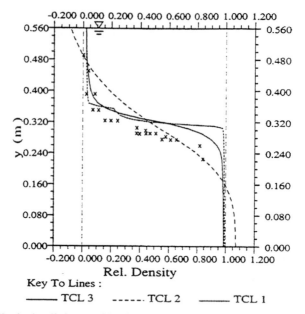

Fig. 3.18 Vertical salinity profiles in stably-stratified mixing layer, Craft *et al.* (1999).
TCL 3: partial third-moment solution;
TCL 2: second-moment closure using GGDH for third-moments;
TCL 1: second-moment closure for dynamic field; GGDH for scalar fluxes.

the third moments from their own transport equation (in particular $\overline{u_2^2\theta}$, which represents the vertical diffusional transport of $\overline{u_2\theta}$), as seen in Fig. 3.18, the levels of Θ remain properly bounded and indeed the Θ profile is in much closer agreement with the experiment.*

Before leaving this case it is worth noting that almost as satisfactory agreement is obtained if, instead of elaborating the second-moment closure to third-moment level, one *simplifies* it by dropping the transport equations for the scalar fluxes, using instead the GGDH formula to obtain them, i.e. $\overline{u_i\theta} = -(c_\theta k\overline{u_i u_k}/\varepsilon)\partial\Theta/\partial x_k$. This is the closure level referred to in Fig. 3.18 as 'TCL 1'. The density profile shows for this case a rather too thin mixing region or, in other words, a somewhat too abrupt switchover from fresh to saline water. Agreement is, however, clearly superior to that of the (complete) second-moment closure. It is further noted that the computations of the downward-directed wall jet shown in Fig. 3.17, Gerasimov (2004) also adopted a 'TCL 1' level of modelling. The questions of strategy which these experiences raise are considered briefly in Section 5.

---

* There is a vertical displacement of the measurements relative to the computations which seems to be a consequence of weak secondary flow in the experiments (which was of course not captured by these two-dimensional computations).

4.5 Other approaches to the modelling of buoyant flows

4.5.1 DNS and LES treatments

Direct and *large-eddy simulations* of turbulent flow are now becoming very widely employed in turbulence research. Direct numerical simulations (DNS) are simply accurate numerical solutions of the time-resolved Navier–Stokes equations and (where appropriate) the corresponding energy or species transport equations at Reynolds or Rayleigh numbers where the flow is turbulent. Because of the huge amount of memory and computing time required, these are still very much the province of the specialist, fundamental research laboratory. The Reynolds numbers that can be resolved are still low and the configurations studied still rather simple. The most extensively studied geometry has been the flow between parallel planes, both forced and buoyantly driven convection being explored. For example, Versteegh & Nieuwstadt (1998) consider the buoyantly driven flow between vertical planes (one heated, one cooled) and provide a detailed quantitative analysis of their results within the context of budgets of the second-moment equations. A similar study was undertaken by Wörner & Grötzbach (1998) for horizontal, differentially heated plates (the Rayleigh–Bénard problem). From this the novel and rather controversial result to emerge was that most of the diffusion of the second moments (for example the turbulence energy) was driven by pressure fluctuations rather than by velocity fluctuations.

While the role of DNS today is very much in the business of helping to improve turbulence modelling at second-moment and simpler levels, large-eddy simulation (LES) is being selectively applied directly to industrial problems that cannot be – or, at least, have not so far been – successfully tackled by second-moment closure. As with DNS, one adopts a three-dimensional time-dependent approximation but acknowledges that one cannot resolve all fine-scale motions responsible for most of the dissipation. Thus a 'subgrid-scale' (sgs) turbulence model is usually added. These models have ranged in complexity from second-moment closure, Deardorff (1973) (though with a turbulent length scale related to the size of the grid) to the mixing-length hypothesis, whose first use within the context of sgs modelling is attributed to Smagorinsky (1963). Moreover, several studies have made no explicit use of a subgrid-scale model. Essentially, these rely on numerical diffusion associated with the numerical discretization scheme to take the place of an sgs model. Such approaches have been adopted by Boris *et al.* (1992), Muramatsu (1993) and Hashiguchi (1996) among others. However, currently the most popular route to subgrid-scale modelling is probably the developed form of Smagorinsky model proposed by Germano *et al.* (1991) in which, essentially, the coefficient of the local subgrid-scale viscosity is obtained by extrapolating statistics of the finest turbulent eddies actually resolved by the grid.

A recent review of LES studies has been provided by Grötzbach & Wörner (1999). Among the many applications cited are several related to problems of *thermal striping*, a problem which necessitates the acquisition of time-resolved information on the large-scale flapping and intermittency associated with the outer regions of turbulent jets, e.g. Voke & Gao (1994). It is certainly such intrinsically time-varying problems where LES is at its most powerful and where the computational penalty of adopting such an approach compared with a time-dependent or stationary RANS approach is less severe.

4.5.2 Unsteady RANS approach

There is an extensive record of periodic or quasi-periodic flows being computed by RANS models in time-dependent numerical schemes. The more challenging situations to examine are those where the unsteadiness arises from some flow-generated instability or other feature, the Karman vortex sheet for example. The level of reliability attainable with this approach relative to that of LES seems to depend on the problem being tackled and on who is doing the computations. Recently, Kenjeres & Hanjalić (1999, 2002, 2003) applied the approach to the study of Rayleigh–Bénard convection. This represented an interesting extension from previous studies because the flow, while certainly dominated by large-scale, buoyancy-driven eddies, did not exhibit the very clear, coherent unsteadiness that arises downstream of a bluff body. The turbulence model adopted was a greatly truncated algebraic form of second-moment closure, similar to that of Hanjalić & Vasić (1993) discussed in Section 3. Figure 3.19 compares an instantaneous 'snapshot' showing the eddy pattern created by the Kenjeres–Hanjalić unsteady RANS (URANS) with the DNS results of Wörner & Grötzbach (1998). Clearly, the eddy structure is similar in the two cases though, as would be expected, the URANS computation lacks the finer scale detail (the role of the turbulence model is to make good the effects of this missing fine-scale structure).

Although the DNS results with which comparison is made are not at the same Rayleigh number, it appears from Fig. 3.20 that satisfactory prediction of the rms temperature fluctuation is achieved since the URANS results at a Rayleigh number of 10^7 lie between the DNS results at 2×10^7 and 6.3×10^5. They are also evidently superior to those obtained with a (steady) two-dimensional RANS treatment employing the same turbulence model which leads to levels of temperature variance at a height mid-way between the upper and lower plate of only about half of that generated by the URANS simulation.

A conclusion seems to be that an unsteady RANS treatment may offer the prospect of simulating various flow phenomena in gravitationally driven turbulent flows which have hitherto been regarded as being predictable, if at all, only with the more costly large-eddy simulations. Indeed, the Kenjeres–Hanjalić team

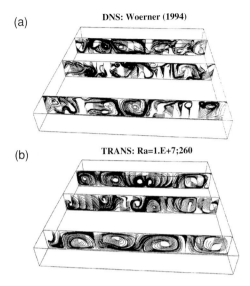

Fig. 3.19 Eddy structure in simulations of Rayleigh–Bénard convection, Kenjeres & Hanjalić (1999). (a) DNS simulation, Wörner & Grötzbach (1998). (b) Unsteady RANS.

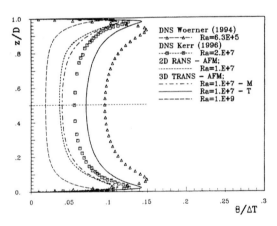

Fig. 3.20 Rms temperature fluctuation profile between upper and lower plates, Kenjeres & Hanjalić (1999).

has continued to press upward in Rayleigh number and in summer 2003 showed impressive results for values up to 10^{14} (a level of *Ra* well beyond what has been simulated with LES) which exhibit a distinct increase in slope of the *Nu–Ra* relationship. This result has greatly interested the experimental physics community as two recent experiments of high-*Ra* R-B convection showed strongly different behaviour at high Rayleigh numbers: one indicated that *Nu* increased as $Ra^{0.31}$ over the whole turbulent-flow regime, while the other found that the exponent changed from 0.30 to 0.38 for $Ra \approx 10^{11}$, a result which the Kenjeres–Hanjalić (2003) computations closely reproduce.

The relative cost of the two strategies (LES or URANS) is still a matter of debate. Hanjalić (personal communication) feels that the cost saving of employing an optimized URANS approach is at least one order of magnitude (which would seem to be more than borne out by the fact that, with the URANS strategy, results have been obtained for Rayleigh–Bénard convection for values of Ra at least two orders of magnitude higher than LES). Grötzbach & Wöhner (1999), by contrast, cite examples where, compared with a URANS treatment employing just a linear k–ε EVM, the LES treatment took, in one case, eight times as long ('but giving much more reliable results and more information') while, for the study of flow through a cross-corrugated heat exchanger, the time penalty was only a factor of two! The matter seems likely to be resolved only by an independent re-examination of this issue. It appears that groups that have so far employed both LES and URANS on the same problem have not adopted gridding and other numerical practices that are equally suited to both schemes. Indeed, Moin (2000) has asserted that really the Kenjeres–Hanjalić study above should be interpreted as a genuine LES because, in his view, any subgrid-scale model should be assigned independently of the grid used (a practice which may well become the adopted norm but which is not at present).

4.5.3 Two-scale modelling

The concept of adopting more than one time or length scale within one's turbulence model was first advanced in the 1970s (Launder & Schiestel, 1978; Hanjalić *et al.*, 1980). The cited papers imagined the energy-containing part of the turbulence spectrum, Fig. 3.21, to be split into a 'production' range (wherein the turbulence energy is produced) and a 'transfer' range (wherein there is little net production or dissipation . . . but where energy is simply broken down into smaller and smaller eddies). Eventually, the energy is dissipated in very fine scale eddies. Thus, instead of transport equations for k and ε, with this split-spectrum concept, k_P, k_T, ε_P and ε_T all appeared as dependent variables.[+] This decomposition added two strengths to the model: firstly, it built in a time delay between new energy entering turbulence and the effects being felt in the dissipation rate; secondly, it provided two new dimensionless parameters, k_P/k_T and $\varepsilon_P/\varepsilon_T$, on which the empirical coefficients could rationally be held to depend. However, the original model was constructed around a linear eddy-viscosity stress-strain hypothesis. The elaborate treatment of the scales seemed, to the writer, to be ill-matched to the rudimentary choice of the

---

[+] Separate equations were not required for k and ε since $k = k_P + k_T$ and ε_T was assumed equal to ε.

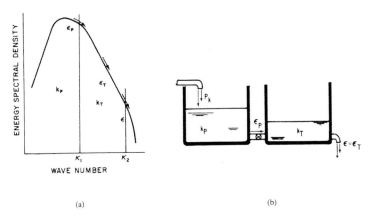

(a)

(b)

Fig. 3.21 Split-spectrum concept (from Hanjalić *et al.*, 1980).
(a) Sketch of energy spectrum and its partitioning.
(b) Plumbing analogy of split-spectrum concept.

stress-strain connection. Schiestel (1987) subsequently proposed a more elaborate second-moment scheme (employing $\overline{u_i u_j}_P$ and $\overline{u_i u_j}_T$ as dependent variables). This level of closure is, personally, seen as a too elaborate approach for other than (numerically) very simple one-dimensional, homogeneous flows (since one needs to solve transport equations for two second-rank tensors). However, Yonemura & Yamamoto (1995) have persisted in this strategy, applying it successfully to a number of (non-buoyant) two-dimensional boundary layers.

The writer believes that it is desirable to pursue further the multi-scale approach but would advocate a different strategy from that proposed above. Firstly, the main need for innovation can be met by simply adopting two transport equations for ε, that is (as before) for the production (ε_P) and transfer (ε_T) ranges (and presuming $\varepsilon_T = \varepsilon$). This feature provides, in principle, both the spectral time lag (between the input and dissipation of turbulence energy) and the opportunity to make the modelling coefficients (weakly) dependent upon $\varepsilon_P/\varepsilon_T$ (which introduces some explicit dependence of the model form on the shape of the energy spectrum). The stress transport equations would be solved either in differential (preferably) or algebraic form adopting (desirably) the TCL approximations introduced in Section 4.1. The time scale for the diffusion processes should be k/ε_P but, in the various dissipation terms, ε_T would be employed. A further feature should be the development of a source term for ε_P which depended on the mean strain in ways other than just the turbulence energy production rate, P_k. So far as the writer knows there has been no work in this direction at present. Indeed, the only application found of the multiple-scale approach to a buoyancy-modified flow was a very preliminary ASM study (Huang *et al.*, 1993) using modelling ideas from the 1970s.

5 Overall conclusions and recommendations

- The effects of gravitational forces on turbulence structure are highly diverse in character and by no means completely understood. It is thus not surprising that there is no sentiment among the turbulence modelling community that a 'universal' treatment for such flows is currently available – or will become available in the foreseeable future.

- As a generalization, it may be said that raising the order of closure modelling improves the likelihood of generating CFD results that correspond closely with reality. Thus, an algebraic second-moment (ASM) approximation may be expected to be more reliable than a standard eddy-viscosity (EVM) treatment; successively refining the ASM model either to a differential second-moment closure (DSM), to a partial third-moment closure or, for unstably-stratified flows, to an unsteady RANS/LES scheme would also be expected to improve the realism of the results.

- Nevertheless, there are known cases that violate this expected link between modelling refinement and computational realism. Thus it is important that workers applying CFD to stratified flows should themselves be familiar with the history of prior applications of CFD in stratified flows, as has been presented in Sections 2–4 of this chapter.

- The standard, very widely used k–ε (linear) eddy-viscosity model is only usable with confidence in buoyancy affected flows adjacent to a wall where the predominant effect of buoyancy is felt in the vertical mean momentum equation rather than in the turbulence equations. The approach has been successfully applied to mixed convection through vertical tubing and related flows.

- Vertical free flows (plumes) affected by buoyancy are not reliably predicted at the EVM level of modelling, but this may be seen as a general weakness of modelling free flows (irrespective of buoyant effects) due, in part, to the greater importance of transport effects in such flow than in flow near walls.

- In refining a linear eddy-viscosity treatment, a simple and reasonably successful strategy, adopted by at least two independent groups, has been to retain the standard linear stress-strain relation but to adopt an improved representation of the scalar flux (heat flux or species flux, according to the application) by using the generalized gradient-diffusion hypothesis or some other more complete approximation. This has the effect of increasing the importance of the buoyant term in the turbulent kinetic energy equation. An advantage of proceeding in this way is that the linear eddy-viscosity formula may be straightforwardly replaced by a non-linear eddy-viscosity model, as discussed in Section 3.4. Such a replacement will ensure a more accurate computation of the stress field in situations where there is appreciable streamline curvature or other 'difficult' strain fields.

- In stably-stratified, horizontally directed shear flows the standard k–ε model gives much too rapid mixing. The main physical reason for this is that vertical velocity fluctuations which are principally responsible for the mixing (since the steepest gradients in mean velocity, temperature or mass fraction are vertical) are directly damped by the stable stratification. Several groups have tried to compensate for this by omitting the buoyant term from the ε equation. This raises the level of energy dissipation, reducing k and hence the turbulent viscosity. However, even among stable, horizontal flows, the success of this practice is substantially case dependent.

- Given the range of flow problems and geometric configurations encountered within industry, a more generally applicable basis for turbulence modelling is desirable. Such a strategy is provided by second-moment closure, a modelling level developed in Section 3. The starting point for modelling at this level is the exact transport equations for the Reynolds stress and scalar flux which explicitly contain the direct effects of buoyant forces on the turbulent field. However, there are also important buoyant effects on the unknown terms in those equations so the overall modelling is not exact ... but, in principle, it should be more secure than closures assuming a linear eddy-viscosity model.

- In simple free flows, if one adopts what is known as the Basic Model and neglects transport effects, the second-moment equations can be reduced to an eddy-viscosity model that includes, more realistically, the effect of buoyant damping on the vertical velocity fluctuations. However, the level of modelling usually adopted requires corrections for the proximity of a wall or a free surface and becomes rather arbitrary if one is trying to account for the effects of surfaces of complex shape. There is, moreover, still no consensus among different groups on what practice to adopt for the buoyant term in the ε equation.

- It is therefore recommended that software should aim to include, as one of its modelling options, what is known as the TCL set of approximations. This modelling strategy applies no wall corrections and, while buoyant terms appear in the closure, they do so without associated empirical coefficients. Moreover, in the ε equation, the most reasonable choice of taking $c_{\varepsilon 1} = c_{\varepsilon 3}$ is uniformly adopted. Over the past decade the model has been successfully applied to a number of horizontal and vertical buoyantly modified turbulent flows exhibiting appreciable improvements in accuracy over alternative simpler schemes. The drawback with the model is that it gives rise to forms which are, algebraically, very complex. However, in partial mitigation, it has been found that convergence is sometimes more rapid, due to the fact that the model respects the constraints of real turbulence as turbulence approaches the two-component limit (TCL).

- Notwithstanding, it has been found that when stable stratification has nearly eliminated turbulent mixing, even the TCL second-moment closure cannot be relied on to provide an entirely accurate representation. In these circumstances it has been shown that the inclusion of third-moment equations greatly improves predictions.

- Finally, it is well to note the progress being made in large-eddy simulation (LES). The most impressive results for a buoyancy-driven flow currently available have been obtained by applying a normal RANS solver containing a k–ε buoyancy-modified closure but operated in time-dependent mode. When applied to the case of Rayleigh–Bénard convection, agreement with the time-averaged properties seemed to be virtually complete and significantly superior to the steady-state RANS approach. Again, the body of evidence is not at present sufficient to justify a switchover to this approach even if the additional computing time (estimated at an order of magnitude more than a time-invariant simulation of an equivalent 3D flow) could be justified. Nevertheless, it is a strategy which, in the future, we may expect to be increasingly applied to the CFD of buoyant flows.

Appendix 1
The basic second-moment closure

A: Computing the Reynolds stresses

$$\frac{\partial \overline{u_i u_j}}{\partial t} + \underbrace{U_k \frac{\partial \overline{u_i u_j}}{\partial x_k}}_{C_{ij}} = d_{ij} + P_{ij} + F_{ij} + G_{ij} + \phi_{ij} - \varepsilon_{ij};$$

$$\frac{\partial \varepsilon}{\partial t} + U_k \frac{\partial \varepsilon}{\partial x_k} = d_\varepsilon + \tfrac{1}{2} c_{\varepsilon 1} (P_{kk} + G_{kk}) \frac{\varepsilon}{k} - c_{\varepsilon 2} \frac{\varepsilon^2}{k} + YC;$$

$$P_{ij} \equiv - \left\{ \overline{u_i u_k} \frac{\partial U_j}{\partial x_k} + \overline{u_j u_k} \frac{\partial U_i}{\partial x_k} \right\};$$

$$F_{ij} \equiv -2\Omega_k \{ \overline{u_j u_m} \, \varepsilon_{ikm} + \overline{u_i u_m} \, \varepsilon_{jkm} \};$$

$$G_{ij} \equiv - (\overline{u_j \theta} \beta_i + \overline{u_i \theta} \beta_j);$$

where

$$\phi_{ij} = \phi_{ij1} + \phi_{ij2} + \phi_{ij3} + (\phi_{ijw});$$

$$\phi_{ij1} = -c_1 \frac{\varepsilon}{k} \left[\overline{u_i u_j} - \frac{1}{3} \delta_{ij} \, \overline{u_k u_k} \right];$$

$$\phi_{ij2} = -c_2 \left[P_{ij} - C_{ij} + F_{ij} - \frac{1}{3} \delta_{ij} \, (P_{kk} - C_{kk}) \right];$$

$$\phi_{ij3} = -c_3 \left[G_{ij} - \frac{1}{3} \delta_{ij} G_{kk} \right];$$

$$\varepsilon_{ij} = \frac{2}{3} \delta_{ij} \varepsilon; \quad d_{ij} = \frac{\partial}{\partial x_k} \left[c_s \frac{k}{\varepsilon} \overline{u_k u_\ell} \frac{\partial \overline{u_i u_j}}{\partial x_\ell} \right];$$

$$d_\varepsilon = \frac{\partial}{\partial x_k} \left[c_\varepsilon \frac{k}{\varepsilon} \overline{u_k u_\ell} \frac{\partial \varepsilon}{\partial x_\ell} \right];$$

$$YC \equiv \text{Yap correction: } 0.83(\varepsilon/k) \left(\frac{k^{3/2}}{c_l y \tilde{\varepsilon}} - 1 \right) \left(k^{3/2} / c_l y \tilde{\varepsilon} \right)$$

Wall flows only

$$\phi_{ijw} = \left[c_1' \frac{\varepsilon}{k} \left[\overline{u_k u_m} \, n_k n_m \, \delta_{ij} - \frac{3}{2} \overline{u_i u_k} \, n_k n_j - \frac{3}{2} \overline{u_k u_j} \, n_k n_i \right] \right.$$

$$+ c_2' \left[\phi_{km2} \, n_k n_m \, \delta_{ij} - \frac{3}{2} \phi_{ik2} \, n_k n_j - \frac{3}{2} \phi_{kj2} \, n_k n_i \right]$$

$$\left. + c_3' \left[\phi_{km3} \, n_k n_m \, \delta_{ij} - \frac{3}{2} \phi_{ik3} \, n_k n_j - \frac{3}{2} \phi_{kj3} \, n_k n_i \right] \right] \frac{k^{3/2}}{c_\ell \varepsilon x_n}.$$

c_1	c_2	c_3	c_s	c_ε	$c_{\varepsilon 1}$	$c_{\varepsilon 2}$	c_1'	c_2'	c_3'	c_ℓ
1.8	0.6	0.5	0.22	0.18	1.44	1.92	0.5	0.3	0	2.5

B: Computing the scalar flux and variance (free shear flow form)

$$\frac{\partial \overline{u_i \theta}}{\partial t} + \underbrace{U_k \frac{\partial \overline{u_i \theta}}{\partial x_k}}_{C_{i\theta}} = d_{i\theta} + P_{i\theta} + F_{i\theta} + G_{i\theta} + \phi_{i\theta} - \varepsilon_{i\theta};$$

$$\frac{\partial \frac{1}{2}\overline{\theta^2}}{\partial t} + U_k \frac{\partial \frac{1}{2}\overline{\theta^2}}{\partial x_k} = d_\theta + P_\theta - \varepsilon_\theta;$$

$$P_{i\theta} \equiv -\overline{u_i u_k} \frac{\partial \Theta}{\partial x_k} - \underbrace{\overline{u_k \theta} \frac{\partial U_i}{\partial x_k}}_{P_{i\theta 2}};$$

$$F_{i\theta} \equiv -2\Omega_k \,\overline{\theta u_m}\, \varepsilon_{ikm};$$

$$G_{i\theta} \equiv -\beta_i \frac{\overline{\theta^2}}{\Theta};$$

$$d_{i\theta} = c_\theta \frac{\partial}{\partial x_k} \left[\overline{u_\ell u_k} \frac{\partial \overline{u_i \theta}}{\partial x_k} \right];$$

$$d_\theta = c_\theta \frac{\partial}{\partial x_k} \left[\overline{u_\ell u_k} \frac{\partial \overline{\theta^2}}{\partial x_k} \right];$$

$$\phi_{i\theta} = \phi_{i\theta 1} + \phi_{i\theta 2} + \phi_{i\theta 3} + (\phi_{i\theta w});$$

$$\phi_{i\theta 1} = -c_{1\theta} \frac{\varepsilon}{k} \,\overline{u_i \theta};$$

$$\phi_{i\theta 2} = -c_{2\theta}(P_{i\theta 2} - C_{i\theta} + F_{i\theta});$$

$$\phi_{i\theta 3} = -c_{3\theta}\, G_{i\theta};$$

$$\varepsilon_{i\theta} = 0;$$

$$P_\theta \equiv -\overline{u_k \theta}\, \partial \Theta / \partial x_k;$$

$$\varepsilon_\theta = \frac{1}{2} \frac{\overline{\theta^2}\varepsilon}{k} R; \quad R^{-1} = 1.5(1 + A_{2\theta});$$

$$A_{2\theta} \equiv \frac{\overline{u_i \theta}\,\overline{u_i \theta}}{\overline{\theta^2}k}.$$

$c_{1\theta}$	$c_{2\theta}$	$c_{3\theta}$	c_θ
3.0	0.5	0.5	0.18

Appendix 2
Models for the TCL pressure correlations
(From Craft *et al.*, 1996)

Table 3.4 *Exact generation terms and pressure-strain model in $\overline{u_i u_j}$ equations*

Generation terms (exact):

$$P_{ij} \equiv -\left(\overline{u_j u_k}\frac{\partial U_i}{\partial x_k} + \overline{u_i u_k}\frac{\partial U_j}{\partial x_k}\right); \quad G_{ij} \equiv -(\beta_i \overline{u_j \theta} + \beta_j \overline{u_i \theta})$$

Pressure-strain model:

$$\phi_{ij} = \phi_{ij1} + \phi_{ij2} + \phi_{ij3},$$

where $\phi_{ij1} = -c_1 \varepsilon[a_{ij} + c_1'(a_{ik}a_{jk} - 1/3 A_2 \delta_{ij})],$

$\phi_{ij2} = -0.6(P_{ij} - 1/3\delta_{ij}P_{kk}) + 0.3\varepsilon a_{ij}(P_{kk}/\varepsilon)$

$$-0.2\left[\frac{\overline{u_k u_j}\,\overline{u_l u_i}}{k}\left(\frac{\partial U_k}{\partial x_l} + \frac{\partial U_l}{\partial x_k}\right) - \frac{\overline{u_l u_k}}{k}\left(\overline{u_i u_k}\frac{\partial U_j}{\partial x_l} + \overline{u_j u_k}\frac{\partial U_i}{\partial x_l}\right)\right]$$

$$-c_2\left[A_2(P_{ij} - D_{ij}) + 3a_{mi}a_{nj}(P_{mn} - D_{mn})\right]$$

$$+c_2'\left[\left(\frac{7}{15} - \frac{A_2}{4}\right)(P_{ij} - 1/3\delta_{ij}P_{kk})\right.$$

$$+0.1\varepsilon[a_{ij} - 1/2(a_{ik}a_{kj} - 1/3\delta_{ij}A_2)](P_{kk}/\varepsilon) - 0.05a_{ij}a_{lk}P_{kl}$$

$$+0.1\left[\left(\frac{\overline{u_i u_m}}{k}P_{mj} + \frac{\overline{u_j u_m}}{k}P_{mi}\right) - 2/3\delta_{ij}\frac{\overline{u_l u_m}}{k}P_{ml}\right]$$

$$+0.1\left[\frac{\overline{u_l u_i}\,\overline{u_k u_j}}{k^2} - 1/3\delta_{ij}\frac{\overline{u_l u_m}\,\overline{u_k u_m}}{k^2}\right]\left[6D_{lk} + 13k\left(\frac{\partial U_l}{\partial x_k} + \frac{\partial U_l}{\partial x_k}\right)\right]$$

$$\left.+0.2\frac{\overline{u_l u_i u_k u_j}}{k^2}(D_{lk} - P_{lk})\right],$$

$$\phi_{ij3} = -\left(\frac{4}{10} - \frac{3A_2}{80}\right)(G_{ij} - 1/3\delta_{ij}G_{kk}) + \frac{1}{4}a_{ij}G_{kk} + \frac{3}{20}\left(\beta_i\frac{\overline{u_m u_j}}{k} + \beta_j\frac{\overline{u_m u_i}}{k}\right)\overline{u_m\theta}$$

$$-\frac{1}{10}\delta_{ij}\beta_k\frac{\overline{u_m u_k}}{k}\overline{u_m\theta} - \frac{1}{4}\beta_k\left(\frac{\overline{u_k u_i}}{k}\overline{u_j\theta} + \frac{\overline{u_k u_j}}{k}\overline{u_i\theta}\right) + \frac{1}{20}\delta_{ij}\beta_k\frac{\overline{u_m u_n}\,\overline{u_m u_k}}{k^2}\overline{u_n\theta}$$

$$-\frac{1}{8}\left(\frac{\overline{u_m u_j}}{k}\overline{u_i\theta} + \frac{\overline{u_m u_i}}{k}\overline{u_j\theta}\right)\frac{\overline{u_m u_k}}{k}\beta_k + \frac{1}{8}\left(\frac{\overline{u_k u_i}\,\overline{u_m u_j}}{k^2} + \frac{\overline{u_k u_j}\,\overline{u_m u_i}}{k^2}\right)\beta_k\overline{u_m\theta}$$

$$+\frac{3}{40}\left(\beta_i\frac{\overline{u_m u_j}}{k} + \beta_j\frac{\overline{u_m u_i}}{k}\right)\frac{\overline{u_m u_n}}{k}\overline{u_n\theta} + \frac{1}{4}\beta_k\frac{\overline{u_m u_k}\,\overline{u_i u_j}}{k^2}\overline{u_m\theta}.$$

Table 3.5 *Exact generation terms and pressure-scalar-flux model in $\overline{u_i\theta}$ equations*

Generation terms (exact):

$$P_{i\theta} \equiv -\left(\overline{u_iu_j}\frac{\partial\Theta}{\partial x_j} + \overline{u_j\theta}\frac{\partial U_i}{\partial x_j}\right); \quad G_{i\theta} \equiv -\beta_i\overline{\theta^2}$$

Pressure-strain gradient model:

$$\phi_{i\theta} = \phi_{i\theta1} + \phi_{i\theta2} + \phi_{i\theta3},$$

where

$$\phi_{i\theta1} = -1.7\left[1 + 1.2(A_2A)^{1/2}\right]R^{1/2}\frac{\varepsilon}{k}[\overline{u_i\theta}(1 + 0.6A_2) - 0.8a_{ik}\overline{u_k\theta}$$

$$+ 1.1a_{ik}a_{kj}\overline{u_j\theta}] - 0.2A^{1/2}Rka_{ij}\frac{\partial\Theta}{\partial x_j},$$

$$\phi_{i\theta2} = 0.8\overline{u_k\theta}\frac{\partial U_i}{\partial x_k} - 0.2\overline{u_k\theta}\frac{\partial U_k}{\partial x_i} + \frac{1}{6}\frac{\varepsilon}{k}\overline{u_i\theta}(P_{kk}/\varepsilon)$$

$$- 0.4\overline{u_k\theta}a_{il}\left(\frac{\partial U_k}{\partial x_l} + \frac{\partial U_l}{\partial x_k}\right)$$

$$+ 0.1\overline{u_k\theta}a_{ik}a_{ml}\left(\frac{\partial U_m}{\partial x_l} + \frac{\partial U_l}{\partial x_m}\right)$$

$$- 0.1\overline{u_k\theta}(a_{im}P_{mk} + 2a_{mk}P_{im})/k$$

$$+ 0.15a_{ml}\left(\frac{\partial U_k}{\partial x_l} + \frac{\partial U_l}{\partial x_k}\right)(a_{mk}\overline{u_i\theta} - a_{mi}\overline{u_k\theta})$$

$$- 0.05a_{ml}\left[7a_{mk}\left(\overline{u_i\theta}\frac{\partial U_k}{\partial x_k} + \overline{u_k\theta}\frac{\partial U_i}{\partial x_l}\right) - \overline{u_k\theta}\left(a_{ml}\frac{\partial U_i}{\partial x_k} + a_{mk}\frac{\partial U_i}{\partial x_l}\right)\right],$$

$$\phi_{i\theta3} = 1/3\beta_i\overline{\theta^2} - \beta_k\overline{\theta^2}a_{ik}.$$

Appendix 3
Third-moment equations and their modelling
(From Kidger, 1999)

A3.1 The triple-moment equations

Transport equations to describe exactly the evolution of the triple-moment terms may be derived in a similar manner as for the second-moment equations, as is shown below in Sections A3.1.1–A3.1.4.

A3.1.1 Equation for triple-velocity $\overline{u_i u_j u_k}$

A transport equation for the triple-velocity term, $\overline{u_i u_j u_k}$, can be produced from the fluctuating velocity equation (Eq. (3.32)), since:

$$\frac{D\overline{u_i u_j u_k}}{Dt} = \overline{u_j u_k \frac{Du_i}{Dt}} + \overline{u_i u_k \frac{Du_j}{Dt}} + \overline{u_i u_j \frac{Du_k}{Dt}}. \tag{3.73}$$

Hence:

$$\frac{D\overline{u_i u_j u_k}}{Dt} = -\overline{u_j u_k u_l}\frac{\partial U_i}{\partial x_l} - \overline{u_j u_k \frac{\partial}{\partial x_l}(u_i u_l - \overline{u_i u_l})} + \frac{\overline{u_j u_k f_i}}{\rho} - \frac{\overline{u_j u_k}}{\rho}\frac{\partial p}{\partial x_i}$$

$$+ \nu\frac{\overline{u_j u_k \partial^2 u_i}}{\partial x_l^2} - \overline{u_i u_k u_l}\frac{\partial U_j}{\partial x_l} - \overline{u_i u_k \frac{\partial}{\partial x_l}(u_j u_l - \overline{u_j u_l})}$$

$$+ \frac{\overline{u_i u_k f_j}}{\rho} - \frac{\overline{u_i u_k}}{\rho}\frac{\partial p}{\partial x_j} + \nu\frac{\overline{u_i u_k \partial^2 u_j}}{\partial x_l^2} - \overline{u_i u_j u_l}\frac{\partial U_k}{\partial x_l}$$

$$- \overline{u_i u_j \frac{\partial}{\partial x_l}(u_k u_l - \overline{u_k u_l})} + \frac{\overline{u_i u_j f_k}}{\rho} - \frac{\overline{u_i u_j}}{\rho}\frac{\partial p}{\partial x_k} + \nu\frac{\overline{u_i u_j \partial^2 u_k}}{\partial x_l^2}.$$

By applying similar equalities to those shown in Eqs. (3.38)–(3.40), the above equation can be re-written as:

$$\frac{D\overline{u_i u_j u_k}}{Dt} = -\left[\overline{u_j u_k u_l}\frac{\partial U_i}{\partial x_l} + \overline{u_i u_k u_l}\frac{\partial U_j}{\partial x_l} + \overline{u_i u_j u_l}\frac{\partial U_k}{\partial x_l}\right]$$

$$+ \left[\overline{u_j u_k \frac{\partial \overline{u_i u_l}}{\partial x_l}} + \overline{u_i u_k \frac{\partial \overline{u_j u_l}}{\partial x_l}} + \overline{u_i u_j \frac{\partial \overline{u_k u_l}}{\partial x_l}}\right]$$

$$+ \frac{1}{\rho}[\overline{u_j u_k f_i} + \overline{u_i u_k f_j} + \overline{u_i u_j f_k}]$$

$$+ \frac{\overline{p}}{\rho}\left[\frac{\partial \overline{u_j u_k}}{\partial x_i} + \frac{\partial \overline{u_i u_k}}{\partial x_j} + \frac{\partial \overline{u_i u_j}}{\partial x_k}\right]$$

$$- \nu\left[\frac{\overline{\partial u_i}}{\partial x_j}\frac{\partial u_j u_k}{\partial x_l} + \frac{\overline{\partial u_j}}{\partial x_l}\frac{\partial u_i u_k}{\partial x_l} + \frac{\overline{\partial u_k}}{\partial x_l}\frac{\partial u_i u_j}{\partial x_l}\right]$$

$$- \left[\frac{\partial}{\partial x_l}\left[\overline{u_i u_j u_k u_l} + \frac{\overline{p u_j u_k}}{\rho}\delta_{il} + \frac{\overline{p u_i u_k}}{\rho}\delta_{jl} + \frac{\overline{p u_i u_j}}{\rho}\delta_{kl}\right.\right.$$

$$\left.\left. - \nu\frac{\partial \overline{u_i u_j u_k}}{\partial x_l}\right]\right], \tag{3.74}$$

or, symbolically:

$$\frac{D\overline{u_i u_j u_k}}{Dt} = P^1_{ijk} + P^2_{ijk} + F_{ijk} + \phi_{ijk} - \varepsilon_{ijk} + d_{ijk}. \tag{3.75}$$

As in the second-moment equations, the production and buoyant processes (P^1_{ijk}, P^2_{ijk} and F_{ijk}) can be implemented exactly, whereas the pressure-correlation, dissipation and diffusion processes (ϕ_{ijk}, ε_{ijk}, d_{ijk}) require some form of modelling.

A3.1.2 Equation for the scalar flux diffusion, $\overline{u_i u_k \theta}$

In a similar manner to that for $\overline{u_i u_j u_k}$, a transport equation for the scalar flux diffusion term $\overline{u_i u_k \theta}$ can be produced from the fluctuating velocity and scalar equations

$$\frac{D\overline{u_i u_k \theta}}{Dt} = \overline{u_k \theta \frac{Du_i}{Dt}} + \overline{u_i \theta \frac{Du_k}{Dt}} + \overline{u_i u_k \frac{D\theta}{Dt}} \tag{3.76}$$

giving a solution in the form:

$$\frac{D\overline{u_i u_k \theta}}{Dt} = P^1_{ik\theta} + P^2_{ik\theta} + F_{ik\theta} + \phi_{ik\theta} - \varepsilon_{ik\theta} + d_{ik\theta}, \tag{3.77}$$

where:

$$P^1_{ik\theta} = -\left[\overline{u_k u_l \theta}\frac{\partial U_i}{\partial x_l} + \overline{u_i u_l \theta}\frac{\partial U_k}{\partial x_l} + \overline{u_i u_k u_l}\frac{\partial \Theta}{\partial x_l} \right], \tag{3.78}$$

$$P^2_{ik\theta} = +\left[\overline{u_i \theta}\frac{\partial \overline{u_k u_l}}{\partial x_l} + \overline{u_k \theta}\frac{\partial \overline{u_i u_l}}{\partial x_l} + \overline{u_i u_k}\frac{\partial \overline{u_i \theta}}{\partial x_l} \right], \tag{3.79}$$

$$F_{ik\theta} = +\left[\frac{\overline{u_k \theta f_i}}{\rho} + \frac{\overline{u_i \theta f_k}}{\rho} \right]. \tag{3.80}$$

Again, the pressure-correlation, dissipation and diffusion processes ($\phi_{ik\theta}$, $\varepsilon_{ik\theta}$, $d_{ik\theta}$) require some form of tubulence closure.

A3.1.3 Equation for the scalar variance diffusion, $\overline{u_k \theta^2}$

In a similar way, a transport equation for the scalar variance diffusion term, $\overline{u_k \theta^2}$, may be formed from:

$$\frac{D\overline{u_k \theta^2}}{Dt} = \overline{2u_k \theta \frac{D\theta}{Dt}} + \overline{\theta^2 \frac{Du_k}{Dt}}, \tag{3.81}$$

which results in a solution in the form:

$$\frac{D\overline{u_k \theta^2}}{Dt} = P^1_{k\theta\theta} + P^2_{k\theta\theta} + F_{k\theta\theta} + \phi_{k\theta\theta} - \varepsilon_{k\theta\theta} + d_{k\theta\theta}, \tag{3.82}$$

where

$$P^1_{k\theta\theta} = -\left[\overline{u_l\theta^2}\frac{\partial U_k}{\partial x_l} + 2\overline{u_k u_l\theta}\frac{\partial \Theta}{\partial x_l}\right],$$
(3.83)

$$P^2_{k\theta\theta} = +\left[\overline{\theta^2}\frac{\partial \overline{u_k u_l}}{\partial x_l} + 2\overline{u_k\theta}\frac{\partial \overline{u_l\theta}}{\partial x_l}\right],$$
(3.84)

$$F_{k\theta\theta} = +\left[\frac{\overline{\theta^2 f_i}}{\rho}\right],$$
(3.85)

and the remaining terms ($\phi_{k\theta\theta}$, $\varepsilon_{k\theta\theta}$, $d_{k\theta\theta}$) require the implementation of a suitable turbulence closure.

A3.1.4 Equation for the triple-scalar $\overline{\theta^3}$

Finally, for the triple-scalar term $\overline{\theta^3}$:

$$\frac{D\overline{\theta^3}}{Dt} = 3\overline{\theta^2}\frac{D\overline{\theta}}{Dt},$$
(3.86)

from which the transport equation for $\overline{\theta^3}$ can be written as:

$$\frac{D\overline{\theta^3}}{Dt} = P^1_{\theta\theta\theta} + P^2_{\theta\theta\theta} - \varepsilon_{\theta\theta\theta} + d_{\theta\theta\theta},$$
(3.87)

where

$$P^1_{\theta\theta\theta} = -3\overline{u_l\theta^2}\frac{\partial \Theta}{\partial x_l},$$
(3.88)

$$P^2_{\theta\theta\theta} = +3\overline{\theta^2}\frac{\partial \overline{u_l\theta}}{\partial x_l}.$$
(3.89)

Note that $\overline{\theta^3}$ contains neither any production due to buoyancy, nor any pressure-correlation terms, leaving just the diffusion and dissipation ($\varepsilon_{\theta\theta\theta}$ and $d_{\theta\theta\theta}$) to be modelled.

A3.2 Closing the triple-moment equations

Various proposals are shown below to model the unknown terms arising in the triple-moment equations.

A3.2.1 Dissipation (ε_{ijk}, $\varepsilon_{ik\theta}$, $\varepsilon_{k\theta\theta}$, $\varepsilon_{\theta\theta\theta}$)

Dekeyser and Launder (1985) (hereafter 'DL-85') proposed closure schemes for the various unknown terms in the triple-moment equations. For the triple-velocity dissipation, ε_{ijk}, they proposed using essentially a GGDH scheme, but simplifying with the presumption of isotropic dissipation (Eq. (3.31)), producing:

$$\varepsilon_{ijk} = C_{\varepsilon uuu}\frac{k}{\varepsilon}\frac{\partial \varepsilon}{\partial x_l}[\overline{u_k u_l}\delta_{ij} + \overline{u_j u_l}\delta_{ki} + \overline{u_i u_l}\delta_{jk}].$$
(3.90)

A similar argument, based on a GGDH, but assuming that $\varepsilon_{i\theta} = 0$, produces a form for

$\varepsilon_{ik\theta}$, proposed by DL-85:

$$\varepsilon_{ik\theta} = C_{\varepsilon uu\theta}\frac{k}{\varepsilon}\frac{\partial\varepsilon}{\partial x_l}\overline{u_l\theta}\delta_{ik}.$$ (3.91)

Similarly, for $\varepsilon_{k\theta\theta}$:

$$\varepsilon_{k\theta\theta} = C_{\varepsilon u\theta\theta}\frac{k}{\varepsilon}\frac{\partial\varepsilon_\theta}{\partial x_l}\overline{u_k u_l}.$$ (3.92)

Dekeyser and Launder (1985) suggested that the three constants in the above take the following values – which arise if the constant in the GGDH model (Eq. (3.18)) is taken as equal to 0.15:

$$C_{\varepsilon uuu} = 0.2; \quad C_{\varepsilon uu\theta} = 0.1; \quad \text{and} \quad C_{\varepsilon u\theta\theta} = 0.3.$$

The dissipation of the triple-scalar (not considered by DL-85) has been noted as a particularly significant process by Wyngaard and Sundarajan (1979), in their studies of $\overline{\theta^3}$, and following Zeman and Lumley (1976), a form based on a normal distribution of fluctuations can be taken as:

$$\varepsilon_{\theta\theta\theta} = 3\left[\varepsilon_{\theta\theta}\frac{\overline{\theta^3}}{\overline{\theta^2}}\right]$$ (3.93)

(with $\varepsilon_{\theta\theta}$ as Eq. (3.41)).

It is worth noting, however, that the inclusion of $\varepsilon_{\theta\theta}$ in Eq. (3.93) raises the issue of how best to describe the turbulent time-scale ratio, R, as discussed in Sections 3.2.1 and 4.2, if $\varepsilon_{\theta\theta}$ is being modelled via an algebraic relation involving R.

A3.2.2 Pressure correlations (ϕ_{ijk}, $\phi_{ik\theta}$, $\phi_{k\theta\theta}$)

Dekeyser and Launder (1985) proposed schemes to account for the non-dispersive role of the pressure fluctuations, based essentially on the 'Basic Model' used for the second moments, with two components: a 'return to isotropy' term and an 'isotropization of production' term. Although DL-85 did not consider flows with buoyant production present, it seems logical to include the buoyant production also into the pressure-redistribution term (as with Eq. (3.34) for ϕ_{ij_3}). For the triple-velocity pressure transport, ϕ_{ijk}, this produces:

$$\phi_{ijk} = -C_{1uuu}\frac{k}{\varepsilon}\overline{u_i u_j u_k} - C_{2uuu}\left[P^1_{ijk} + P^2_{ijk} + F_{ijk}\right],$$ (3.94)

with $C_{1uuu} = (1/0.075)$ and $C_{2uuu} = 0.5$.

A similar form can also be derived for $\phi_{ik\theta}$. However, as with the pressure-scalar gradient, $\phi_{i\theta}$ (e.g. Eq. (3.38), and Launder, 1975), the pressure redistribution only arises from the Du_i/Dt terms, not the $D\theta/Dt$ terms. Thus:

$$\phi_{ik\theta} = -C_{1uu\theta}\frac{k}{\varepsilon}\overline{u_i u_k\theta} - C_{2uu\theta}\left[P^{1*}_{ik\theta} + P^{2*}_{ik\theta} - F_{ik\theta}\right],$$ (3.95)

where

$$P^{1*}_{ik\theta} = -\left[\overline{u_k u_l\theta}\frac{\partial U_i}{\partial x_l} + \overline{u_i u_l\theta}\frac{\partial U_k}{\partial x_l}\right],$$ (3.96)

$$P^{2*}_{ik\theta} = +\left[\overline{u_i\theta}\frac{\partial\overline{u_k u_l}}{\partial x_l} + \overline{u_k\theta}\frac{\partial\overline{u_i u_l}}{\partial x_l}\right],$$ (3.97)

and likewise, $C_{1uu\theta} = (1/0.075)$ and $C_{2uu\theta} = 0.5$ (DL-85).

DL-85 proposed that the $P^{1*}_{ik\theta}$ term could be dropped since, certainly for the flows they considered, in $P^1_{ik\theta}$, the production from gradients of the mean velocity is insignificant in comparison to that from the gradients of the mean scalar. Hence it could be presumed that the effect of the $P^{1*}_{ik\theta}$ contribution to the pressure redistribution is negligible in comparison to the total $P^1_{ik\theta}$ production term.

A similar modelling approach for $\phi_{k\theta\theta}$ yields:

$$\phi_{k\theta\theta} = -C_{1u\theta\theta}\frac{k}{\varepsilon}\overline{u_i\theta^2} - C_{2u\theta\theta}\left[P^{1*}_{k\theta\theta} + P^{2*}_{k\theta\theta} + F_{k\theta\theta}\right], \tag{3.98}$$

where

$$P^{1*}_{k\theta\theta} = -\left[\overline{u_l\theta^2}\frac{\partial U_k}{\partial x_l}\right], \tag{3.99}$$

$$P^2_{k\theta\theta} = +\left[\overline{\theta^2}\frac{\partial \overline{u_k u_l}}{\partial x_l}\right], \tag{3.100}$$

and again, $C_{1u\theta\theta} = (1/0.075)$ and $C_{2u\theta\theta} = 0.5$ (DL-85).

A3.2.3 Diffusion (d_{ijk}, $d_{ik\theta}$, $d_{k\theta\theta}$, $d_{\theta\theta\theta}$)

There are several ways in which the diffusion process for the triple moments can be considered. Following Millionshtchikov (1941), Dekeyser and Launder (1985) assumed a Gaussian distribution for the quadruple-velocity correlation, in which case the fourth-rank tensor can be written as:

$$\overline{\alpha\beta\gamma\delta} = \overline{\alpha\beta}\cdot\overline{\gamma\delta} + \overline{\alpha\gamma}\cdot\overline{\beta\delta} + \overline{\alpha\delta}\cdot\overline{\beta\gamma}. \tag{3.101}$$

Hence, a general triple-moment diffusion process, which consists of gradients of $\overline{\alpha\beta\gamma u_l}$, can be written as:

$$d_{\alpha\beta\gamma} = -\frac{\partial(\overline{\alpha\beta\gamma u_l})}{\partial x_l} = -\frac{\partial[\overline{\alpha\beta}\cdot\overline{\gamma u_l} + \overline{\alpha\gamma}\cdot\overline{\beta u_l} + \overline{\alpha u_l}\cdot\overline{\beta\gamma}]}{\partial x_l}, \tag{3.102}$$

where α, β and γ may represent either u_i or θ.

In the course of the studies described in this thesis, attempts were made to solve the transport equations for the triple-moment terms, with the closures shown above, using the GGDH for the diffusion of the triple moments. However, the performance was found to be disappointing, with coarse, grid-dependent fluctuations, particularly for the scalar fluxes. Attempts to use the Gaussian redistribution for the diffusion process (Eq. (3.102)) – which modified the P^2 production terms – also proved disappointing, since the region where the triple moments were most influential (in this case in the most strongly stably-stratified layer) was also the region where the quadruple moment (3.102) was the most significant term, yet arguably least accurate. Kawamura et al. (1995) also observed this, noting that the combination of the production and the diffusion processes to form a new effective production term was inadequately predicting the triple moments. Thus Kawamura et al. (1995) proposed that Eq. (3.102) should be *subtracted* from the production (P^2) term, and *added* to the diffusion term. Thus, from Eq. (3.102):

$$D_{ijk} = \frac{\partial}{\partial x_l}[\overline{u_i u_j}\cdot\overline{u_k u_l} + \overline{u_j u_k}\cdot\overline{u_i u_l} + \overline{u_i u_k}\cdot\overline{u_j u_l}], \tag{3.103}$$

subtract from P^2,

$$P^{2(new)}_{ijk} = P^2_{ijk} - D_{ijk}; \tag{3.104}$$

add to d_{ijk},

$$d_{ijk}^{(new)} = d_{ijk} + D_{ijk}. \tag{3.105}$$

Hence

$$P_{ijk}^{2(new)} = -\overline{u_i u_l}\frac{\partial \overline{u_j u_k}}{\partial x_l} - \overline{u_j u_l}\frac{\partial \overline{u_i u_k}}{\partial x_l} - \overline{u_k u_l}\frac{\partial \overline{u_i u_j}}{\partial x_l}, \tag{3.106}$$

and

$$d_{ijk}^{(new)} = -\frac{\partial}{\partial x_l}\left[\overline{u_i u_j u_k u_l} - (\overline{u_i u_j}\cdot\overline{u_k u_l} + \overline{u_j u_k}\cdot\overline{u_i u_l} + \overline{u_i u_k}\cdot\overline{u_j u_l})\right]. \tag{3.107}$$

The production term, $P_{ijk}^{2(new)}$ can be implemented exactly, and the GGDH can be used to model Eq. (3.107), thus:

$$d_{ijk}^{(new)} = -C_{dijk}\frac{\partial}{\partial x_l}\left[\frac{k}{\varepsilon}\overline{u_l u_m}\frac{\partial(\overline{u_i u_j u_k})}{\partial x_m}\right]. \tag{3.108}$$

Likewise, for the other triple-moment terms.
   For $\overline{u_i u_k \theta}$:

$$P_{ik\theta}^{2(new)} = -\overline{u_k u_l}\frac{\partial \overline{u_i \theta}}{\partial x_l} - \overline{u_i u_l}\frac{\partial \overline{u_k \theta}}{\partial x_l} - \overline{u_l \theta}\frac{\partial \overline{u_i u_k}}{\partial x_l}, \tag{3.109}$$

$$d_{ik\theta}^{(new)} = -C_{dik\theta}\frac{\partial}{\partial x_l}\left[\frac{k}{\varepsilon}\overline{u_l u_m}\frac{\partial(\overline{u_i u_k \theta})}{\partial x_m}\right]. \tag{3.110}$$

For $\overline{u_k \theta^2}$:

$$P_{k\theta\theta}^{2\ (new)} = -\overline{u_k u_l}\frac{\partial \overline{\theta^2}}{\partial x_l} - 2\overline{u_l \theta}\frac{\partial \overline{u_k \theta}}{\partial x_l}, \tag{3.111}$$

$$d_{k\theta\theta}^{(new)} = -C_{dk\theta\theta}\frac{\partial}{\partial x_l}\left[\frac{k}{\varepsilon}\overline{u_l u_m}\frac{\partial(\overline{u_k \theta^2})}{\partial x_m}\right]. \tag{3.112}$$

Finally $\overline{\theta^3}$:

$$P_{\theta\theta\theta}^{2(new)} = -3\overline{u_l \theta}\frac{\partial \overline{\theta^2}}{\partial x_l}, \tag{3.113}$$

$$d_{\theta\theta\theta}^{(new)} = -C_{d\theta\theta\theta}\frac{\partial}{\partial x_l}\left[\frac{k}{\varepsilon}\overline{u_l u_m}\frac{\partial(\overline{\theta^3})}{\partial x_m}\right]. \tag{3.114}$$

The constants in the above (C_{dijk}, $C_{dik\theta}$, $C_{dk\theta\theta}$, $C_{d\theta\theta\theta}$) are all taken as equal, $C_{d\alpha\beta\gamma} = 0.11$.

A3.3 An algebraic solution for the triple moments

Full transport equations for the triple moments (Eqs. (3.75), (3.77), (3.82) and (3.87)) could be solved with the inclusion of the closures shown in Sections A3.2.1–A3.2.3. However, it can be seen that this would be no mean feat, computationally, For a two-dimensional buoyancy affected flow, it requires the solution of twelve simultaneous differential equations for the triple moments (one for $\overline{\theta^3}$, two for $\overline{u_k \theta^2}$, three for $\overline{u_i u_k \theta}$ and six for $\overline{u_i u_j u_k}$) in addition to the seven for the second moments (one for $\overline{\theta^2}$, two for $\overline{u_i \theta}$

and four for $\overline{u_i u_j}$). Earlier investigations (e.g. Craft, Ince & Launder, 1996) have, understandably, chosen to use simplified *algebraic* forms for the triple moments, retaining the production and dissipative terms, but neglecting the transport effects. The Dekeyser and Launder (1985) paper was, in fact, aimed at producing such algebraic forms. Suggestions for closure of the equations were proposed at various levels of complexity, ranging from retaining only the production from second-moment gradients ($P^2_{\alpha\beta\gamma}$), diffusion ($d_{\alpha\beta\gamma}$) terms, and the 'return to isotropy' model for the turbulent-turbulent pressure interactions ($\phi^1_{\alpha\beta\gamma} = -C\dfrac{k}{\varepsilon}\overline{\alpha\beta\gamma}$), to the inclusion of all the production, pressure-interaction and dissipative processes.

The general format for the algebraic truncations proposed by DL-85 is the same for all the triple-moment equations (except $\overline{\theta^3}$, not considered by DL-85). If one regards the flow as steady $\left[\dfrac{D\overline{\alpha\beta\gamma}}{Dt} = 0\right]$ and moves the turbulent-turbulent pressure-interaction term $\left(\phi^1_{\alpha\beta\gamma} = -C\dfrac{k}{\varepsilon}\overline{\alpha\beta\gamma}\right)$ to the left hand side of the equation, then divides through by the time scale (k/ε), one arrives at the expression:

$$\overline{\alpha\beta\gamma} = C^{-1}\frac{\varepsilon}{k}\left[P^{-1}_{\alpha\beta\gamma} + P^{2\,(\text{new})}_{\alpha\beta\gamma} + d^{(\text{new})}_{\alpha\beta\gamma} + F_{\alpha\beta\gamma} + \phi^2_{\alpha\beta\gamma} - \varepsilon_{\alpha\beta\gamma}\right] \tag{3.115}$$

(where the modified $P^{(\text{new})}$ and $d^{(\text{new})}$ arise from the Kawamura redistribution – see Section A3.2.3).

Hence, for example, the algebraic form for the triple-velocity term, $\overline{u_i u_j u_k}$, is (including all pressure and buoyant contributions):

$$\overline{u_i u_j u_k} = -C\frac{k}{\varepsilon}\left[\begin{array}{c} 0.5\left(\begin{array}{c}+\overline{u_j u_k u_l}\dfrac{\partial U_i}{\partial x_l} + \overline{u_i u_k u_l}\dfrac{\partial U_j}{\partial x_l} + \overline{u_i u_j u_l}\dfrac{\partial U_k}{\partial x_l} \\[2mm] -\dfrac{\overline{u_j u_k f_i}}{\rho} - \dfrac{\overline{u_i u_k f_j}}{\rho} - \dfrac{\overline{u_i u_j f_k}}{\rho} \\[2mm] +\overline{u_k u_l}\dfrac{\partial\overline{u_i u_j}}{\partial x_l} + \overline{u_j u_l}\dfrac{\partial\overline{u_i u_k}}{\partial x_l} + \overline{u_i u_l}\dfrac{\partial\overline{u_j u_k}}{\partial x_l}\end{array}\right) \\[6mm] \underbrace{}_{P^1+P^2+F+\phi^2} \\[2mm] -C_{\varepsilon uuu}\dfrac{k}{\varepsilon}\dfrac{\partial}{\partial x_l}\left(\overline{u_k u_l}\delta_{ij} + \overline{u_j u_l}\delta_{ik} + \overline{u_i u_l}\delta_{jk}\right) \\[2mm] \underbrace{}_{\varepsilon} \\[2mm] +C_{duuu}\dfrac{\partial}{\partial x_l}\left(\dfrac{k}{\varepsilon}\overline{u_l u_m}\dfrac{\partial(u_i u_j u_k)}{\partial x_m}\right) \\[2mm] \underbrace{}_{d} \end{array}\right] \tag{3.116}$$

It is perhaps worth highlighting a consequence of applying the Gaussian redistribution for the diffusion term (see Eq. (3.101)). Combining this with the production term arising from second-moment gradients (P^2), the resultant term can be written in a similar form to P^2, but with the gradient operator acting on the other second-moment than it does in P^2 (i.e., obtaining $+\overline{\gamma\delta}\cdot\dfrac{\partial\overline{\alpha\beta}}{\partial x_l}$, when $-\overline{\alpha\beta}\cdot\dfrac{\partial\overline{\gamma\delta}}{\partial x_l}$ appears in P^2).

A3.4 Clipping the triple-moment equations

There is a risk that the closures above for the triple-moment equations will produce unphysical values for the triple moments and that therefore, when applied to the diffusion process for the second moments, will result in the second-moment equations no longer satisfying the ordinary Schwarz inequalities. To prevent that happening, André *et al.* (1979) proposed a clipping algorithm for the triple moments which, upon further investigation, they report as being realizable, in that non-physical values for the triple moments are prevented.

If, as previously, we consider a general triple moment $\overline{\alpha\beta\gamma}$, where α, β and γ may represent either the fluctuating velocity or scalar, (u_i or θ), the clipping algorithm defines:

$$\left|\overline{\alpha\beta\gamma}\right| \leq \min \left\{ \begin{array}{l} \left[\overline{\alpha^2}(\overline{\beta^2} \cdot \overline{\gamma^2} + \overline{\beta\gamma}^2)\right]^{1/2} \\[4pt] \left[\overline{\beta^2}(\overline{\alpha^2} \cdot \overline{\gamma^2} + \overline{\alpha\gamma}^2)\right]^{1/2} \\[4pt] \left[\overline{\gamma^2}(\overline{\alpha^2} \cdot \overline{\beta^2} + \overline{\alpha\beta}^2)\right]^{1/2} \end{array} \right\}. \tag{3.117}$$

In practice, for this present study, it was found that Eq. (3.117) made virtually no difference to the converged solution (based on *transport* equations for the triple moments), but that during the convergence process, its inclusion proved to be necessary to ensure stability.

Nomenclature

a_{ij}	dimensionless anisotropic Reynolds stress, Eq. (3.33)
A_2	second dimensionless invariant of the stress tensor, $a_{ij}a_{ji}$
A_3	third dimensionless invariant of stress tensor, $a_{ij}a_{jk}a_{ki}$
A	flatness factor of stress tensor; $1 - 9\,(A_2 - A_3)/8$
c_μ	coefficient appearing in turbulent viscosity formula
$C_{ij}, C_{i\theta}$	convective transport of $\overline{u_i u_j}$, $\overline{u_i \theta}$
$d_{ij}, d_{i\theta}$	diffusive transport of $\overline{u_i u_j}$, $\overline{u_i \theta}$
f_μ, f_ε	viscous damping functions
G_k	generation rate of turbulent kinetic energy by the gravitational field, $\frac{1}{2}\,G_{kk}$ (see Eq. (3.14))
G_{ij}	generation rate of $\overline{u_i u_j}$ by the gravitational field
$G_{i\theta}$	generation rate of $\overline{u_i \theta}$ by the gravitational field
g_i	gravitational acceleration vector
k	turbulent kinetic energy, $\overline{u_i^2}/2$
l	turbulent length scale, $k^{3/2}/\varepsilon$
P_k	production rate of turbulence energy by mean shear, $\frac{1}{2}\,P_{kk}$
P_{ij}	production rate of $\overline{u_i u_j}$ by mean shear
$P_{i\theta 1}$	production of $\overline{u_i \theta}$ by mean scalar gradients
$P_{i\theta 2}$	production of $\overline{u_i \theta}$ by mean velocity gradients
$P_{\theta\theta}$	production rate of $\overline{\theta^2}$ by mean scalar gradients
Re_t	local turbulent Reynolds number, $k^2/\nu\varepsilon$
Ri	gradient Richardson number
R_f	flux Richardson number, $-G_k/P_k$

S	dimensionless strain parameter, $(S_{ij}S_{ij})^{1/2}$
S_{ij}	dimensionless strain tensor $\dfrac{k}{\varepsilon}\left(\dfrac{U_i}{\partial x_j} + \dfrac{\partial U_i}{\partial x_j}\right)$
S_Y	the Yap correction to ε equation (Eq. (3.10a))
t	time
U	mean velocity in x direction (the principal component in a simple shear flow)
$\overline{uv}$	turbulent shear stress
$\overline{u_i u_j}$	turbulent Reynolds stress
$\overline{u_j \theta}$	turbulent flux of scalar Θ
y	distance normal to a wall
y_v	conceptual thickness of the viscous sublayer
y^*	normalized distance from wall, $c_\mu^{1/4} y k^{1/2}/\nu$
α	dimensionless volumetric expansion coefficient, Eq. (3.24)
β	conventional volumetric expansion coefficient
β_i	$\beta\, g_i$
δ_{ij}	Kronecker delta ($=1$ if $i=j$; $=0$ otherwise)
ε	dissipation rate of kinetic energy
$\tilde{\varepsilon}$	homogeneous part of ε $\left(\tilde{\varepsilon} \equiv \varepsilon - 2\nu \left(\dfrac{\partial k^{1/2}}{\partial x_j}\right)^2\right)$
ε_{ij}	dissipation rate of $\overline{u_i u_j}$
ε_{ijk}	alternating third-rank unit tensor ($=1$ if i, j, k all different and in cyclic order; $=-1$ if i, j, k all different and in anti-cyclic order; $=0$ otherwise)
$\varepsilon_{i\theta}$	dissipation rate of $\overline{u_i \theta}$
$\varepsilon_{\theta\theta}$	dissipation rate of $\overline{\theta^2}$
ϕ_{ij}	pressure-strain correlation, $\dfrac{p}{\rho}\left(\dfrac{\partial u_i}{\partial x_j} + \dfrac{\partial u_j}{\partial x_i}\right)$
$\phi_{ij1}, \phi_{ij2}, \phi_{ij3}$	'turbulence', 'mean-strain' and 'gravitational' contributions to ϕ_{ij}
$\phi_{i\theta}$	pressure-scalar-gradient correlation, $\dfrac{p}{\rho}\left(\dfrac{\partial \theta}{\partial x_i}\right)$ (likewise decomposed into $\phi_{i\theta 1}, \phi_{i\theta 2}, \phi_{i\theta 3}$)
λ	thermal diffusivity
ν	molecular kinematic viscosity
ν_t	turbulent kinematic viscosity
Θ	mean scalar (denoting temperature, mass fraction, etc.)
θ	fluctuating scalar
ρ	fluid density
ρ'	turbulent fluctuations in density
σ_ϕ	turbulent Prandtl number for diffusive transport of ϕ (ϕ may denote Θ, $\overline{\theta^2}, k, \varepsilon$)
τ_w	wall shear stress

References

Abrahamsson, H., Johansson, B. and Löfdahl, L., 1997, 'An investigation of the turbulence field in the fully developed three-dimensional wall-jet', Int. Rep 97/1, Chalmers University of Technology, Sweden.

Addad, Y., Laurence, D. and Benhamadouche, S., 2003, 'The negatively buoyant wall jet. Part 1: The LES database', *Fourth Int. Symp. on Turbulence, Heat & Mass Transfer*, Antalya, Turkey.

André, J. C., De Moor, G., Lacarrère, L., Therry, G. and du Vachat, R., 1979, 'The clipping approximation and inhomogeneous turbulence simulations', in *Turbulent Shear Flows* 1 (eds. F. Durst, B. E. Launder, F. W. Schmidt and J. H. Whitelaw), pp. 307–318, Berlin, Springer Verlag.

el Baz, A., Craft, T. J., Ince, N. Z. and Launder, B. E., 1993, 'On the adequacy of the thin-shear-flow equations for computing turbulent jets in stagnant surroundings', *Int. J. Heat Fluid Flow*, **14**, 164–169.

Bertoglio, J.-P., 1982, 'Homogeneous turbulence within a rotating reference frame', *AIAA Journal*, **20**, 1175–1178.

Betts, P. L. and D'Affa Alla, 1986, 'Turbulent buoyant air flow in a tall rectangular cavity', ASME Heat Transfer Division, Volume HTD-60, pp. 83–91, ASME, Winter Annual Meeting.

Boris, J. P., Grinstein, F. F., Oran, E. S. and Kolbe, R. L., 1992, 'New insights into large eddy simulation', *Fluid Dynamics Research*, **10**, 199–228.

Bradshaw, P., 1973, 'Effects of streamline curvature in turbulent flow', AGARDograph 196, NATO.

Cheah, S. C., Cheng, L., Cooper, D. and Launder, B. E., 1993, 'On the structure of turbulent flow in spirally fluted tubes', *Proc. Fifth IAHR Conf. on Refined Flow Modelling and Turbulence Measurements*, pp. 293–300, Paris, Presses Ponts et Chausées.

Chen, C. J. and Rodi, W., 1980, *Turbulent Buoyant Jets – a Review of Experimental Data*, HMT Series, Vol. 4, Oxford, Pergamon Press.

Chieng, C. C. and Launder, B. E., 1980, 'On the calculation of turbulent heat transport downstream from an abrupt pipe expansion', *Numerical Heat Transfer*, **3**, 189–207.

Choi, Y. D., Iacovides, H. and Launder, B. E., 1989, 'Numerical computation of turbulent flow in a square-sectioned 180-degree bend', *ASME J. Fluids Engineering*, **111**, 59–68.

Chou, P.-Y., 1945, 'On velocity correlations and the solutions of the equations of turbulent fluctuation', *Quart. Appl. Math.*, **3**, 38–54.

Cooper, D., Jackson, D. C., Launder, B. E. and Liao, G. X., 1993, 'Impinging jet studies for turbulence model assessment. I. Flow-field experiments', *Int. J. Heat & Mass Transfer*, **36**, 2675–2684.

Cordier, H., 1999, 'Simulation numérique de l'interaction en milieu confiné d'un écoulement de convection forcée avec un panache thermique', Doctoral Thesis, Université d'Aix-Marseille-II, Marseille.

Cotton, M. A. and Ismael, J. O., 1995, 'Some results for homogeneous shear flows computed using a strain-parameter model of turbulence', *Proc. Tenth Symp. Turbulent Shear Flows*, pp. 26.7–26.12, Pennsylvania State University.

Cotton, M. and Jackson, J. D., 1987, 'Calculation of turbulent mixed convection using a low-Reynolds number k-ε model', *Proc. Sixth Turbulent Shear Flow Symposium*, Paper 9-6, Toulouse.

Craft, T. J., 1991, 'Second-moment modelling of turbulent scalar transport', Ph.D. Thesis, Faculty of Technology, University of Manchester, Manchester.

Craft, T. J. and Launder, B. E., 1989, 'A new model for the pressure/scalar-gradient correlation and its application to homogeneous and inhomogeneous free shear flows',

Paper 17-1, *Proc. Seventh Symposium on Turbulent Shear Flows*, Stanford, California.

1992, 'A new model of wall-reflection effects on the pressure-strain correlation and its application to the turbulent impinging jet', *AIAA Journal*, **30**, 2970–2972.

1999, 'The self-similar three-dimensional wall jet', in *Turbulence and Shear-Flow Phenomena – 1* (eds. S. Banerjee and J. K. Eaton), pp. 1129–1134, New York, Begell.

2001, 'On the spreading mechanism of the three-dimensional wall jet', *J. Fluid Mech.* **435**, 305–326.

2002, 'Application of TCL modelling to stratified flow', in *Closure Strategies for Turbulent and Transitional Flows* (eds. B. E. Launder and N. D. Sandham), Cambridge, Cambridge University Press.

Craft, T. J., Launder, B. E. and Suga, K., 1993, 'Extending the applicability of eddy-viscosity models through the use of deformation invariants and non-linear elements', *Proc. Fifth IAHR Conf. on Refined Flow Modelling and Measurements*, pp. 125–132, Paris, Presses Ponts et Chaussées.

Craft, T. J., Ince, N. Z. and Launder, B. E., 1996a, 'Recent developments in second-moment closure for buoyancy-affected flows', *Dynamics of Atmospheres and Oceans*, **23**, 99–114.

Craft T. J., Launder, B. E. and Suga, K., 1996b, 'Development and application of a cubic eddy-viscosity model of turbulence', *Int. J. Heat Fluid Flow*, **17**, 590–596.

Craft, T. J., Kidger, J. W. and Launder, B. E., 1997a, 'The importance of third-moment modelling in horizontal, stably-stratified flow', *Proc. Eleventh Symp. Turbulent Shear Flows*, pp. 20.13–20.18, Grenoble.

Craft, T. J., Launder, B. E. and Suga, K., 1997b, 'The prediction of turbulent transitional phenomena with a non-linear eddy-viscosity model', *Int. J. Heat Fluid Flow*, **18**, 15–28.

Craft, T. J., Kidger, J. W. and Launder, B. E., 1999, 'The development of three-dimensional, turbulent surface jets', *Turbulence & Shear Flow Phenomena – I* (eds. S. Banerjee and J. K. Eaton), pp. 1117–1122, New York, Begell House.

Craft, T. J., Gant, S. E., Iacovides, H. and Launder, B. E., 2001, 'Development and application of a new wall function for complex turbulent flows', *Proc. ECCOMAS CFD Conference*, Swansea.

Craft, T. J., Gant, S. E., Gerasimov, A. V., Iacovides, H. and Launder, B. E., 2002a, 'Wall-function strategies for use in turbulent flow CFD', Keynote Paper, *Proc. Twelfth International Heat Transfer Conference*, Grenoble.

Craft, T. J., Gerasimov, A. V., Iacovides, H. and Launder, B. E., 2002b, 'Progress in the generalization of wall-function treatments', *Int. J. Heat Fluid Flow*, **23**, 148–160.

Craft, T. J., Gerasimov, A. V., Iacovides, H., Kidger, J. and Launder, B. E., 2003, 'The negatively buoyant turbulent wall jet: performance of alternative options in RANS modelling', *Fourth Int. Symp. on Turbulence, Heat & Mass Transfer*, Antalya, Turkey.

Craft, T. J., Gant, S. E., Iacovides, H. and Launder, B. E., 2004, 'A new wall-function strategy for complex turbulent flows', *Numerical Heat Transfer*, **45B**, 301–318.

Cresswell, R., 1988, 'An experimental study of a turbulent jet in which buoyancy acts against initial momentum', Ph.D. Thesis, Council for National Academic Awards (CNAA).

Cresswell, R., Haroutunian, V., Ince, N. Z., Launder, B. E. and Szczepura, R. T., 1989, 'Measurement and modelling of buoyancy-modified elliptic turbulent flows', *Proc. Seventh Symp. Turbulent Shear Flows*, Stanford.

Daly, B. J. and Harlow, F. H., 1970, 'Transport equations in turbulence', *Physics of Fluids*, **13**, 2634.

Deardorff, J. N., 1973, 'The use of sub-grid transport equations in a three-dimensional model of atmospheric turbulence', *J. Fluids Engineering*, **95**, 429–438.

Dekeyser, I. and Launder, B. E., 1985, 'A comparison of triple-moment temperature–velocity correlations in the asymmetric heated jet with alternative closure models', in *Turbulent Shear Flows 4* (eds. L. J. S. Bradbury, F. Durst, B. E. Launder, F. W. Schmidt and J. H. Whitelaw), pp. 102–117, Berlin, Springer Verlag.

Durbin, P. A., 1993, 'Application of a near-wall turbulence model to boundary layers and heat transfer', *Int. J. Heat & Fluid Flow*, **14**, 316–323.

  1995, 'Separated flow computations with the k-ε-v^2 model', *AIAA Journal*, **33**, 659–664.

Ellison, T. H. and Turner, J. S., 1959, 'Turbulent entrainment in stratified flows', *J. Fluid Mech.*, **6**, 423–448.

Forstall, W. and Shapiro, A. H., 1950, 'Momentum and mass transfer in coaxial gas jets', *J. Applied Mechanics*, **17**, 399–408.

Fu, S., Launder, B. E. and Tselipidakis, D. P., 1987, 'Accommodating the effects of high strain rates in modelling the pressure-strain correlation', Mechanical Engineering Dept. Rep. TFD/87/5, UMIST, Manchester.

Gant, S. E., 2002, 'Development and application of a new wall function for complex turbulent flows', Ph.D. Thesis, Dept. Mech., Aerosp. & Manuf. Eng, UMIST, Manchester.

Gatski, T. B. and Rumsey, C. L., 2002, 'Linear and non-linear eddy viscosity models', in *Closure Strategies for Laminar and Turbulent Flows* (eds. B. E. Launder and N. D. Sandham), pp. 9–46, Cambridge, Cambridge University Press.

Gerasimov, A., 2004, 'Development and validation of an analytical wall-function strategy for modelling forced, mixed and natural convection flows' Ph.D. Thesis, Dept. Mech., Aerospace & Manufacturing Engineering, UMIST.

Germano, M., Piomelli, U., Moin, P. and Cabot, W. H., 1991, 'A dynamic, sub-grid-scale eddy-viscosity model', *Phys. Fluids*, **A3**, 1760–1765.

Gibson, M. M. and Launder, B. E., 1976, 'On the calculation of horizontal free shear flows under gravitational influence', *ASME J. Heat Transfer*, **98**, 379–386.

Gibson, M. M. and Launder, B. E., 1978, 'Ground effects on pressure fluctuations in the atmosphere boundary layer', *J. Fluid Mech.*, **86**, 491–511.

Grötzbach, G. and Wörner, M., 1999, 'Direct numerical and large-eddy simulations in nuclear applications', *Int. J. Heat Fluid Flow*, **20**, 222–240.

Hanjalić, K. and Launder, B. E., 1980, 'Sensitizing the dissipation equation to irrotational strains', *ASME J. Fluids Eng.*, **102**, 34–40.

Hanjalić, K. and Musemić, 1997, 'Modelling the dynamics of double diffusive scalar fields at various stability conditions', *Int. J. Heat Fluid Flow*, **18**, 360–367.

Hanjalić, K. and Vasić, S., 1993, 'Some further exploration of turbulence models for buoyancy driven flows', in *Turbulent Shear Flows – 8* (eds. F. J. Durst *et al.*), Heidelberg, Springer Verlag.

Hanjalić, K., Launder, B. E. and Schiestel, R., 1980, 'Multiple-time-scale concepts in turbulent transport modelling', in *Turbulent Shear Flows – 2* (eds. L. J. S. Bradbury *et al.*), Heidelberg, Springer Verlag.

Haroutunian, V. and Launder, B. E., 1988, 'Second-moment modelling of free buoyant shear flows: a comparison of parabolic and elliptic solutions', in *Stably-Stratified Flow and Dense Gas Dispersion* (ed. J. S. Puttock), pp. 409–430, Oxford, Oxford University Press.

Hashiguchi, M., 1996, 'Turbulence simulations in the Japanese automobile industry', in *Proc. Engineering Turbulence Modelling & Experiments – 3* (eds. W. Rodi and G. Bergeles), pp. 291–308, Amsterdam, Elsevier.

Hinze, J. O., 1973, 'Experimental investigation on secondary currents in the turbulent flow through a straight conduit', *Appl. Sci. Res.*, **28**, 453.

Hossain, M. S. and Rodi, W., 1982, 'A turbulence model for buoyant flows and its application to vertical buoyant jets', in *Turbulent Buoyant Jets and Plumes* (ed. W. Rodi), pp. 121–178, Oxford, Pergamon.

Huang, P. G., 1986, 'The computation of elliptic turbulent flows with second-moment closure models', Ph.D. Thesis, Faculty of Technology, University of Manchester.

Huang, G.-W., Kawahara, Y. and Tamai, N., 1993, 'Numerical investigation of a buoyant shear flow with a multiple-time-scale model', *Proc. Fifth Symp. Refined Flow Modelling and Turbulence Measurements*, pp. 769–776, Paris, Presses Ponts et Chaussées.

Iacovides, H. and Raisee, M., 1997, 'Computation of flow and heat transfer in 2-D rib-roughened passages', in *Addendum to Second Int. Symp. on Turbulence, Heat and Mass Transfer* (eds. K. Hanjalic and T. W. J. Peeters), pp. 21–30, The Netherlands, Delft University Press.

Iacovides, H. and Raisee, M., 1999, 'Recent progress in the computation of flow and heat transfer in internal cooling passages of turbine blades', *Int. J. Heat & Fluid Flow*, **20**, 320–328.

Iacovides, H., Launder, B. E. and Li, H.-Y., 1996, 'The computation of flow development through stationary and rotating U-ducts of strong curvature', *Int. J. Heat & Fluid Flow*, **17**, 22–33.

Ilyushin, B. B., 2002, 'Higher moment diffusion in stable stratification', in *Closure Strategies for Laminar and Turbulent Flows* (eds. B. E. Launder and N. D. Sandham), pp. 424–448, Cambridge, Cambridge University Press.

Ince, N. Z. and Launder, B. E., 1989, 'On the computation of buoyancy-driven turbulent flows in rectangular enclosures', *Int. J. Heat Fluid Flow*, **10**, 110–117.

Ince, N. Z. and Launder, B. E., 1995, 'Three-dimensional and heat-loss effects on turbulent flow in a nominally two-dimensional cavity', *Int. J. Heat & Fluid Flow*, **16**, 171–177.

Jackson, J. D., Cotton, M. A. and Axcell, B. P., 1989, 'Studies of mixed convection in vertical tubes', *Int. J. Heat & Fluid Flow*, **10**, 2–15.

Johnson, R. W., 1984, 'Turbulent convecting flow in a square duct with a 180 degree bend: an experimental and numerical study', Ph.D. thesis, Faculty of Technology, University of Manchester.

Jones, W. P. and Launder, B. E., 1972, 'The prediction of laminarization with a two-equation model of turbulence', *Int. J. Heat Mass Transfer*, **15**, 301–314.
　1973, 'The calculation of low-Reynolds-number phenomena with a two-equation model of turbulence', *Int. J. Heat Mass Transfer*, **16**, 1119–1130.

Kawamura, H., Sasaki, J. and Kobayashi, K., 1995, 'Budget and modelling of triple-moment velocity correlations in a turbulent channel flow based on DNS', *Tenth Int. Symp. on Turbulent Shear Flows*, Pennsylvania State University.

Kenjeres, S. and Hanjalić, K., 1999, 'Transient analysis of Rayleigh–Bénard convection with a RANS model', *Int. J. Heat & Fluid Flow*, **20**, 329–340.

Kenjeres, S. and Hanjalić, K., 2002, 'A numerical insight into flow structure in ultra-turbulent thermal convection', *Phys. Rev. E*, **66**, 1–5.

Kenjeres, S. and Hanjalić, K., 2003, 'Numerical contribution to insight into the elusive, *ultimate* state of turbulent thermal convection', EUROMECH Coll. 443, Lorentz Centre Leiden, The Netherlands.

Kidger, J. W., 1999, 'Turbulence modelling for stably stratified flows and free-surface jets', Ph.D. Thesis, Dept. of Mechanical Engineering, UMIST, Manchester.

Kim, J., Moin, P. and Moser, R., 1987, 'Turbulence statistics in fully developed channel flow at low Reynolds number', *J. Fluid Mech.*, **177**, 133–166.

Launder, B. E., 1975, 'On the effects of a gravitational field on the turbulent transport of heat and momentum', *J. Fluid Mech.*, **67**, 569–581.

1976, 'Heat Transfer', in *Topics in Appl. Phys., Vol 12: Turbulence* (ed. P. Bradshaw), Berlin, Springer Verlag.

1985, 'Progress and problems in phenomenological turbulence models', Chapter 7 in *Theoretical Approaches to Turbulence* (eds. D. L. Dwoyer, M. Y. Hussaini and R. G. Voigt), New York, Springer Verlag.

1986, 'Low Reynolds number turbulence near walls', UMIST Mechanical Engineering Department Report TFD/86/4.

1988, 'On the computation of convective heat transfer in complex turbulent flows', *ASME J. Heat Transfer*, **110**, 1112–1128.

1989a, 'Second-moment closure: present . . . and future?', *Int. J. Heat Fluid Flow*, **10**, 282–300.

1989b, 'The prediction of force-field effects on turbulent shear flows via second-moment closure', *Advances in Turbulence – 2* (eds. H. Fernholz and H. E. Fieldler), pp. 338–358, Berlin, Springer.

Launder, B. E. and Li, S.-P., 1994, 'On the elimination of wall-topography parameters from second-moment closure', *Phys. Fluids*, **6**, 999–1000.

Launder, B. E. and Sandham, N., 2002 (eds.), *Closure Strategies for Turbulent and Transitional Flows*, Cambridge, Cambridge University Press.

Launder, B. E. and Schiestel, R., 1978, 'Sur l'utilisation d'échelles temporelles multiples en modélisation des écoulements turbulents', *Comptes Rendues Acad. Sciences, Paris*, **A288**, 709–712.

Launder, B. E. and Sharma, B. I., 1974, 'Application of the energy-dissipation model of turbulence to the calculation of flow near a spinning disc', *Letters in Heat Mass Transfer*, **1**, 131–138.

Launder, B. E. and Spalding, D. B., 1974, 'The numerical computation of turbulent flows', *Computational Methods in Applied Mechanics and Engineering*, **3**, 269–289.

Launder, B. E. and Tselipidakis, D. P., 1992, 'Contribution to the modelling of near-wall turbulence', *Turbulence Shear Flows – 8* (eds. F. Durst *et al.*), pp. 81–96, Heidelberg, Springer Verlag.

Launder, B. E., Morse, A. P., Rodi, W. and Spalding, D. B., 1973, 'Prediction of free shear flows: a comparison of the performance of six turbulence models', *Proc. NASA Conf. On Free Turbulent Shear Flows*, NASA SP-321, pp. 361–426.

Launder, B. E., Reece, G. J. and Rodi, W., 1975, 'Progress in the development of a Reynolds stress turbulence closure', *J. Fluid Mech.*, **68**, 537–566.

Launder, B. E., Priddin, C. H. and Sharma, B. I., 1977, 'The calculation of turbulent boundary layers on curved and spinning surfaces', *ASME J. Fluids Engineering*, **99**, 231–239.

Lumley, J. L., 1978, 'Computational modeling of turbulent flows', *Advances in Applied Mechanics*, **18**, 123.

McGuirk, J. J. and Papadimitriou, C., 1985, 'Buoyant surface layers under fully-entraining and internal-hydraulic-jump conditions', *Proc. Fifth Symp. Turbulent Shear Flows*, pp. 22.30–22.41, Cornell University.

Mellor, G. L., 1973, 'Analytic prediction of the properties of stratified planetary boundary layers', *J. Atmos. Sci.*, **30**, 1061–1069.

Millionshtchikov, M. D., 1941, 'On the role of the third moments in isotropic turbulence', *C. R. Acad. Sci. SSSR*, **32**, 619.

Moin, P., 2000, Oral remarks at open-forum discussion held at Burgers Day 2000, TU Delft, The Netherlands.

Muramatsu, T., 1993, 'Intensity and frequency evaluations of sodium temperature fluctuations related to thermal striping phenomena based on numerical methods', *Fifth Int. Symp. Refined Flow Modelling and Turbulence Measurements*, pp. 351–358, Paris, Presses Ponts et Chaussées.

Naot, D., Shavit, A. and Wolfshtein, M., 1970, 'Interactions between components of the turbulent velocity correlation tensor due to pressure fluctuations', *Israel J. Tech.*, **8**, 259–269.

Oberlack, M., 1995, 'Closure of the dissipation tensor and the pressure-strain tensor based on the two-point correlation equation', in *Turbulent Shear Flows – 9* (eds. F. Durst *et al.*), pp. 33–52, Berlin, Springer.

Patel, V. C., Rodi, W. and Scheuerer, G., 1985, 'Turbulence models for near-wall and low-Reynolds-number flows: a review', *AIAA J.*, **23**, 1308.

Pope, S. B., 1975, 'A more general effective viscosity hypothesis', *J. Fluid Mech.*, **72**, 331–340.

   1978, 'An explanation of the turbulent round-jet/plane-jet anomaly', *AIAA J.*, **16**, 279–281.

Prandtl, L., 1925, 'Bericht über Untersuchungen zur ausgebildeten Turbulenz', *ZAMM*, **5**, 136.

Rajaratnam, N. and Humphreys, J. A., 1984, 'Turbulent, non-buoyant surface jets', *J. Hydraulic Res.*, **22**, 103–115.

Robinson, C., 2001, 'Advanced CFD modelling of road vehicle aerodynamics', Ph.D. Thesis, Department of Mechanical Engineering, UMIST, Manchester.

Rodi, W., 1972, 'The prediction of free turbulent boundary layers by use of a two-equation model of turbulence', Ph.D. Thesis, Faculty of Engineering, University of London.

   1976, 'A new algebraic relation for calculating the Reynolds stresses', *ZAMM*, **56**, T219–T221.

Rotta, J. C., 1951, 'Statistische Theorie nichthomogener Turbulenz', *Zeitschr. Physik*, **129**, 547–572.

Savill, A. M., 1991, 'Predicting transition induced by free stream turbulence', *First European Fluid Mechanics Conference*.

Schiestel, R., 1987, 'Multiple-time-scale modelling of turbulent flows in one-point closures', *Phys. Fluids*, **30**, 722.

Schumann, U., 1976, 'Numerical simulation of the transition from three- to two-dimensional turbulence under a uniform magnetic field', *J. Fluid Mech.*, **74**, 31–58.

Schumann, U., 1977, 'Realizability of Reynolds-stress turbulence models', *Phys. Fluids*, **20**, 721–723.

Shih, T. H. and Lumley, J. L., 1985, 'Modelling of pressure-correlation terms in Reynolds-stress and scalar-flux equations', Sibley School of Mechanical & Aerospace Eng. Rep. FDA-85-3, Cornell University, Ithaca, New York.

Shir, C. C., 1973, 'A preliminary numerical study of atmospheric turbulent flows in the idealized planetary boundary layer', *J. Atmos. Sciences*, **30**, 1327–1339.

Smagorinsky, J. S., 1963, 'General circulation experiments with the primitive equations: 1. The basic experiment', *Mon. Weather Rev.*, **91**, 99–164.

Suga, K., 1995, 'Development and application of a non-linear eddy-viscosity model sensitized to stress and strain invariants', Ph.D. Thesis, Faculty of Technology, University of Manchester.

Uittenbogaard, R., 1992, 'Measurement of turbulent fluxes in a steady-stratified mixing layer', *Proc. Third Int. Symp. on Refined Flow Modelling and Turbulence Measurement*, Tokyo.

Versteegh, T. A. M. and Nieuwstadt, F. T. M., 1998, 'Turbulent budgets of natural convection in an infinite, differentially heated vertical channel', *Int. J. Heat Fluid Flow*, **19**, 135–149.

Viollet, J.-P., 1980, 'Turbulent mixing in a two-layer stratified shear flow', *Second Int. Symp. on Stratified Flows*, Trondheim, Norway.

  1987, 'The modelling of turbulent recirculating flows for the purpose of reactor thermal-hydraulic analysis', *Nuclear Eng. Design*, **99**, 365–377.

Voke, P. R. and Gao, S., 1994, 'Large eddy simulation of heat transfer from an impinging plane jet', *Int. J. Num. Methods Eng.*

Webster, C. A. G., 1964, 'An experimental study of turbulence in a density-stratified shear flow', *J. Fluid Mech.*, **19**, 221–245.

Wilcox, D., 1993, *Turbulence modeling for CFD*, La Cañada, CA, DCW Industries.

Woodburn, P., 1995, 'CFD simulation of fire-generated flows in tunnels and corridors', Ph.D. Thesis, Cambridge University.

Wörner, M. and Grötzbach, G., 1998, 'Pressure transport in direct numerical simulations of turbulent natural convection in horizontal fluid layers', *Int. J. Heat Fluid Flow*, **19**, 150–158.

Wyngaard, J. C. and Sundarajan, A., 1979, 'The temperature skewness budget in the lower atmosphere and its implications for turbulence modelling', in *Turbulent Shear Flows 1* (eds. F. Durst, B. E. Launder, J. H. Schmidt and J. H. Whitelaw), pp. 319–326, Berlin, Springer Verlag.

Yap, C., 1987, 'Turbulent heat and momentum transfer in recirculating and impinging flows', Ph.D. Thesis, Faculty of Technology, University of Manchester.

Yonemura, A. and Yamamoto, M., 1995, 'Proposal of a multiple-time-scale Reynolds stress model for turbulent boundary layers', *Proc. Int. Symp. Math. Modelling Turb. Flows*, pp. 381–386, Tokyo, Japan.

Young, S. T. B., 1975, 'Turbulence measurements in a stably-stratified shear flow', Rep QMC-EP6018, Queen Mary College, London.

Zeman, O. and Lumley, J. L., 1976, 'Modeling buoyancy driven mixing layers', *J. Atmos. Sci.*, **33**, 1974–1988.

4

Turbulent flames

W. P. Jones

Imperial College London

1 Introduction

Over 80% of the world's energy is generated by the combustion of hydrocarbon fuels, and this is likely to remain the case for the foreseeable future. In addition to the release of heat, this combustion is accompanied by the emission, in the exhaust stream, of combustion generated pollutants such as carbon monoxide, unburnt hydrocarbons and oxides of nitrogen, NO_x. The former two quantities arise as a result of incomplete combustion, whereas NO_x is formed from the reaction of nitrogen present in the air or fuel with oxygen, usually at high temperatures. An unavoidable outcome of the burning of hydrocarbon fuels is the formation of carbon dioxide, CO_2, which is a 'greenhouse' gas that may contribute to global warming. While the amount of carbon dioxide generated depends on the fuel burnt, any improvements which can be achieved to combustion efficiency will clearly contribute to an overall reduction in the emissions of CO_2. Because of the growing need to reduce the emissions of combustion generated pollutants and improve combustion efficiencies, there is increased interest in accurate methods for predicting the properties of combustion systems. The combustion in the vast majority of practical systems is turbulent and this poses a number of difficulties for prediction. The development of accurate methods for predicting turbulent combusting flows remains a largely unresolved problem which continues to attract a large number of researchers.

The categorisation of combusting flows into premixed and non-premixed cases serves as an aid to analysis and constitutes an idealisation of reality. For example, combustion in spark ignition engines occurs in the premixed mode, however the degree to which the fuel/air mixture has homogenised prior to entry into the

Prediction of Turbulent Flows, eds. G. F. Hewitt and J. C. Vassilicos.
Published by Cambridge University Press. © Cambridge University Press 2005.

combustion chamber may not be complete and so a totally premixed characterisation would be a simplification. Combustion in equipment such as furnaces, diesel engines and gas turbine combustors is classified as non-premixed. However, in such flames the presence of local extinction pockets can lead to fuel/air mixtures becoming partially premixed. Consequently a flame front would propagate into this partially premixed region. Several reviews and texts on various aspects of turbulent combustion are available, e.g. Bilger (1980, 1989), Bray (1980, 1986), Heitor *et al.* (1996), Jones and Whitelaw (1982), Libby and Williams (1993), Peters (2000) and Pope (1985, 1987).

The provision of methods for computing the properties of turbulent flames and combustion systems requires that several distinct phenomena be modelled, with the main two being the representation of turbulent transport (diffusion) and the interaction between chemical reaction and turbulent fluctuations. The former aspect is covered in other contributions and here attention will be focused on chemical reaction. The rates of reaction and heat release are invariable highly non-linear functions of temperature and composition and are 'fast' compared to mean flow time scales. Consequently, an accurate and practicably feasible description of reaction in a turbulent environment presents severe modelling difficulties. The present chapter attempts to summarise the current 'state of the art' in the modelling of turbulent combustion. Emphasis is placed on the ideas behind closure principles, and after a brief presentation of the governing equations and basic assumptions, combustion in non-premixed and premixed configurations is considered. This is then followed by a separate section in which assumptions about the particular mode of combustion are avoided. The main emphasis in this latter section is transported probability density function (pdf) equation methods; a formulation that is not, in principle, restricted to non-premixed or premixed flame calculations.

2 The governing equations

The governing equations describing the flow of a fluid can be derived from the principles of conservation of mass, momentum and energy. Unless otherwise stated, they are written below using Cartesian tensor notation:

Continuity

$$\frac{\partial \varrho}{\partial t} + \frac{\partial \varrho u_j}{\partial x_j} = 0, \tag{4.1}$$

where u_j is the instantaneous velocity vector, and ϱ is the density.

Navier–Stokes equation

$$\varrho \frac{\partial u_i}{\partial t} + \varrho u_j \frac{\partial u_i}{\partial x_j} = -\frac{\partial p}{\partial x_i} + \frac{\partial \tau_{ij}}{\partial x_j} + \varrho g_i, \tag{4.2}$$

where τ_{ij} is the viscous stress tensor defined by:

$$\tau_{ij} = \mu \left(\frac{\partial u_i}{\partial x_j} + \frac{\partial u_j}{\partial x_i} - \frac{2}{3} \delta_{ij} \frac{\partial u_k}{\partial x_k} \right), \tag{4.3}$$

and where p is the pressure, μ is the mixture viscosity, δ_{ij} is the Kronecker delta and g_i is the gravitational acceleration vector.

Energy

$$\varrho \frac{\partial h}{\partial t} + \varrho u_j \frac{\partial h}{\partial x_j} = \frac{\partial p}{\partial t} + \frac{\partial}{\partial x_i} \left\{ \frac{\mu}{\sigma} \frac{\partial h}{\partial x_i} + \mu \left(\frac{1}{Sc} - \frac{1}{\sigma} \right) \sum_{\alpha=1}^{n} h_\alpha \frac{\partial Y_\alpha}{\partial x_i} \right\}, \tag{4.4}$$

where h is the specific enthalpy[1] of the mixture, Y_α is the mass fraction, Sc is the Schmidt number of the mixture and σ is the mixture Prandtl number. Low Mach number flows ($Ma \ll 1$) are considered, and consequently the terms associated with the kinetic energy of the mixture and the viscous dissipation are neglected. Radiative heat losses are not considered. Fourier's and Fick's laws for molecular fluxes are employed and work done against body forces is assumed to be negligible in comparison to thermal energy changes associated with chemical reaction. The term $\frac{\partial p}{\partial t}$ is assumed to be negligible in many practically occurring situations, e.g. jets, gas turbines and furnaces, but cannot be neglected in the flows occurring in internal combustion engines. The implication of $\frac{\partial p}{\partial t} \approx 0$ and the low Mach number assumption is that the density and the source term for reaction (in the species conservation equation, to be described later on) are only functions of the scalar fields involving the specific enthalpy and the mass fraction of the species. The effect of chemical reaction appears in the energy equation through the definition of h:

$$h = \sum_{\alpha=1}^{n} Y_\alpha h_\alpha, \quad \text{where} \quad h_\alpha = c_{p,\alpha} T + \Delta_\alpha. \tag{4.5}$$

The Δ_α represent the heats of formation for the species involved in the reactions. The inclusion of this term accounts for the heat release (and absorption) effects due to the chemical reactions taking place. Finally, the Lewis number, defined as the

---

[1] Strictly, h is the stagnation enthalpy, but for low Mach number flows the kinetic energy contribution to this is negligible.

ratio of the Schmidt number to the Prandtl number is close to unity in combusting flows, and it is a useful assumption to take it as exactly unity (since the third term on the rhs of Eq. (4.4) disappears). This simplifies the energy equation to:

$$\varrho \frac{\partial h}{\partial t} + \varrho u_j \frac{\partial h}{\partial x_j} = \frac{\partial}{\partial x_i}\left(\frac{\mu}{\sigma}\frac{\partial h}{\partial x_i}\right). \tag{4.6}$$

Species conservation

$$\varrho \frac{\partial Y_\alpha}{\partial t} + \varrho u_j \frac{\partial Y_\alpha}{\partial x_j} = -\frac{\partial J_{i,\alpha}}{\partial x_i} + \dot{\omega}_\alpha \quad \text{for} \quad \alpha = 1, \ldots, n, \tag{4.7}$$

where $\dot{\omega}_\alpha$ is the net formation rate of species α per unit volume and $J_{i,\alpha}$ is the diffusional flux of species α. Fick's law of diffusion, $J_{i,\alpha} = -\frac{\mu}{Sc}\frac{\partial \phi_\alpha}{\partial x_i}$, is strictly applicable only to binary mixtures, and so the use of Fickian diffusion in the context of multi-component mixtures is only justified if the Schmidt number is the same for all species (in order to ensure that the mass fractions sum to unity). The assumption of uniform Sc (equal diffusivity) is known to be poor when species with widely differing molar masses are being considered. In the case of turbulent flows the discussion presented later demonstrates that equal (and constant) diffusivity is invariably invoked. To obtain a general expression for $\dot{\omega}_\alpha$, consider the system involving n chemical species and r reaction steps, expressed in the form (Libby and Williams, 1980):

$$\sum_{\alpha=1}^{n} v'_{\alpha\beta} M_\alpha \rightleftharpoons \sum_{\alpha=1}^{n} v''_{\alpha\beta} M_\alpha, \quad \beta = 1, \ldots, r, \tag{4.8}$$

where the rate constants for the forward and backward reactions are $k_{f\beta}$ and $k_{b\beta}$, respectively, and M_α denotes the chemical symbol of species α. If species α does not participate in a particular reaction then $v'_{\alpha\beta}$ and $v''_{\alpha\beta}$ are zero as appropriate. The rate constants can be expressed as:

$$k_{f\beta} = B_\beta T^{a_\beta} \exp\left(-\frac{E_\beta}{RT}\right), \tag{4.9}$$

where B_β is the pre-exponential factor, a_β is a constant exponent, E_β is the activation energy and R is the universal gas constant. Clearly the rate of formation of species α is the sum of the formation rates of species α in all the individual reaction steps. In any particular reaction step, the rate of formation of species α is the net rate of formation of species α due to the forward and backward processes in reaction β.

Mathematically, this can be expressed as:

$$
\dot{\omega}_\alpha = W_\alpha \sum_{\beta=1}^{r} (v''_{\alpha\beta} - v'_{\alpha\beta}) \left\{ k_{f\beta} \varrho^{m_\beta} \prod_{\alpha=1}^{n} \left(\frac{Y_\alpha}{W_\alpha} \right)^{v'_{\alpha\beta}} - k_{b\beta} \varrho^{l_\beta} \prod_{\alpha=1}^{n} \left(\frac{Y_\alpha}{W_\alpha} \right)^{v''_{\alpha\beta}} \right\},
$$

(4.10)

where $m_\beta \equiv \sum_{i=1}^{n} v'_{\alpha\beta}$ and $l_\beta \equiv \sum_{i=1}^{n} v''_{\alpha\beta}$ are the orders of the forward and reverse reactions respectively. In order to close the system of equations, an expression for the density is required. This is given by the equation of state for an ideal gas:

$$
p = \varrho RT \sum_{\alpha=1}^{n} \frac{Y_\alpha}{W_\alpha}.
$$

(4.11)

With the provision of a suitable reaction mechanism and rate constants the equations described so far constitute a closed set and providing appropriate initial and boundary conditions are available they can, in principle, be solved (numerically). However, it can be shown that typically the CPU requirements for a direct numerical simulation (DNS) of inert mixing are $O(Re^3)$. It is quite clear that for the high Reynolds number flows of engineering interest, DNS is computationally prohibitive (Reynolds, 1990). The computational requirements for DNS of combusting turbulent flow are even more severe. These arise because burning will occur in thin layers. Typically, the thickness of the reaction zone in, say, methane–air flame burning at atmospheric pressure will be of order 10^{-1} mm and the number of grid nodes required to resolve this region, at minimum, will be $O(10)$. Since the orientation and position of the flame is a random function of time with respect to any fixed grid, the implication is that a DNS of turbulent combusting flows would require of $O(10^6)$ grid nodes per mm³ of flame. The situation may be expected to become even more restrictive at higher pressures as the thickness of the reaction zone decreases as pressure is increased; for methane–air the burning zone thickness typically reduces to $O(10^{-2})$ mm at 30 bar and even finer meshes would then be required. However, it should not be implied that DNS is of little value; it is simply that DNS does not represent a solution to the vast majority of engineering problems and is unlikely to become so in the foreseeable future. For 'low' Re flows and in simple configurations, DNS is feasible and can lead to extremely useful results that are an invaluable aid to understanding and to the provision of information for constructing simpler representations of turbulence.

The traditional approach for treating high Re turbulent flows has been to average the conservation equations, with the consequence that all the fine details of the flow do not require spatial and temporal resolution. Different types of averaging can be applied, but they all share a common feature: the closed set of instantaneous equations are transformed by averaging into an 'open' set of equations. Terms appear which cannot be accounted for in terms of the dependent variables of the flow field. Modelling serves to approximate these terms, and the purpose of this chapter is to address the principles behind the representation of such terms. Two common types of averaging are the Reynolds (unweighted) and Favre (density-weighted) approach, defined as follows:

$$\bar{A}(\mathbf{x}, t) \equiv \lim_{N_p \to \infty} \frac{1}{N_p} \sum_{i=1}^{N_p} A^{(i)}(\mathbf{x}, t), \qquad \text{Reynolds averaging} \qquad (4.12)$$

$$\tilde{A}(\mathbf{x}, t) \equiv \frac{1}{\bar{\varrho}} \lim_{N_p \to \infty} \frac{1}{N_p} \sum_{i=1}^{N_p} \varrho^{(i)} A^{(i)}(\mathbf{x}, t), \quad \text{Favre averaging} \qquad (4.13)$$

where $A^{(i)}(\mathbf{x}, t)$ is the ith realisation. For statistically stationary flows the averages defined by Eqs. (4.12) and (4.13) are identical to Reynolds and Favre (or density-weighted) time averages respectively. The standard practice is to decompose the instantaneous equations into mean and fluctuating parts and then to average them. With density-weighted averaging the velocity and scalars (but not the density and pressure) are decomposed according to $u_i = \tilde{u}_i + u_i''$, $\phi_\alpha = \tilde{\phi}_\alpha + \phi_\alpha''$. With the definition of Favre averaging in Eq. (4.13) it can easily be shown that $\overline{\varrho \phi_\alpha''} = \overline{\varrho u_i''} = 0$. For Reynolds averaging, $u_i = \bar{u}_i + u_i'$, $\phi_\alpha = \bar{\phi}_\alpha + \phi_\alpha'$ and where $\overline{u_i'} = \overline{\phi_\alpha'} = 0$. From the basic definitions, the following results can be established between weighted and unweighted statistics:

$$\bar{u}_i = \tilde{u}_i + \overline{u_i''}; \qquad \bar{\phi}_\alpha = \tilde{\phi}_\alpha + \overline{\phi_\alpha''}; \qquad (4.14)$$

$$\overline{u_i''} = -\frac{\overline{\varrho' u_i'}}{\bar{\varrho}}; \qquad \overline{\phi_\alpha''} = -\frac{\overline{\varrho' \phi_\alpha'}}{\bar{\varrho}}; \qquad (4.15)$$

$$\overline{\varrho' u_i''} = \overline{\varrho' u_i'}; \qquad \overline{\varrho' \phi_\alpha''} = \overline{\varrho' \phi_\alpha'}. \qquad (4.16)$$

The main reason for employing Favre averaging is that the resulting conservation equations do not include terms involving correlations of density fluctuations. Consequently, the resulting equations are easier to interpret (and hence model) than their Reynolds averaged counterparts. Furthermore, Favre averaging leads to equations describing the means of conserved quantities, e.g., the resulting momentum equation is in terms of $\overline{\varrho u_i} \equiv \bar{\varrho} \tilde{u}_i$ rather than $\bar{\varrho} \bar{u}_i$ which is obtained from Reynolds averaging and is not conserved. Despite the advantages of adopting Favre averaging over Reynolds averaging for calculation, different instruments may measure values close to density-weighted or unweighted averages (Jones and Whitelaw, 1982). Clearly the ability to calculate both unweighted and density-weighted quantities is desirable. But $\overline{\varrho u_i} \equiv \bar{\varrho}\tilde{u}_i = \bar{\varrho}\bar{u}_i + \overline{\varrho' u'_i}$, and $\overline{\varrho' u'_i}$ can be very difficult to calculate accurately; in terms of the pdf, unweighted and weighted statistics can be used interchangeably without approximation (Bilger, 1980). Applying Favre averaging to Eqs. (4.1), (4.2), (4.6) and (4.7) yields:

$$\frac{\partial \bar{\varrho}}{\partial t} + \frac{\partial \bar{\varrho}\tilde{u}_j}{\partial x_j} = 0, \tag{4.17}$$

$$\frac{\partial}{\partial t}(\bar{\varrho}\tilde{u}_i) + \frac{\partial}{\partial x_j}(\bar{\varrho}\tilde{u}_j\tilde{u}_i) = -\frac{\partial \bar{p}}{\partial x_i} - \frac{\partial}{\partial x_j}(\bar{\varrho}\widetilde{u''_j u''_i}) + \bar{\varrho}g_i, \tag{4.18}$$

$$\frac{\partial}{\partial t}(\bar{\varrho}\tilde{h}) + \frac{\partial}{\partial x_j}\bar{\varrho}\tilde{u}_j\tilde{h} = \frac{\partial}{\partial x_j}(\bar{\varrho}\widetilde{u''_j h''}), \tag{4.19}$$

$$\frac{\partial}{\partial t}(\bar{\varrho}\tilde{Y}_\alpha) + \frac{\partial}{\partial x_j}\bar{\varrho}\tilde{u}_j\tilde{Y}_\alpha = \frac{\partial}{\partial x_j}(\bar{\varrho}\widetilde{u''_j Y''_\alpha}) + \bar{\dot{w}}_\alpha \quad \text{for} \quad \alpha = 1, \ldots, n. \tag{4.20}$$

The above equations have been written for high Re, so that the averaged molecular fluxes can be considered as negligible in comparison with the turbulent fluxes $\bar{\varrho}\widetilde{u''_i u''_j}$ (Reynolds stress) and $\bar{\varrho}\widetilde{u''_j \phi''_\alpha}$ (scalar flux). Consequently the equations have a form identical to their constant density counterparts. The unknowns appearing in the equations are the Reynolds stress, scalar flux, mean density and the mean formation rate due to chemical reaction. This chapter is concerned mainly with the methods of determining $\bar{\varrho}$ and $\bar{\dot{w}}_\alpha$. However, turbulence models are needed to calculate $\bar{\varrho}\widetilde{u''_i u''_j}$ and $\bar{\varrho}\widetilde{u''_j \phi''_\alpha}$. They form a crucial part of any calculation method, and various approaches to turbulence models are discussed elsewhere in this volume. In what follows ϕ_α is used to denote the scalar field, i.e. species mass fractions and mixture enthalpy.

3 Non-premixed combustion

In the many conventional combustors and furnaces the fuel and air streams enter the combustion chamber in separate streams and the flames that result are diffusion or non-premixed flames. If in these circumstances it is presumed that the reaction is 'fast', that the flow is adiabatic and the pressure is thermodynamically constant (low speed flow) then the instantaneous major species composition, temperature and density can be related to a strictly conserved scalar quantity, the mixture fraction ξ. The mixture fraction can be viewed as a normalised chemical element mass fraction defined so that it takes the value of unity in the fuel stream and zero in the air stream. The equation for ξ contains no reaction source terms and is only affected indirectly by reaction through its influence on fluid density. This conserved scalar approach to non-premixed combustion is described in detail in Bilger (1980). To determine the dependence of composition, temperature and density on ξ a thermo-chemical model is needed.

3.1 Thermo-chemical models

3.1.1 The chemical equilibrium approach

If the mixture is presumed to be in chemical equilibrium then the mixture state can be determined by minimising the free energy (Gordon and McBride, 1971). The major shortcoming of the approach is related to CO levels in hydrocarbon–air flames. For hydrocarbon–air mixtures it is known, e.g. Jones (1980), that in the case of lean mixtures, negligible quantities of CO concentrations are predicted under equilibrium conditions. The proportion of equilibrium CO increases progressively with mixture strength, and for equivalence ratios greater than about 1.2 the CO mole fraction is greater than that of CO_2; at equivalence ratios of around three, CO mole fractions of about 20% can be achieved. These levels are wholly unrealistic in laminar and turbulent hydrocarbon flames. It is also not appropriate for any species involving relatively slow chemistry such as NO_x or soot. Nevertheless, the chemical equilibrium approach has been widely used and reasonable temperature and velocity field characteristics can be obtained. A major advantage is that a reaction mechanism is not required, knowledge of ξ and the appropriate thermo-chemical data are sufficient.

3.1.2 The laminar flamelet approach

In the laminar flamelet approach the turbulent flame is presumed to comprise an ensemble of laminar flames. The formulation is restricted to 'thin' flame burning where the thickness of the burning zone is less than the Kolmogorov length scale – this fundamental requirement for the existence of flamelets (i.e. laminar-like burning zones in a turbulent medium) is referred to as the Klimov–Williams criterion. For a more detailed review of the theory, the papers by Williams (1975) and Peters (1986) may be consulted. Laminar flame properties are usually obtained from computations of counterflow diffusion flames (Fig. 4.1)

The results of the computations are then used to construct flamelet 'libraries' on a once and for all basis and expressed as a function of mixture fraction, ξ, and some measure of the flame stretch rate usually taken to be the mixture fraction variance 'dissipation' rate, i.e. the scalar dissipation rate. One practical difficulty of the flamelet technique is that for many fuels of engineering interest, the details of the reaction mechanisms are not adequately known, e.g. kerosene in gas turbines and the variety of petrols used in internal combustion engines. Nevertheless the method

(a)   (b)

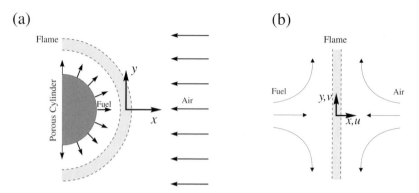

Fig. 4.1 (a) Schematic of a laminar counterflow diffusion flame geometry. The width of the flame is exaggerated for clarity. (b) Locally the flame is treated as planar and the flow field is well represented by a potential flow solution.

does generate realistic CO levels in the burning zone and often yields plausible results even under circumstances when its use cannot be strictly justified.

3.2 Chemistry–turbulence interactions: the presumed-shape pdf method

Because of the non-linearity of the state relationships relating composition, temperature and density to mixture fraction etc., a pdf is needed to obtain mean values. If the state relationships depend only on the mixture fraction (for example, when a flamelet obtained at a specific value of strain is selected a priori) then density-weighted averages can be evaluated from:

$$\tilde{\phi}_\alpha(\mathbf{x}, t) = \int_0^1 \phi_\alpha(\hat{\xi}) \tilde{P}(\hat{\xi}) d\hat{\xi}, \tag{4.21}$$

where the density-weighted pdf is defined as:

$$\tilde{P}(\hat{\xi}) = \frac{\varrho(\hat{\xi}) P(\hat{\xi})}{\bar{\varrho}(\mathbf{x}, t)} \quad \text{and} \quad \bar{\varrho} = \left\{ \int_0^1 \frac{\tilde{P}(\hat{\xi})}{\varrho(\hat{\xi})} d\hat{\xi} \right\}^{-1}. \tag{4.22}$$

In the context of the conserved scalar formalism the usual approach for determining the pdf is the presumed-shape method. In general the joint pdf of mixture fraction and the scalar dissipation rate is needed. However, if only the mixture fraction is used then a two-parameter form of the pdf is often presumed in terms of the mixture fraction mean and variance. The values of these are then obtained from the solution of their respective modelled transport equations. Since for combusting flows these equations are usually solved in density-weighted form, the resulting pdf that is constructed from these moments will be the density-weighted pdf. The modelled forms of these equations are:

$$\bar{\varrho}\frac{\partial\tilde{\xi}}{\partial t} + \bar{\varrho}\tilde{u}_j\frac{\partial\tilde{\xi}}{\partial x_j} = \frac{\partial}{\partial x_j}\left(\frac{\mu_t}{\sigma_t}\frac{\partial\tilde{\xi}}{\partial x_j}\right),$$ (4.23)

$$\bar{\varrho}\frac{\partial\widetilde{\xi''^2}}{\partial t} + \bar{\varrho}\tilde{u}_j\frac{\partial\widetilde{\xi''^2}}{\partial x_j} = \frac{\partial}{\partial x_j}\left(\frac{\mu_t}{\sigma_t}\frac{\partial\widetilde{\xi''^2}}{\partial x_j}\right) + 2\frac{\mu_t}{\sigma_t}\left(\frac{\partial\widetilde{\xi''^2}}{\partial x_j}\right)^2 - \bar{\varrho}C_d\frac{\varepsilon}{k}\widetilde{\xi''^2}.$$ (4.24)

The most popular form for the pdf is the beta pdf:

$$\tilde{P}(\hat{\xi}) = \frac{\hat{\xi}^{a-1}(1-\hat{\xi})^{b-1}}{\int\limits_0^1 x^{a-1}(1-x)^{b-1}dx},$$ (4.25)

where

$$a = \tilde{\xi}\left[\tilde{\xi}\frac{(1-\tilde{\xi})}{\widetilde{\xi''^2}} - 1\right] \quad \text{and} \quad b = (1-\tilde{\xi})\left[\tilde{\xi}\frac{(1-\tilde{\xi})}{\widetilde{\xi''^2}} - 1\right].$$ (4.26)

This form of pdf has been widely used for engineering calculations of reacting turbulent flow. The implementation of Eq. (4.21) is particularly straightforward in the context of the β-pdf. The instantaneous relations, $\phi_\alpha(\xi)$, can usually be curve-fitted using a least-squares polynomial (of order ≈ 20) and the relations can then be expressed in the form

$$\phi_\alpha = \sum_{j=0}^N A_{j,\alpha}\xi^j.$$ (4.27)

By substituting (4.27) into (4.21) and by employing the relations pertaining to the β- and Γ-function:

$$\beta(a, b) = \frac{\Gamma(a)\Gamma(b)}{\Gamma(a+b)} \quad \text{and} \quad \Gamma(a+1) = a\Gamma(a),$$ (4.28)

the following analytical expression for the mean of a dependent scalar variable is obtained:

$$\tilde{\phi}_\alpha = \sum_{j=0}^N A_{j,\alpha}\frac{\prod\limits_{i=0}^{j-1}(a+i)}{\prod\limits_{i=0}^{j-1}(a+b+i)}.$$ (4.29)

The numerator and denominator are defined for $j \geq 1$.

If the state relationships are presumed to depend on mixture fraction and some measure of the stretch rate then a joint pdf is needed. If the instantaneous scalar dissipation rate evaluated at stoichiometric mixture fraction, χ_{st}, is used for this purpose, then the pdf required is $P(\xi, \chi_{st}, \mathbf{x}, \mathbf{t})$. The necessary state relationships – flamelet 'libraries' – may be constructed on a once and for all basis and expressed in the instantaneous form, $\phi_\alpha(\xi, \chi_{st})$. Further details of the method are described in Jones and Kakhi (1996) and Peters (2000). However, it now appears that the advantage to be gained from such an approach is small; away from extinction the effects of flame stretch are relatively small and the local stretch rate provides insufficient information to characterise local extinction phenomena. The conditional moment closure (CMC) (Klimenko and Bilger, 1999), and the unsteady flamelet formulation (Mauss *et al.*, 1990; Peters, 2000), represent alternative methods of accurately incorporating flame stretch and local extinction effects. In the CMC approach fluctuations in the thermo-chemical state are presumed to arise predominantly from fluctuations in mixture fraction, and fluctuations in conditionally averaged quantities are then neglected so that the chemical source terms depend only on (known) conditionally averaged mean quantities. In the unsteady flamelet approach the time dependent forms of the flamelet equations are solved and a Lagrangian time defined as:

$$t = \int_0^x \frac{1}{\tilde{u}(x|\tilde{\xi} = \xi_{st})} \mathrm{d}x \tag{4.30}$$

is used to describe flamelet evolution. There are a number of similarities between the two approaches and in both cases quite detailed kinetics can be incorporated. Both models have led to excellent predictions in a range of jet flames (TNF4, TNF5, TNF6).

3.2.1 Nitric oxide formation

The formation of NO at high temperatures is known to be controlled by the extended Zeldovich mechanism:

$$O + N_2 \rightleftharpoons NO + N,$$
$$N + O_2 \rightleftharpoons NO + O,$$
$$N + OH \rightleftharpoons NO + H.$$

The overall NO formation rate is much slower than those of the main heat release reactions, and the concentrations of NO involved are such that neither the temperature nor the main species compositions are affected by its presence. In most combustion systems the NO concentration will be well below its equilibrium levels. In order to close the system of the above equations, so that it is completely specified in terms of ξ, a steady state assumption for the N atom concentration and a partial equilibrium assumption for the O atom concentration are usually made.

Consequently the mean rate of formation of NO per unit mass of mixture, $\bar{\dot{S}}_{NO}(\mathbf{x})$, is:

$$\bar{\dot{S}}_{NO}(\mathbf{x}) = \int_0^1 \dot{S}_{NO}(\hat{\xi}) P(\hat{\xi}; \mathbf{x}) d\hat{\xi}.$$

The results of Jones and Priddin (1978) suggest that the above formulation is adequate at conventional aircraft gas turbine operating pressures and inlet temperatures, but the method does not produce accurate results at atmospheric conditions. Furthermore, the approach is not adequate in low NO_x combustion systems, particularly land based devices, where prompt NO_x and other reaction paths need to be included. In these circumstances the CMC and unsteady flamelet methods both represent viable alternatives.

4 Premixed combustion

4.1 Introduction

Premixing implies that the composition of the fuel is essentially uniform – the fuel and oxidant have mixed down to the molecular scale prior to combustion. As a consequence a strictly conserved scalar (such as the mixture fraction, ξ) will take a uniform value set by the initial fuel/air ratio. Therefore the closure problem associated with mean formation rates of non-conserved scalars cannot be avoided. The representation of a premixed flame propagating into the cold unburned fuel/air mixture is another difficulty for modelling and is still the subject of research.

The structure of premixed hydrocarbon–air flames is known to comprise three regions: a preheat zone which is essentially chemically inert, a thin reaction zone where the fuel reacts to form CO and H_2 etc. and an oxidisation layer in which the products CO_2 and H_2O are formed. At atmospheric pressure the reaction zone occupies typically one-tenth of the total flame thickness. The various combustion regimes that can arise as a result of the influence of turbulence can be characterised in terms of various dimensionless quantities such as the Damköhler (Da), Karlovitz (Ka) and Reynolds (Re) numbers and has been conveniently summarised in the Borghi diagram (Borghi, 1985).

Peters (1999) has extended the work of Borghi through consideration of an additional regime in which the fine Kolmogorov sized eddies are supposed smaller than the flame thickness but are larger than the reaction zone width. In this regime the reaction zone will remain thin and will be essentially unaffected by turbulent mixing. To characterise the regime a Karlovitz number (Ka_δ) based on the reaction

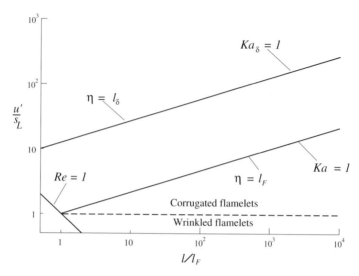

Fig. 4.2 Phase diagram showing the different regimes of combustion in premixed flames. u' – characteristic velocity of energy-containing motions, S_L – laminar burning velocity (strained or unstrained), l – integral length scale, l_F – flame thickness, l_δ – reaction zone thickness, η – Kolmogorov length scale.

zone thickness is introduced. Figure 4.2 serves to illustrate the different regimes of turbulent combustion which can then arise.

Turbulent combustion is characterised by $Re > 1$, and for large Da and small Ka ($u' < u_L$), the effect of turbulence is to wrinkle the flame surface, increasing its area and hence the effective flame speed (Damköhler, 1940). Increases in Ka ($Ka < 1$), such that $u' > u_L > v_k$ (where v_k denotes the Kolmogorov velocity scale, characteristic of the smallest dissipative motions in the flow), imply that the large eddies (that scale with u') will cause substantial convolution of the flame surface. However, the smaller eddies cannot corrugate the flame front by virtue of their smaller velocity; this regime is referred to as corrugated flamelets. To characterise this regime, Peters (1986) introduced the Gibson length scale, $L_G \equiv u_L^3/\varepsilon$ such that an eddy of 'size' L_G has a velocity of the order of the local flame velocity. L_G is a measure of the size of burnt gas pockets that move into the unburnt mixture, try to grow there due to the advancement of the flame front normal to itself, but are reduced in size again by newly arriving eddies of size L_G. It is argued that flame stretch effects are most effective at this particular scale. The region corresponding to $Re > 1$, $Ka > 1$ and $Ka_\delta < 1$ is described as the thin reaction regime (Peters, 1999). Here the fine scale turbulent motions distort the overall flame structure but not the thin reaction zone. Small scale Kolmogorov sized eddies will interact with the preheat zone, thereby increasing scalar mixing while at the same time destroying the quasi-steady flamelet structure that prevails in the corrugated flamelet regime.

Typically $Ka_\delta \approx 100Ka$ (Poinsot *et al.*, 1991). As the ratio of the reaction zone thickness and the Kolmogorov length scale increases ($Ka_\delta > 1$), the distributed reaction regime prevails. Under these circumstances the turbulent flame can no longer be envisaged as an ensemble of laminar flamelets (quasi-steady or unsteady), but instead, the instantaneous region of combustion is 'distributed' throughout the time-averaged combustion zone (Summerfield *et al.*, 1955). Further reduction in the turbulent time scale ($Da < 1$) moves the combustion into the well-stirred reactor regime, where kinetic control significantly influences the evolution of the flow.

4.2 Spalding's eddy break-up (EBU) model

Early concepts concerning premixed flames centred around the prediction of the turbulent flame speed and its relation to dimensionless parameters (e.g. the Reynolds number and the ratio of a characteristic turbulence velocity to the laminar flame speed u'/S_L). Such work has been reviewed by Bray (1980). For the purpose of predicting turbulent premixed flames in realistic geometries, knowledge of the flame speed is not sufficient. An appropriate starting point in the mathematical modelling of turbulent premixed flames is the work of Spalding (1971), who employed the averaged form of the conservation equations to predict a bluff body stabilised flame spreading obliquely across a confined duct into the unburned stream until the whole duct cross-section was covered by the flame. A major motivation for the study was to obtain an accurate prediction of the angle of spread where experimental findings showed that at any downstream position, the distance from the symmetry plane to the flame edge was approximately one-tenth of the distance from the baffle, regardless of the mixture strength, the approach velocity and the initial temperature of the mixture.

Prandtl's mixing length hypothesis was used for turbulent transport and two combustion models were employed. The first assumed kinetic control, with a bimolecular reaction rate (one-step irreversible chemistry) employing the time averaged concentrations and temperatures. Comparison of concentration profiles with experimental data showed that the predicted concentration curves, albeit qualitatively correct given the simplicity of the turbulence/chemistry model, were too steep, implying that the chemical reaction was too densely concentrated in the high temperature part of the flame. Comparison with flame spread data showed that the predictions were quite poor since an increase in flame spread with increasing approach temperature was predicted while a slight decrease was measured. Spalding recognised that the reaction rate had to be formulated in a way which increased the influence of the turbulence field, and diminished that of the chemical kinetic constants. As a consequence the eddy break-up (EBU) model was formulated using ideas analogous to the energy cascade arguments of turbulence; namely that the rate controlling step is the breaking up of large eddies of burned and unburned gas into successively smaller ones. At sufficiently small scales, the flame zone (which can be identified as the interface between burned and unburned fragments of gas) can consume reactants at an effective rate. In the EBU model it is presumed that the rate of species consumption/production is proportional to the rate of turbulent mixing of the unburnt and burnt gases,

$$\bar{\omega}_{pr} \propto \frac{\left(\overline{Y'^2_{pr}}\right)^{\frac{1}{2}}}{\tau}. \tag{4.31}$$

Due to the assumption of one-step chemistry (involving fuel, oxidant and products) and adiabatic flow, knowledge of the product mass fraction, Y_{pr}, characterises the thermo-chemistry. The time scale τ is associated with an integral time scale representative of the energy-containing motions and is consistent with the idea of turbulent mixing being the rate controlling parameter. Cast in terms of k and ε, $\tau \approx k/\varepsilon$, so that

$$\bar{\omega}_{pr} \propto \frac{\varepsilon}{k}\left(\overline{Y'^2_{pr}}\right)^{\frac{1}{2}}. \tag{4.32}$$

The EBU model, based on conventional modelling practice (as for Eqs. (4.23) and (4.24) in Section 3.2) can be summarised as:

$$\bar{\varrho}\frac{\partial \bar{Y}_{pr}}{\partial t} + \bar{\varrho}\bar{u}_j\frac{\partial \bar{Y}_{pr}}{\partial x_j} = \frac{\partial}{\partial x_j}\left\{\frac{\mu_t}{\sigma_t}\frac{\partial \bar{Y}_{pr}}{\partial x_j}\right\} + \bar{\omega}_{pr}, \tag{4.33}$$

$$\bar{\varrho}\frac{\partial \overline{Y'^2_{pr}}}{\partial t} + \bar{\varrho}\bar{u}_j\frac{\partial \overline{Y'^2_{pr}}}{\partial x_j} = \frac{\partial}{\partial x_j}\left\{\frac{\mu_t}{\sigma_t}\frac{\partial \overline{Y'^2_{pr}}}{\partial x_j}\right\} + 2\frac{\mu_t}{\sigma_t}\left(\frac{\partial \bar{Y}_{pr}}{\partial x_j}\right)^2 - C_d\bar{\varrho}\frac{\varepsilon}{k}\overline{Y'^2_{pr}}, \tag{4.34}$$

where $\bar{\omega}_{pr} = \min[\dot{\mathcal{R}}, C_{EBU}\bar{\varrho}(\varepsilon/k)(\overline{Y'^2_{pr}})^{1/2}]$, $\dot{\mathcal{R}} = a\bar{\varrho}\bar{Y}_{fu}\bar{Y}_{ox}\exp[-E/(R\bar{T})]$ and the model 'constant', $C_{EBU} \approx 0.35$.

The reaction source term in Eq. (4.33) includes some kinetic control. For example, in cold regions of a flow $\dot{\mathcal{R}} \approx 0$, while the EBU rate term is generally finite everywhere, so that $\bar{\omega}_{pr} = 0$. However, this constitutes an ad hoc, though in some cases an essential, addition to the modelling in order to account for kinetic control. Furthermore, a potentially large term representing the correlation between the instantaneous reaction rate and the scalar fluctuation, $\overline{Y'_{pr}\omega_{pr}}$, has been omitted from Eq. (4.34). Application of the model to the previously described test problem yielded concentration profiles in much better qualitative agreement with experiment than those resulting from the kinetically controlled combustion model. The predictions of the flame spread were also appreciably better, and in the case of approach temperature the slight drop in flame spread was correctly predicted. The ideas behind the EBU concept, although plausible, are essentially intuitive – no formal derivations were performed. Because of its reliance on the assumption of combustion being effectively mixing-controlled, the model is consistent with the concept of thin flame burning.

4.3 The premixed flame model of Bray and Moss

A further, important, development in the theory of premixed combustion was the work of Bray and Moss (1977, 1974) (also reviewed in terms of Favre averaging in Bray, 1980). A key assumption in the work was that combustion occurred through a global, one-step irreversible chemical reaction. It was argued that since the chemical kinetics of combustion rarely conforms to such a global description, the analysis would be most successful when the combustion was rate-limited by turbulent mixing rather than chemical kinetics. In what follows only the main features of the model are presented. Two reactive chemical species are considered, namely 'reactant' (fuel and oxidant) which form 'product'; the mixture may also include diluents. A progress variable, c, for the global combustion reaction is defined as the mass fraction of product normalised so that $c = 1$ when combustion is complete (i.e. the fuel and/or oxidant is exhausted), while $c = 0$ in the unburned mixture which has no product. So $Y_{pr} = Y_3 = cY_{3,\infty}$, where $Y_{3,\infty}$ is the mass fraction of product in the fully burned gas whose temperature, T_∞, is the adiabatic flame temperature. From the species transport equation (4.7) for Y_3, a transport equation for c can be easily derived:

$$\frac{\partial}{\partial t}(\varrho c) + \frac{\partial}{\partial x_k}(\varrho u_k c) = \frac{\partial}{\partial x_k}\left(\varrho D \frac{\partial c}{\partial x_k}\right) + \frac{\dot{r}_3}{Y_{3,\infty}}. \qquad (4.35)$$

From the definition of element mass fraction, $Z_\alpha \equiv \sum_{\beta=1}^{N} \mu_{\alpha\beta} Y_\beta$, where $\mu_{\alpha\beta}$ denotes the number of Kg of the αth element in 1 Kg of the βth species,

$$Y_1 = \frac{1}{\mu_{11}}(Z_1 - \mu_{13}Y_{3,\infty}c), \qquad Y_2 = \frac{1}{\mu_{22}}(Z_2 - \mu_{23}Y_{3,\infty}c), \qquad (4.36)$$

and Y_4 is clearly independent of c. In the present notation, $\alpha = 1, 2, 3$ and 4 represent fuel, oxidant, products and diluents, respectively. Denoting T_o and ϱ_o as the temperature and density of the reactants, it can be shown (Bray, 1980) that:

$$T = T_o(1 + \tau c) \quad \text{and} \quad \varrho = \frac{\varrho_o}{(1 + \tau c)}, \quad \text{where} \quad \tau = \frac{T_\infty}{T_o} - 1, \qquad (4.37)$$

and τ can vary from 4 to 9. In general, $\dot{r}_3 = \dot{r}_3(\varrho, T, Y_1, Y_2)$, but because ϱ, T, Y_1 and Y_2 are all functions of c, this can be replaced by $\dot{r}_3 = \dot{r}_3(c)$. The statistics of the flame are calculated by way of a pdf (similar in principle to the presumed-shape pdf method described earlier). It is argued that the pdf $P(c; \mathbf{x})$, must represent gas which is unburned and gas which has burned completely, as well as gas mixture which is in the process of burning. It is convenient then to separate these three contributions such that:

$$P(c; \mathbf{x}) = \alpha(\mathbf{x})\delta(c) + \beta(\mathbf{x})\delta(1 - c) + [H(c) - H(c - 1)]\gamma(\mathbf{x})f(c; \mathbf{x}), \quad (4.38)$$

where $\delta(c)$ and $H(c)$ are Dirac delta and Heaviside step functions, respectively. The coefficients α, β and γ are non-negative functions of position and $f(c)$ satisfies the normalisation condition, $\int_0^1 f(c)dc = 1$. α, β and γ may be regarded as the probabilities that unburned mixture, fully burned product, and burning mode, respectively, exist at the point in question – it can be shown by integrating $P(c)$ with respect to c that $\alpha + \beta + \gamma = 1$. In this a priori pdf method, $f(c)$ is to be represented empirically and the functional form for $P(c)$ is assumed to have at most two parameters, $\tilde{c}$ and c''^2.

An approximation which drastically simplifies the model is that $\gamma \ll 1$. This represents, in the limiting case, a flow which fluctuates between unburned and fully burned mixtures. Bray (1980) expressed the transport equations for $\tilde{c}$ and c''^2 in non-dimensional form and in the expression for the source terms identified the grouping γDa. It was argued that γDa was of order unity and thus the approximation $\gamma \ll 1$ was accompanied by $Da \gg 1$; this implies that the characteristic chemical time is much smaller than the characteristic turbulence time and so the rate of combustion is controlled mainly by turbulent mixing. In fact it can be shown that with the condition $\gamma \ll 1$, an expression for $\dot{r}_3/Y_{3,\infty}$ which is very similar to the form of Spalding's EBU model is obtained. The condition $\gamma \ll 1$ leads to a pdf $P(c)$ consisting mainly of two delta functions with $\alpha + \beta \approx 1$. Using the relation for the density, Eq. (4.37), in averaged form and the equation of state expressed as $\varrho_\infty T_\infty = \varrho_0 T_0$ there results:

$$\beta = \frac{(1 + \tau)\tilde{c}}{1 + \tau\tilde{c}} \quad \text{and so} \quad \alpha = 1 - \beta = \frac{1 - \tilde{c}}{1 + \tau\tilde{c}}. \quad (4.39)$$

The only term remaining in the specification of the pdf $P(c)$ is the burning mode pdf. Consistent with the $\gamma \ll 1$ approximation is the idea that the burning mode pdf makes little contribution to the pdf $P(c)$; under these circumstances $c''^2 \approx \tilde{c}(1 - \tilde{c})$ and, as shown in Eq. (4.38), to construct $P(c)$ only α and β would be required, which are functions of $\tilde{c}$ only. It can be shown (Bray, 1980), however, that $f(c)$ does influence $\bar{\dot{\omega}}$ and $\overline{c''\dot{\omega}}$. Furthermore, the source terms are related to $f(c)$ through the ratio of its moments, and in this respect it is argued that the results are insensitive to the details of the specification of $f(c)$.

Many of the applications of the Bray–Moss model have been confined to an infinite plane one-dimensional reaction zone which is swept relative to an incoming stream of reactants. Further details of the application and modelling can be found in Bray and Libby (1976), Libby and Bray (1981) and Bray *et al.* (1981). Turbulent transport was modelled initially through an eddy viscosity concept, and subsequently through the joint pdf of velocity and reaction progress variable. The model has been used to study the effects on the turbulence kinetic energy, k, as the flow passed through the reaction zone. In such a situation the effects to be accounted for include dilatation associated with the heat release (which tends to bring about a reduction in the turbulence kinetic energy, k), and interaction between the mean motion and the Reynolds stresses which generates k; this latter effect was neglected (on the basis of an order of magnitude argument involving an angular deflection of a typical streamline passing through a planar flame) in comparison to the production of k from buoyancy effects associated with the mean pressure gradient. The use of the joint pdf implies that closure of the turbulent transport terms can be achieved without invoking an eddy viscosity assumption, and at the same time it allows the validity of such a hypothesis to be investigated in turbulent combustion. The second moment closure of Bray *et al.* (1981) focused attention on the solution of the equations governing the evolution of $\overline{\varrho u'' u''}$ and $\overline{\varrho u'' c''}$. The numerical treatment was facilitated by replacing the x-coordinate as the independent variable with the mean product concentration $\tilde{c}$; this manipulation also had the advantage of eliminating $\bar{\omega}_{pr}$ from the formulation and thereby obviating the need for its specification (unless the spatial structure in the form of $x(\tilde{c})$ was required). The results demonstrated that as τ (a measure of the heat release) was increased, extensive regions within the flame showed signs of counter-gradient diffusion – i.e diffusion of opposite sign to that required by gradient transport. The physical process responsible for this effect was due to a 'buoyancy' mechanism in which packets of low-density burned gas were preferentially accelerated relative to packets of reactants by the mean pressure gradient acting across the flame. The resulting relative motion between burned and unburned packets took place in the direction opposite to that predicted by gradient diffusion. The predictions of the model based on the joint pdf reproduced with good accuracy the changes in the intensity of the velocity and scalar fluctuations across the flame. The mean pressure gradient provided the dominant term which augmented k throughout the interior of the flame, but its effect diminished at the boundaries of the flame where the effects of dilatation and viscous dissipation (both sinks of k) became relatively more dominant. Despite the success of the above treatment, the generalisation of this approach to two- and three-dimensional flows is likely to require further effort, firstly because of the increased dimensionality of the joint pdf (more velocity components have to be accounted for) and secondly the Reynolds stress terms require more detailed modelling. Some progress has been made in this respect by Bailly and Champion (1997).

The original Bray–Moss model has been extended to situations which involve more realistic chemistry. Within the framework of multi-step chemistry, a redefinition of the reaction progress variable is required. Clearly the normalised mass fraction of product is unsuitable since with complex chemistry there is no single product. In addition some species may have been depleted where others just begin to appear and therefore the state of any particular reactive species is not (necessarily) indicative of the state of the reaction. c is therefore redefined in terms of temperature:

$$c = \left(\frac{T - T_o}{T_\infty - T_o} \right),$$
(4.40)

where T_o and T_∞ are the temperatures upstream and downstream of the reaction zone, respectively. One of the most notable applications of complex chemistry is the work of Champion et al. (1978) in which the chemistry is assumed to proceed sequentially. The semi-global chemical mechanism of Edelman and Fortune (described in Champion et al., 1978) was employed to investigate a reacting mixture of propane and air in a ducted, stationary and one-dimensional flow. It was assumed that the complete chemical scheme could be partitioned into two distinct zones (not necessarily spatially distinct). Firstly a delay zone which was represented by the global reaction for the propane break down. This reaction led to small heat release and had to be the fastest of the reactions for sequential chemistry to be postulated. In the next combustion zone, the chemical description included all the remaining reactions in the semi-global scheme (involving slower radical recombinations). Thermo-chemical closure was achieved by expressing conservation of species in terms of c, invoking partial equilibrium assumptions and by taking the concentration of oxygen to hydrogen radicals as a constant.

The Bray–Moss model (1974, 1977) has drawn upon ideas from laminar flamelet theory; this is related to the specification of $f(c)$. For example, if the distribution of product or temperature through a laminar flame can be represented by $c = c(x/\delta_f)$, where $c(x/\delta_f)$ is a monotonically increasing function from $c = 0$ to $c = 1$, then it can be shown that:

$$f(c) = \left\{ \frac{dc}{d(x/\delta_f)} \right\}^{-1}.$$
(4.41)

Therefore, given a representative laminar flame profile, it is possible to determine $f(c)$. The original theory makes no further use of the laminar flamelet idea thereafter. However, more recently, Bray (1986) has extended the flamelet pdf model to include effects of flame stretch. The motivation for this line of approach is that the basic EBU-type expression involves the time scale k/ε. This implies that $\bar{\dot{\omega}}$ will increase without limit if the turbulence time scale is reduced, since no allowance is made for flamelet quenching. It is argued that in a turbulent flame burning may be represented by a number of strained laminar flames with

Karlovitz numbers $Ka_0, Ka_1, \ldots, Ka_n$ which are assumed to occur with probabilities $\wp_0(\mathbf{x}), \wp_1(\mathbf{x}), \ldots, \wp_n(\mathbf{x})$ such that $\sum_{s=0}^{n} \wp_s(\mathbf{x}) = 1$. Under these circumstances, the mean reaction rate can be written as:

$$\bar{\omega} = \gamma \sum_{s=0}^{n} \left\{ \wp_s \int_0^1 \dot{\omega}_s(\hat{c}) f_s(\hat{c}) d\hat{c} \right\}, \tag{4.42}$$

where $\dot{\omega}_s$ is the source term for the flamelet with Karlovitz number Ka_s and $f_s(c)$ is given by Eq. (4.41) with the conditioning on the Ka_s, which affects the flame structure and thickness. The mean species concentration then becomes:

$$\bar{\phi}_i(c) = \alpha\phi_i(0) + \beta\phi_i(1) + \gamma \sum_{s=0}^{n} \left\{ \wp_s \int_0^1 \phi_{i,s}(\hat{c}) f_s(\hat{c}) d\hat{c} \right\}, \tag{4.43}$$

where $\phi_{i,s}(c)$ is determined from a flamelet library – in order to use such a thermochemical specification, the same ideas invoked in the work of Champion *et al.* (1978) are required, i.e. a redefinition of c in terms of temperature and sequential chemistry. Note that even if $\gamma \ll 1$, the last term on the rhs of Eq. (4.43) must be retained, particularly if reaction intermediates are to be calculated, since these species are predominantly created in the reaction zone. Bray (1986) assumed that γ was independent of flamelet quenching and controlled by the scalar dissipation mechanism. The effects of stretch are necessarily coupled to the position of the flamelet and thus to the progress variable c. It was assumed that Ka was uncorrelated with c, which is similar in principle to an assumption of statistical independence. The implication of such an assumption is that the integral $\int_0^1 \dot{\omega}_s(\hat{c}) f_s(\hat{c}) d\hat{c}$ can be taken out of the summation in Eq. (4.42), and in doing so, the subscript 's', characterising the degree of stretch, can be replaced by the unstrained flamelet subscript 'o'. By invoking a log-normal distribution for the viscous dissipation rate, relating this to the Karlovitz number under quench conditions and assuming that the quenched flamelets make no contribution to $\bar{\omega}$, $\sum_{s=0}^{n} \wp_s$ can be replaced by the probability of unquenched flamelets, $\wp_{uq}$. The final result is:

$$\bar{\omega} = \bar{\varrho} \left(\frac{\varepsilon}{k} \right) \frac{\tilde{c}(1 - \tilde{c})}{(2c_m - 1)} \wp_{uq}, \quad \text{where} \quad c_m = \frac{\overline{c\dot{\omega}_o}}{\bar{\dot{\omega}}_o}. \tag{4.44}$$

This is a basic EBU-type expression multiplied by the proportion of flamelets that are unquenched.

4.4 Further developments in premixed flame theory

At large Damköhler numbers the source term for chemical reaction has been shown to be proportional to the scalar dissipation rate in premixed (Bray, 1978) and non-premixed (Bilger, 1976) configurations; in the premixed situation,

$\bar{\omega} \propto \tilde{\chi} \equiv (\bar{\varrho})^{-1} \frac{\mu}{Sc} \overline{(\nabla c'')^2}$. The main difficulty in exploiting this relation for the purpose of calculating the mean reaction rate is in the representation of $\tilde{\chi}$. However, in this section we are concerned with a reactive scalar, c, and the appearance of the time scale (k/ε) as the controlling parameter is questionable since it ignores the distribution of scalar time scales in the reaction zone, which can be very different to (k/ε).

Recently, EBU-type combustion models, as embodied in the original Bray–Moss theory, have come under careful scrutiny, particularly with regards to their prediction of flame propagation. For example, Weller *et al.* (1990) pointed out that the use of such a combustion model led to problems of high flame speed predictions near a wall, whereas observations show the flame to slow down in such regions. Catlin and Lindstedt (1991) studied flame propagation through a uniform turbulence field in a one-dimensional planar open-ended tube geometry where the turbulent reaction rate was quenched near the cold front of the flame. The computations demonstrated that the burning velocity and the reaction zone thickness increased rapidly as the quench point approached the cold front and, although the burning velocity remained bounded, the reaction zone thickness became indefinitely large. The conclusion was that cold front quenching was necessary for the prediction of experimentally observed flame behaviour in tubes using mixing-controlled reaction models.

It has been noted (Bray *et al.*, 1984) that in the Bray–Moss theory length scale information, particularly with reference to the local scalar field, is not present. An attempt to remedy this situation is described in Bray *et al.* (1984) and Bray and Libby (1986). The formalism centres around the representation of the mean reaction rates as $\bar{\omega}(\mathbf{x}) = \dot{\omega}_f(\mathbf{x})\nu(\mathbf{x})$, where ν represents the mean frequency at which flamelets cross the spatial location under consideration and where $\dot{\omega}_f$ is the mean reaction rate per crossing. The main task is the determination of ν, which is achieved through the solution of a modelled transport equation at the one-point, two-time level for the autocorrelation of the progress variable, c, with time advance τ, i.e. $\overline{c(\mathbf{x}, t)c(\mathbf{x}, t + \tau)} \equiv \mathcal{P}_{11}(\mathbf{x}, \tau)$. It was assumed that $c(\mathbf{x}, t)$ observes a zero–one function or square-wave time series behaviour, by virtue of the $\gamma \ll 1$ approximation. In the modelled equation for $\mathcal{P}_{11}(\mathbf{x}, \tau)$, terms involving molecular diffusion were neglected (high Reynolds number assumption) and correlations such as $\overline{c\dot{\omega}}$ were treated by analysing the covariance between a delta function (simulating $\dot{\omega}$ as a pulse train) and a square-wave time series (for c). For the remaining terms, the limiting behaviour for $\tau = 0$ and $\tau \to \infty$ was either known or postulated, and consequently closure was achieved by assuming that an exponential function could be employed to interpolate between these limits. By assuming a pdf for the passage times for the reactant and product modes at a point, an expression for ν was obtained. In a one-dimensional planar flame geometry it was found that the predictions of mean passage times for reactants and products as well as the normalised power spectral density of c were in very good agreement with the measurements of Shepherd and Moss (1983). It must be stressed, however, that for the validation of the model, data from the measurements were employed to evaluate some of the constants associated with the model. Although this is one of the advantages of the theory (i.e. the quantities appearing in the model are directly measurable), it is not clear how easily the experimental input can be supplemented by theory. In order to calculate $\bar{\omega}$, some means of prescribing $\dot{\omega}_f$ is necessary, and this can be achieved through the use of a flamelet library (Bray, 1986). In this approach a term, $\bar{V}_{n,s}$, representing the mean speed normal to a flamelet with a characteristic Karlovitz number, Ka_s, is required. This was modelled by assuming that this mean speed is independent of stretch effects, and then related to the mean flow velocity (one-dimensional situation) with $\bar{V}_n \approx \bar{V}/\mathcal{K}$, where $\mathcal{K}$ is an empirical constant. No predictions of $\bar{\omega}$ have yet been obtained to demonstrate whether the flamelet theory incorporated in the model behaves correctly. Furthermore, the extension of this model to two-dimensional and three-dimensional situations requires further effort.

Recent work on premixed combustion has attempted to characterise the burning rate through the evaluation of a flame area and burning velocity. The work of Darabiha *et al.* (1989) serves as an example of an approach to premixed flame modelling based on the solution of an equation for the flame area per unit volume (also referred to as the flame surface density). The principle of the flame area technique is to obtain closure for the mean reaction rate in terms of the mean flame area per unit volume, $\tilde{\Sigma}$. Instantaneously, this quantity is defined as $\Sigma = \lim_{\delta V \to 0}(\partial S/\partial V)$, where S denotes a material surface element with area δS. A balance equation for Σ can be derived:

$$\frac{\partial}{\partial t}(\varrho\Sigma) + \frac{\partial}{\partial x_k}(\varrho u_k \Sigma) = -\frac{1}{2}\varrho n_i n_j \left(\frac{\partial u_i}{\partial x_j} + \frac{\partial u_j}{\partial x_i}\right), \tag{4.45}$$

where n_i is the unit vector normal to the surface. The Favre-averaged form of this equation requires closure. The turbulent flux of Σ, $-\bar{\varrho}(\widetilde{u_k''\Sigma''})$, was closed using a conventional gradient transport model including a turbulent Schmidt number for the surface density. The Favre-averaged form of the rhs of Eq. (4.45) was modelled as a function of the strain rate (assumed $\propto \varepsilon/k$), $\bar{\varrho}$ and $\tilde{\Sigma}$. In order to complete the closure, further modelled terms had to be added, to account for the reduction in flame area brought about by the interaction of two flamelets separated by fresh reactant and removal of flame area by quenching when the flamelets are situated in a highly-strained field. Intuitive arguments were employed to represent these complicated processes. The mean consumption rates were calculated from $\bar{\omega}_i = -\bar{\varrho}V_{r,i}\tilde{\Sigma}$, where $V_{r,i}$ is the volume rate of consumption per unit flame area of the ith species (presumably based on mean concentrations). For the purpose of calculating a turbulent flame, a flamelet library was constructed based on the calculation of a laminar premixed counterflow flame of fresh reactants and burnt products, using a detailed propane–air mechanism. The flamelet model was characterised by the strain rate, temperature of the unburnt gas stream and the equivalence ratio of the incoming reactant stream. A baffle-stabilised ducted turbulent premixed flame was investigated, supported by measurements of the local heat release rate, obtained by light scattering techniques. For the purpose of prediction, the empirical constants emerging from the modelling were assigned the value of unity (without further optimisation/calibration). The reacting flow calculations were initiated by introducing a temperature zone with a Gaussian shape and a localised distribution of flame surface density in a region of recirculation of the cold flow. How sensitive the predictions were to variations in the specification of the reaction zone is not clear. Qualitatively, the predictions followed the correct trends; for example a reduction in the equivalence ratio brought about the observed reduction in the length of the reaction zone. However, the model predicted heat release in regions not observed in the measurements. Despite the extensive modelling employed to close the Σ equation, the shortcomings of the prediction were attributed to the use of the gradient transport hypothesis.

Cant *et al.* (1990) also investigated the surface-to-volume ratio equation in premixed combustion. They considered the flame surface to be defined by all points $\underline{X}$ such that $c(\underline{X}, t) = c^*$, where c^* was a specified and fixed value of the progress variable c. In this respect the flame surface was a constant-property surface. An exact equation for the mean surface area per unit volume was proposed, although comparison with the equation appearing in the work of Darabiha *et al.* (1989) was rendered difficult due to the use of surface averaging in the former case. Cant *et al.* identified realisability conditions which had to be satisfied and modelled the unclosed terms appearing in their formulation of the Σ-equation. In particular, the strain rate in the tangent plane of the surface was modelled to be consistent with the limiting behaviour of a material surface and a fixed surface in isotropic turbulence. The mean curvature, influenced by small scale turbulence effects and instabilities in laminar flame propagation, was represented as a linear function of Σ and the local mean surface laminar flame speed. A one-dimensional test case with open boundaries was considered. At time $t = 0$ the flame was assumed to exist in a field of statistically homogeneous isotropic turbulence with zero mean flow and no density variations. The results yielded a flame which quickly settled down to a steady propagation speed and a self-preserving structure. In addition the effect on the turbulent flame speed as a function of the laminar flame speed and the Reynolds number (based on the Taylor micro-scale), showed the correct trends. Cant (1993) has extended the formulation to the variable-density case. The basic modelling follows on from Cant *et al.* (1990). Certain results from the Bray–Moss theory were employed together with a flamelet library for CH_4–air combustion in order to investigate the same simplified test case as described above. The results once again yielded a plausible evolution for the flame profiles.

The performance of various flame surface density models has been investigated by Prasad and Gore (1999) and Prasad *et al.* (1999). In Prasad and Gore (1999) the results obtained with four different flame surface density model formulations and the Bray–Moss model are compared with measurements in a simple premixed jet flame. All of the models predicted the increase in flame brush thickness with downstream distance and reasonable agreement with experiment. The different models did, however, display differing sensitivities to the inlet turbulence characteristics. The models investigated in Prasad and Gore (1999) have also been applied to non-isenthalpic premixed flames in Prasad *et al.* (1999). Here it was found that the Mantel and Borghi (1994) and Bray–Moss (1977) models gave best results, but that the other FSD models with slight adjustments of the model constants could provide similar agreement.

Peters (1992) investigated premixed turbulent combustion in the flamelet regime by a level set method (Sethian, 1998) where the instantaneous flame contour is described in terms of an iso-scalar surface of the scalar field $G(x, t)$. The condition for the flame front $G(x, t) = G_o$ (an arbitrary constant) divides the flow field into two regions where $G > G_o$ is the region of burnt gas and $G < G_o$ that of the unburnt mixture. The scalar difference $G - G_o$ was interpreted as the distance from the flame surface. The Markstein length, $\mathcal{L}$, which is a characteristic length scale for flame response and proportional to the flame thickness, was introduced into the G-equation by expressing the laminar burning velocity (subjected to stretch) in terms of the unstretched burning velocity and terms associated with flame curvature and flow divergence. The result was:

$$\frac{DG}{Dt} \equiv \frac{\partial G}{\partial t} + \underline{v} \cdot \nabla G = u_{l,o}|\nabla G| + \mathcal{D}_\mathcal{L}\nabla^2 G - \mathcal{L}\left(\frac{D|\nabla G|}{Dt}\right)_{\mathcal{L}\to 0}, \quad (4.46)$$

where $\mathcal{D}_\mathcal{L} = u_{l,o}\mathcal{L}$, with $\mathcal{D}_\mathcal{L}$ being referred to as the Markstein diffusivity. $u_{l,o}$ and $\mathcal{L}$ were defined with respect to the unburnt mixture, so that $\underline{v}$ was the velocity conditioned ahead of the flame. Peters then constructed (through Reynolds decomposition and ensemble-averaging) transport equations for $\bar{G}$ and $\overline{G'^2}$ which contained several terms requiring closure. In particular, unclosed terms were identified in the $\overline{G'^2}$ equation which were associated with the smoothing of the flame front and which were argued to be most effective at different scales of turbulence. Guidance for the modelling of the $\overline{G'^2}$ equation was obtained from consideration of constant-density homogeneous isotropic turbulence. The existence of an equilibrium range for the scalar field under consideration was hypothesised, and consequently between wavenumbers (κ) representing the reciprocal of the integral and Gibson scale, the scalar spectrum function $\Gamma(\kappa, t)$ assumed a universal form. $\Gamma(\kappa, t) = 4\pi\kappa^2\widehat{g^2}(\kappa, t)$, where $\widehat{g^2}(\kappa, t)$ is the Fourier transform of $g^2(r, t) \equiv \overline{G'(r, t)G'(\mathbf{x} + r, t)}$. Introduction of a gradient transport hypothesis permitted closure of the $\Gamma(\kappa, t)$ equation, thereby constraining the modelled terms to be a function of the local wavenumber. The resulting linear differential equation was solved exactly by the method of characteristics, and integration of the differential equation over wavenumber space and comparison with the constant-density homogeneous analogue of the $\overline{G'^2}$ equation completed the closure. Modelling of the $\bar{G}$ equation remained. A term representing the product of the flame surface area per unit volume and the unstretched laminar flame speed was modelled using intuitive arguments and dimensional analysis, with the result that the closed term involved k/ε as a characteristic time scale. The turbulent transport and Markstein diffusion terms were combined and modelled as a curvature term incorporating a turbulent diffusivity. Calculations were performed for plane and oblique flames for which the curvature term disappeared, and while no direct comparison with measurements was made, the predicted flame brush thickness and position agreed qualitatively with experimental observations.

An alternative flamelet formulation for premixed turbulent combustion has been adopted by Bradley et al. (1998b), see also Bradley (2001). Combustion is described in terms of temperature, used to define a reaction progress variable, and the flame stretch rate, s. A joint pdf, $P(c, s)$, is then introduced to evaluate the mean volumetric heat release rate that appears in the transport equation for mean temperature, viz:

$$\bar{q} = \int_{s^-}^{s^+} \int_0^1 q_l(c, s)P(c, s)\mathrm{d}c\,\mathrm{d}s, \qquad (4.47)$$

where $q_l(c, s)$ is the heat release rate in a stretched laminar flame and where s^- and s^+ are the flame extinction limits under negative and positive stretch rates. The volumetric heat release rate, $q_l(c, s)$, is then obtained from:

$$q_l(c, s) = f(s)q_l(c, 0), \qquad (4.48)$$

where the function $f(s)$ depends on s and the Markstein numbers of the mixture. For turbulent flow the function is written in terms of the dimensionless strain rate, $\hat{s} = s\tau_K$, where τ_K is the Kolmogorov time scale and the results of computations of spherically propagating laminar flames are used to construct suitable forms for $f(\hat{s})$. The variables c and $\hat{s}$ are assumed to be uncorrelated so that Eq. (4.47) becomes:

$$\bar{q} = \int_{s^-}^{s^+} f(\hat{s})P(\hat{s})\mathrm{d}s \int_0^1 q_l(c, 0)P(c)\mathrm{d}c, \qquad (4.49)$$

where $\int_{s^-}^{s^+} f(\hat{s})P(\hat{s})ds$ represents the 'probability of burning', P_b. In Bradley *et al.* (1998a,d) the functional forms of the two pdfs, $P(\hat{s})$ and $P(c)$, are assumed a priori to be a near-Gaussian distribution and a beta function, with the former being suggested by DNS results.

5 Partially premixed, premixed and non-premixed combustion

The combustion models described so far are restricted to either premixed or non-premixed turbulent combustion and the availability of models that are equally applicable to all regimes of burning is severely limited. The model perhaps most widely used in commercial CFD codes is the so-called 'eddy dissipation' model. In this approach, together with the presumption of a single step reaction, the mean fuel consumption rate is given, typically, by:

$$\tilde{\omega}_F = -C_B \bar{\varrho} \frac{\varepsilon}{k} \min \left(\frac{\tilde{Y}_P}{1 + r_F}, \min \left(\tilde{Y}_F, \frac{\tilde{Y}_O}{r_F} \right) \right). \tag{4.50}$$

While the formulation has some similarities to Spalding's eddy break-up model for premixed combustion, it appears to have little rigorous basis in general and is based on ad hoc assumptions. From a practical point of view the constant C_B is far from universal, and a very wide range of values have been used – and results are extremely sensitive to the value of C_B. Further ad hoc modifications are also required for burning near solid surfaces.

The premixed turbulent flame model of Bradley *et al.* (1998b) has been extended to partially premixed combustion, Bradley *et al.* (1998c). It is assumed that burning takes place in the form of *premixed* flamelets of varying mixture strength, and the model is augmented by the addition of mixture fraction and the associated joint pdf, $P(\xi, c, s)$. This joint pdf is expressed as:

$$P(\xi, c, s) = P(s)P(c|\xi)P(\xi), \tag{4.51}$$

where the conditional pdf, $P(c|\xi)$, is assumed to be a beta function. The model has been successfully applied to the calculation of flame lift-off heights and blow-off velocities for methane jet flames and the results were in excellent agreement with the available experimental data. In a related approach, Bondi and Jones (2002) have formulated a model, based on the flame surface density and mixture fraction, for premixed flames with varying stoichiometry. The model led to reasonable agreement with measurements in a natural gas fuelled burner.

The G-equation method of Peters (1992) has also been extended (Chen *et al.*, 2000), to partially premixed combustion by the addition of the variables $\tilde{\xi}$, ξ''^2 and

$\widetilde{h}$ for which transport equations are solved. The approach led to good agreement with measured lift-off heights, and further details of the method can be found in Peters (2000).

5.1 The transported pdf approach

If the assumption of 'thin' flame burning or 'fast' chemistry, invoked in practically all of the combustion models described so far, is not appropriate then the problem of evaluating averaged species net formation rates is unavoidable. For example, the successful prediction of CO and unburnt hydrocarbon *emissions* from combustion systems requires account to be taken of finite-rate chemical processes. In addition, the low NO_x combustion chambers currently being developed involve partial pre-mixing of the fuel and air prior to combustion, and here also averaged net formation must be determined. To achieve this the joint probability density function for the temperature and mass fractions of the necessary chemical species must be known. One and perhaps the only method of obtaining this is from solutions of the evolution equation for the probability density function of the relevant scalar quantities. The joint pdf of a set of scalars, $P(\psi_1, \ldots, \psi_N; \mathbf{x}, t)$, contains all single point, single time moments of the scalar field. Consequently, in the exact transport equation for $P(\psi_1, \ldots, \psi_N; \mathbf{x}, t)$ the chemical source terms associated with reaction appear in closed form, though, as in all statistical approaches, the equation itself is unclosed. The unknown terms represent micro-mixing in the 'scalar' space and turbulent transport in physical space, and it is the former of these which represents a central and only partially resolved difficulty. Nevertheless, providing sufficiently accurate approximations can be devised, then the inclusion of chemical reaction should in principle present little further difficulty. Pdf transport equation modelling has been investigated for well over two decades, but it is only relatively recently that the methodology has become a potential tool for engineering calculations. Whilst this is in part due to the successful development of Monte Carlo solution algorithms, it is mainly because of computer hardware developments which provide the increased resource necessary to obtain solutions to the pdf equation. Essentially for the reasons above there is currently considerable interest in the pdf transport approach to turbulent combustion; reviews of the topic can be found in O'Brien (1980), Pope (1985), Kollmann (1990, 1992), Jones and Kakhi (1996) and Jones (2001).

5.1.1 The governing equations

The terms representing chemical source terms appear in closed form in the balance equation for the joint pdf of the relevant scalar quantities. For combusting flows the density-weighted pdf defined by:

$$\tilde{P}(\psi_1, \ldots, \psi_N; \mathbf{x}, t) = \frac{\rho(\psi_1, \ldots, \psi_N)}{\bar{\rho}} P(\psi_1, \ldots, \psi_N; \mathbf{x}, t) \tag{4.52}$$

is appropriate and an equation for this may be derived from the exact conservation equations by standard methods. For high Reynolds number turbulent flow and with the assumption of equal molecular diffusivities this may be written as:

$$\bar{\rho}\frac{\partial \tilde{P}(\underline{\psi})}{\partial t} + \bar{\rho}\tilde{u}_k \frac{\partial \tilde{P}(\underline{\psi})}{\partial x_k} + \bar{\rho}\sum_{\alpha=1}^{N} \frac{\partial}{\partial \psi_\alpha}\{\dot{\omega}_\alpha(\underline{\psi})\tilde{P}(\underline{\psi})\}$$

$$= +\frac{\partial}{\partial x_k}\{\bar{\rho}\langle u_k''|\underline{\phi} = \underline{\psi}\rangle \tilde{P}(\underline{\psi})\}$$

$$+\sum_{\alpha=1}^{N}\sum_{\beta=1}^{N} \frac{\partial^2}{\partial \psi_\alpha \partial \psi_\beta}\left\{\left\langle \frac{\mu}{Sc}\frac{\partial \phi_\alpha}{\partial x_i}\frac{\partial \phi_\beta}{\partial x_i}\middle|\underline{\phi} = \underline{\psi}\right\rangle \tilde{P}(\underline{\psi})\right\}. \tag{4.53}$$

The pdf equation contains two types of unknown terms for which closure is required; that representing turbulent transport and that describing micro-mixing – the final term in Eq. (4.53). In this latter term the quantity

$$\left\langle \frac{\mu}{Sc}\frac{\partial \phi_\alpha}{\partial x_i}\frac{\partial \phi_\beta}{\partial x_i}\middle|\underline{\phi} = \underline{\psi}\right\rangle$$

represents the scalar dissipation conditioned on $\underline{\phi} = \underline{\psi}$.

For the turbulent flux of pdf a gradient transport model is usually adopted so that:

$$-\bar{\rho}\langle u_k''|\underline{\phi} = \underline{\psi}\rangle \tilde{P}(\underline{\psi}) = \frac{\mu_t}{\sigma_t}\frac{\partial \tilde{P}(\underline{\psi})}{\partial x_k} \tag{4.54}$$

where the value of σ_t is assigned the value of 0.7 and where μ_t is obtained from the k–ε turbulence model, Jones and Launder (1973). An alternative is to use the joint pdf for velocity and the necessary scalars, e.g. Wouters *et al.* (2001), in which the turbulent transport terms are closed, though there are other terms that then require modelling.

The micro-mixing term is of central importance and its closure represents a major difficulty in the pdf transport equation approach. Though there have been a number of proposals, it appears that none are entirely satisfactory, even in inert flows. For combusting flows modelling represents an even greater challenge for, as is evident, the dominant contribution to micro-mixing is from the fine scale motions and this is precisely where combustion normally occurs. Micro-mixing and combustion are thus intimately coupled and this has clear implications for modelling. Of the models available at present, it appears that the modified coalescence-dispersion model of Janicka *et al.* (1979) and Dopazo (1979), see also Jones and Kollman (1987), represents an acceptable balance of complexity and realism. It may be written as:

$$\sum_{\alpha=1}^{N}\sum_{\beta=1}^{N}\frac{\partial^2}{\partial\psi_\alpha\partial\psi_\beta}\left\{\left\langle\frac{\mu}{Sc}\frac{\partial\phi_\alpha}{\partial x_i}\frac{\partial\phi_\beta}{\partial x_i}\middle|\underline{\phi}=\underline{\psi}\right\rangle\tilde{P}(\underline{\psi})\right\}$$
$$=-3C_d\bar{\rho}\frac{\epsilon}{k}\left\{\int_\varphi d\varphi\int_\psi d\Psi\,\tilde{P}(\varphi)\tilde{P}(\Psi)T(\varphi,\Psi|\psi)-\tilde{P}(\psi)\right\}. \qquad (4.55)$$

The constant C_d is assigned the value 2.0 so that the predicted variance decay rate of a passive scalar is in conformity with measurements. To complete the description the mean velocity, $\tilde{U}_i$ is needed. This is normally obtained from solution of the mean flow equations in which the k–ε model, written in terms of density-weighted quantities, is used to represent turbulent transport.

The terms in the pdf equation which involve chemical reaction appear in closed form, and in principle reaction mechanisms of arbitrary complexity may be incorporated. However, to keep the method tractable it is necessary that chemical reaction be described in terms of a small number of scalar quantities – that is quantities such as species mass fraction etc. for which conservation equations must be solved. Hence the use of reduced reaction mechanisms appears essential with the method. In addition it is usual to tabulate the chemistry on a once and for all basis in order to keep computational requirements manageable. The method has been applied successfully to both premixed and non-premixed combustion, e.g. Anand and Pope (1987), Jones and Prasetyo (1996), and Jones and Kakhi (1997, 1998). More recently realistic and quite detailed chemistry has been incorporated with quite impressive results, Xu and Pope (2000), Tang et al. (2000) and Lindstedt et al. (2000). The pdf transport equation approach is without doubt a powerful method with considerable potential for representing a wide range of combustion phenomena. However, while there are a number of modelling issues which remain to be resolved, the main limitation to its use in engineering applications is the high computational cost involved. This arises because of the large number of independent variables involved in the pdf equation ($\mathbf{x}$, t and the scalars needed to describe reaction) which in turn implies that stochastic solution methods must be used. The resulting computation cost of solution is thus substantially greater than that required for the majority of other combustion models described elsewhere in this chapter; for three-dimensional flows, particularly if realistic chemistry is used, the computational requirements may well be beyond current capabilities.

6 Conclusions

This chapter has attempted to provide a review of the mathematical models available for predicting combustion in turbulent flows. The processes involved in the various regimes are quite different, and so the modelling of non-premixed (diffusion), premixed and partially premixed flames has been considered separately. The modelling of non-premixed flames is perhaps the most firmly based and has reached the most satisfactory state. If only heat release rate is of concern then the single laminar flamelet/conserved scalar/presumed shape often provides a sufficiently accurate description of burning. This should not be taken to imply that agreement with experiment will always be good; an accurately predicted velocity field – which may not be forthcoming because of limitations in turbulence models – is clearly a necessary prerequisite for an overall accurate prediction. If local extinction effects are important or NO_x formation, particularly at ambient pressures, is of interest then the simple flamelet model is clearly not appropriate and the conditional moment closure and unsteady flamelet model represent viable alternatives. There are many similarities between the two approaches. Both involve a flamelet (thin flame) description of burning, allow fairly detailed chemistry to be incorporated and account for flame history effects. Furthermore, the computational costs of the two approaches appear to be closely similar.

The fundamental assumption underlying practically all of the useful models of premixed combustion is that burning occurs in the thin flame regime with the consequence that the flame can be treated much as a propagating material surface, with a propagation speed being dependent on flame properties such as the stretch and curvature etc. Of those available the BML model, the *G*-equation model, the flame surface density (FSD) model and the model of Bradley *et al.* appear to be the most well developed and all are capable of producing good results in the appropriate circumstances. The main differences between the various formulations appear to rest mainly on the method adopted to account for the influences of flame stretch and curvature. It is also possible, however, that some of the differences in the results obtained with the various models arise because of differences in model calibration and burning velocity data. At the present time it does not appear possible to conclude which, if any, of the above models is optimum. To some degree the model to be selected is a matter of taste though it does also depend on the application to be considered.

The situation is much less satisfactory in circumstances where the thin flame burning assumption is not appropriate or where there is a degree of partial premixing, or where burning in both the premixed and non-premixed regimes occurs within the same combustion chamber. The partially premixed models of Bradley *et al.* (1998c) and Chen *et al.* (2000) both allow the calculation of lifted diffusion flames and blow-out. While the agreement achieved is good, an essential assumption is that burning occurs in the form of premixed flamelets. While this may be appropriate in lifted flames it is inconsistent with the fast chemistry limit described in Section 3 and this will undoubtedly limit the generality of the methods. Currently there appears to be no model capable of accurately describing all regimes of turbulent burning that is *computationally* tractable in three-dimensional engineering applications. The pdf transport equation method is potentially capable of being such a method, and the results obtained with it to date are impressive. However, the computational cost of the method is large, particularly if realistic chemistry is to be used. At present this restricts its use to relatively simple configurations, though this situation is likely to change with current rapid increases in computing power.

Acknowledgements

It is a pleasure to acknowledge the helpful and constructive comments of Professor Ken Bray, Dr Mike Fairweather, Professor Norbert Peters and Dr Chris Priddin on a draft of this review.

References

Anand, M. S. and S. B. Pope, 1987. Calculations of premixed turbulent flames by pdf methods. *Combustion and Flame*, **67**: 127.

Bailly, P., M. Champion and D. Garréton, 1997. Counter-gradient diffusion in a confined turbulent premixed flame. *Physics of Fluids*, **9**: 766–775.

Bilger, R. W., 1976. The structure of diffusion flames. *Combustion Science and Technology*, **13**: 155–170.

1980. Turbulent flows with non-premixed reactants. In P. A. Libby and F. A. Williams, eds., *Turbulent Reacting Flows*, pp. 65–113. 'Topics in Applied Physics', vol. 44. Springer-Verlag.

1989. Turbulent diffusion flames. *Annual Review of Fluid Mechanics*, **21**: 101–135.

Bondi, S. and W. P. Jones, 2002. A combustion model for premixed flames with varying stoichiometry. *Proceedings of the Combustion Institute*, **29**, 2123–2129.

Borghi, R., 1985. On the structure and morphology of turbulent premixed flames. In C. Casci, ed., *Advances in Aerospace Science*, pp. 117–138. Plenum Press.

Bradley, D., 2001. Problems of predicting turbulent burning rates. In *Plenary Lecture, Eighteenth International Colloquium on the Dynamics of Explosions and Reactive Systems, Seattle, USA*, pp. 1–25.

Bradley, D., P. H. Gaskell and X. J. Gu, 1998a. Application of a Reynolds stress, stretched flamelet, mathematical model to computations of turbulent burning velocities and comparisons with experiment. *Combustion and Flame*, **96**: 221–248.

1998b. The modelling of aerodynamic strain rate and flame curvature effects in premixed turbulent combustion. In *Twenty-Seventh Symposium (International) on Combustion*, The Combustion Institute, pp. 849–856.

1998c. The modelling of liftoff and blowoff of turbulent non-premixed methane jet flames at high strain rates. In *Twenty-Seventh Symposium (International) on Combustion*, The Combustion Institute, pp. 1199–1206.

Bradley, D., P. H. Gaskell, X. J. Gu, M. Lawes and M. J. Scott, 1998d. The measurement of laminar burning velocities and Markstein numbers for iso-octane-air and iso-octane-n-heptane-air mixtures at elevated temperatures and pressures in an explosion bomb. *Combustion and Flame*, **115**: 515.

Bray, K. N. C., 1978. The interaction between turbulence and combustion. In *Seventeenth Symposium (International) on Combustion*, The Combustion Institute, pp. 223–233.

1980. Turbulent flows with premixed reactants. In P. A. Libby and F. A. Williams, eds., *Turbulent Reacting Flows*, pp. 115–183. 'Topics in Applied Physics', vol. 44. Springer-Verlag.

1986. Methods of including realistic chemical reaction mechanisms in turbulent combustion models. In *Second Workshop on Modelling of Chemical Reaction Systems*, pp. 356–375. Meeting held in Heidelberg, August 1986.

Bray, K. N. C. and P. A. Libby, 1976. Interaction effect in turbulent premixed flames. *Physics of Fluids*, **19(11)**: 1687–1701.

1986. Passage times and flamelet crossing frequencies in premixed turbulent combustion. *Combustion Science and Technology*, **47**: 253–274.

Bray, K. N. C. and J. B. Moss, 1974. A unified statistical model of the premixed turbulent flame. AASU Report 335, University of Southampton.

1977. A unified statistical model of the premixed turbulent flame. *Acta Astronautica*, **4**: 291–320.

Bray, K. N. C., P. A. Libby, G. Masuya and J. B. Moss, 1981. Turbulence production in premixed turbulent flames. *Combustion Science and Technology*, **25**: 127–140.

Bray, K. N. C., P. A. Libby and J. B. Moss, 1984. Flamelet crossing frequencies and mean reaction rates in premixed turbulent combustion. *Combustion Science and Technology*, **41**: 143–172.

Cant, R. S., 1993. Turbulent reaction rate modelling using flamelet surface area. In *Proceedings of the Anglo-German Combustion Symposium*, p. 48. The Combustion Institute, British Section of the Combustion Institute, April 1993. Meeting held at Queen's College, Cambridge.

Cant, R. S., S. B. Pope and K. N. C. Bray, 1990. Modelling of flamelet surface-to-volume ratio in turbulent premixed combustion. In *Twenty-Third Symposium (International) on Combustion*, The Combustion Institute, pp. 809–815.

Catlin, C. A. and R. P. Lindstedt, 1991. Premixed turbulent burning velocities derived from mixing controlled reaction models with cold front quenching. *Combustion and Flame*, **85**: 427–439.

Champion, M., K. N. C. Bray and J. B. Moss, 1978. The turbulent combustion of a propane-air mixture. *Acta Astronautica*, **5**: 1063–1077. Previously presented at the Sixth Colloquium on Gas Dynamics of Explosions and Reacting Systems, Stockholm 1977.

Chen, M., M. Herrmann and N. Peters, 2000. Flamelet modelling of lifted turbulent methane/air and propane/air flames. In *Twenty-Eighth Symposium (International) on Combustion*, The Combustion Institute, pp. 167–174.

Damköhler, G., 1940. Der einflußder Turbulenz auf die Flammengeschwindigkeit in Gasgemischen. *Z. Elecktrochem.*, **46**: 601–626. English translation: NACA Technical Memorandum, 1112, 1941.

Darabiha, N., V. Giovangigli, A. Trouvé, S. M. Candel and E. Esposito, 1989. Coherent flame description of turbulent premixed ducted flames. In R. Borghi and S. N. B. Murthy, eds., *Turbulent Reactive Flows*, pp. 591–637. Springer-Verlag.

Dopazo, C., 1979. Relaxation of initial probability density functions in the turbulent convection of scalar fields. *Physics of Fluids*, **22(1)**: 20–30.

Gordon, S. and B. J. McBride, 1971. Computer program for calculation of complex equilibrium composition, rocket performance, incident and reflected shocks and Chapman–Jouget detonations. SP-273, KASA.

Heitor, M. V., F. Culick and J. H. Whitelaw, eds., 1996. *Unsteady Combustion.* Kluwer Academic Publishers.

Janicka, J., W. Kolbe and W. Kollmann, 1979. Closure of the equations for the probability density function of turbulent scalar fields. *Journal of Non-Equilibrium Thermodynamics*, **4**: 47–66.

Jones, W. P., 1980. Models for turbulent flows with variable density and combustion. In W. Kollmann, ed., *Prediction Methods for Turbulent Flows*, pp. 380– 421. Hemisphere Publ. Corp.

    2001. The joint scalar probability density function. In B. E. Launder and N. D. Sandham, eds., *Closure Strategies for Turbulent and Transitional Flows.* Cambridge University Press.

Jones, W. P. and M. Kakhi, 1996. Mathematical modelling of turbulent flames. In M. V. Heitor, F. Culick and J. H. Whitelaw, eds., *Unsteady Combustion*, pp. 411–491. Kluwer Academic Publishers.

    1997. Application of the transported pdf approach to hydro-carbon turbulent jet diffusion flames. *Combustion Science and Technology*, **129**: 393–430.

    1998. Pdf modelling of finite-rate chemistry effects in turbulent non-premixed jet flames. *Combustion and Flame*, **115**: 210–229.

Jones, W. P. and W. Kollmann, 1987. Multi-scalar pdf transport equations for turbulent diffusion flames. In F. Durst, B. E. Launder, J. L. Lumley, F. W. Schmidt and J. H. Whitelaw, eds., *Turbulent Shear Flows 5*, pp. 296–309. Springer-Verlag.

Jones, W. P. and B. E. Launder, 1972. The prediction of laminarisation with a two-equation model of turbulence. *International Journal of Heat and Mass Transfer*, **15**: 301–314. Also AIAA selected reprint series XIV, pp. 119–132, 1973.

Jones, W. P. and Y. Prasetyo, 1996. Probability density function modelling of premixed turbulent opposed jet flames. In *Twenty-Sixth Symposium (International) on Combustion*, The Combustion Institute, pp. 275–282.

Jones, W. P. and C. H. Priddin, 1978. Predictions of the flow field and local gas composition in gas turbine combustors. In *17th Symposium (International) on Combustion*, The Combustion Institute, pp. 399–409.

Jones, W. P. and J. H. Whitelaw, 1982. Calculation methods for reacting turbulent flows. *Combustion and Flame*, **48**: 1–26.

Klimenko, A. Y. and R. Bilger, 1999. Conditional moment closure for turbulent combustion. *Progress in Energy and Combustion Science*, **25**: 595.

Kollmann, W., 1990. The pdf approach to turbulent flow. *Theoretical and Computational Fluid Dynamics*, **1**: 249–285.

    1992. Pdf transport modelling. Modelling of combustion and turbulence, Von Karman Institute for Fluid Dynamics, lecture series.

Libby, P. A. and K. N. C. Bray, 1981. Countergradient diffusion in premixed turbulent flames. *AIAA Journal*, **19(2)**: 205–213. Originally presented as Paper 80-0013 at the AIAA meeting, Jan. 1980, Pasadena, California.

Libby, P. A. and F. A. Williams, 1980. Fundamental aspects. In P. A. Libby and F. A. Williams, eds., *Turbulent Reacting Flows*, pp. 1–43. 'Topics in Applied Physics', vol. 44. Springer-Verlag.

    eds., 1993. *Turbulent Reacting Flows.* Academic Press.

Lindstedt, R. P., S. A. Louloudi and E. M. Váos, 2000. Joint scalar pdf modelling of pollutant formation in piloted turbulent jet diffusion flames with comprehensive chemistry. In *Twenty-Eighth Symposium (International) on Combustion*, The Combustion Institute, pp. 149–156.

Mantel, T. and R. Borghi, 1994. A new model of premixed wrinkled flame propagation based on a scalar dissipation equation. *Combustion and Flame*, **96**: 443–457.

Mauss, F., D. Keller and N. Peters, 1990. A Lagrangian simulation of flamelet extinction and re-ignition in turbulent jet diffusion flames. In *Twenty-Third Symposium (International) on Combustion*, The Combustion Institute, pp. 693–698.

O'Brien, E. E., 1980. The probability density function (pdf) approach to reacting turbulent flows. In P. A. Libby and F. A. Williams, eds., *Turbulent Reacting Flows*, pp. 185–218. 'Topics in Applied Physics', vol. 44. Springer-Verlag.

Peters, N., 1986. Laminar flamelet concepts in turbulent combustion. In *Twenty-First Symposium (International) on Combustion*, The Combustion Institute, pp. 1231–1250.

  1992. A spectral closure for premixed turbulent combustion in the flamelet regime. *Journal of Fluid Mechanics*, **242**: 611–629.

  1999. The turbulent burning velocity for large-scale and small-scale turbulence. *Journal of Fluid Mechanics*, **384**: 107–132.

  2000. *Turbulent Combustion*. Cambridge University Press.

Poinsot, T. J., D. Veynante and S. Candel, 1991. Quenching processes and premixed turbulent combustion diagrams. *Journal of Fluid Mechanics*, **228**: 561–606.

Pope, S. B., 1985. Pdf methods for turbulent reactive flows. *Progress in Energy and Combustion Science*, **11**: 119–192.

  1987. Turbulent premixed flames. *Annual Review of Fluid Mechanics*, **19**: 237–270.

Prasad, R. O. S. and J. P. Gore, 1999. An evaluation of flame surface density models for turbulent premixed jet flames. *Combustion and Flame*, **116**: 3–14.

Prasad, R. O. S., R. N. Paul, Y. R. Sivathanu and J. P. Gore, 1999. An evaluation of combined flame surface density and mixture fraction models for nonisenthalpic premixed turbulent flames. *Combustion and Flame*, **117**: 514–528.

Reynolds, W. C., 1990. The potential and limitations of direct and large eddy simulation. In J. L. Lumley, ed., *Whither Turbulence? Turbulence at the Crossroads*, pp. 313–343. Lecture Notes in Physics, 357. Springer-Verlag. Proceedings of a workshop held at Cornell University, Ithaca, March 1989.

Sethian, J. A., 1998. *Level Set Methods*. Cambridge University Press.

Shepherd, I. G. and J. B. Moss, 1983. Characteristic scales for density fluctuations in turbulent premixed flames. *Combustion Science and Technology*, **33**: 231.

Spalding, D. B., 1971. Mixing and chemical reaction in steady confined turbulent flames. In *Thirteenth Symposium (International) on Combustion*, The Combustion Institute, pp. 649–657.

Summerfield, M., S. H. Reiter, V. Kebely and R. W. Mascolo, 1955. The structure and propagation mechanism of turbulent flames in high speed flow. *Jet Propulsion*, **25(8)**: 377–384.

Tang, Q., Jun Xu and S. B. Pope, 2000. Pdf calculations of local extinction and NO production in piloted-jet turbulent methane/air flames. In *Twenty-Eighth Symposium (International) on Combustion*, The Combustion Institute, pp. 133–139.

TNF4. Proceedings of TNF4 workshop, 1999. URL www.ca.sandia.gov/TNF/4thWorkshop/TNF4.html. Darmstadt, Germany.

TNF5. Proceedings of TNF5 workshop, 2000. URL www.ca.sandia.gov/TNF/5thWorkshop/TNF5.html. Delft, The Netherlands.

TNF6. Proceedings of TNF6 workshop, 2002. URL www.ca.sandia.gov/TNF/6thWorkshop/TNF6.html. Sapporo, Japan.

Weller, H. G., C. J. Marooney and A. D. Gosman, 1990. A new spectral method for calculation of the time-varying area of a laminar flame in homogeneous turbulence.

In *Twenty-Third Symposium (International) on Combustion*, The Combustion Institute, pp. 629–636.

Williams, F. A., 1975. Recent advances in the theoretical descriptions of turbulent diffusion flames. In S. N. B. Murthy, ed., *Turbulent Mixing in Non-reactive and Reactive Flows*. Plenum Press.

Wouters, H. A., Peeters, T. W. J. and Roekaerts, D. Joint velocity-scalar PDF methods, In B. E. Launder and N. D. Sandham, eds., *Closure Problems for Turbulent and Transitional Flows*, p. 626. Cambridge University Press.

Xu, Jun and S. B. Pope, 2000. Pdf calculations of turbulent nonpremixed flames with local extinction. *Combustion and Flame*, **123(3)**: 281–307.

5

Boundary layers under strong distortion: an experimentalist's view

J. F. Morrison

Imperial College, London

Abstract

The problems concerning the understanding and prediction of strongly-distorted turbulent boundary layers are reviewed. Some of the views expressed emanate from the Isaac Newton Institute (INI) programme on turbulence in 1999; others are an experimentalist's view of current modelling techniques and the requirements for future work. The purpose of this chapter is, first, to pinpoint research that has already been carried out in this area and, second, to highlight gaps in our knowledge where there is a need for specific experiments to enable the subsequent development of existing turbulence models. Attention is not restricted to Reynolds-averaged Navier–Stokes (RANS) solvers but also considers the problems regarding the modelling for large-eddy simulation (LES). The subject of this chapter is vast, and therefore ample use is made of existing reviews by authors who are experts in specific subject areas. These include 'extra rates of strain', changes in boundary conditions, as well as even more complex phenomena such as shock/boundary-layer interaction and boundary layers with a variety of embedded vortices. While we have many of the numerical tools required for design and prediction, it is clear that physical understanding of the dominant mechanisms is lacking and therefore our ability to predict them is also. As far as our existing knowledge is concerned, emphasis is placed on an empirical approach for reasons of pragmatism. Some 'application challenges' presented during the course of the INI programme on turbulence are addressed. It is clear that more experiments are required: these should be specific and well defined, and have very clear objectives.

Prediction of Turbulent Flows, eds. G. F. Hewitt and J. C. Vassilicos.
Published by Cambridge University Press. © Cambridge University Press 2005.

1 Introduction

The subject of this chapter necessarily involves a wide range of flows that are ubiquitous in both nature and engineering. It is treated as that class of turbulent boundary layers subjected to the imposition of a length or velocity scale (or both) in addition to those necessarily imposed in order that the boundary layer may exist, namely the viscous length scale, v/u_τ (where v is the kinematic viscosity and the wall friction velocity $u_\tau = \sqrt{\tau_w/\rho}$) and the boundary layer thickness, δ. The aim of this chapter is to provide a *physical* description for those interested in the prediction of, and modelling for, distorted boundary layers. Owing to its uncertain definition, δ is a representative length scale: more pragmatic definitions exist such as the Rotta–Clauser boundary layer thickness. The velocity scale, u_τ, is problematical owing to the uncertainties associated with its accurate measurement. However, as both the log law and the velocity defect law indicate, it provides the best description of the mean velocity profile, so that there need to be very good reasons for not using it as the key velocity scale. Additional scales can be introduced in a variety of ways, essentially through modification of the boundary conditions, such as through changes in pressure gradient, or by roughness or free-stream turbulence. Bradshaw's (1971) article, 'Variations on a theme of Prandtl', was the first of a sequence of articles outlining a definition of a class of 'complex' flows: for a fuller treatment see also Bradshaw (1973). Our starting point is therefore Prandtl's boundary layer approximation. Here the emphasis is on the prediction of distorted boundary layers, where the addition of further imposed scale(s) will lead inevitably to further complexity of the physical processes, and unfortunately, to a greater level of empiricism in their modelling. As such, the subject of this chapter is a slightly more formal, but physical, definition of complex flows – those that cannot be predicted with acceptable accuracy by methods developed for thin shear layers.

Formally, imposition of an additional scale implies that a distorted boundary layer cannot be self-preserving, immediately raising questions concerning the suitability or otherwise of simple eddy-viscosity/mixing-length turbulence models. Effects due only to the changes in the ratio of the two imposed length scales, $\delta u_\tau/v$, we take to be those of an *un*distorted boundary layer, and these are usually considered as 'Reynolds number effects'. However, as is well known, Reynolds number effects imply the use of a variable length scale in the formulation of eddy viscosity, as in the viscous sublayer. The case of a rough surface offers further imposed length scales such as roughness height, roughness element spacing or other measures of the roughness topology. Taking the effect of roughness to be parameterised *only* using a roughness height, k, a roughness Reynolds number, ku_τ/v, now represents the ratio of two imposed length scales and, using the definition above, is a distorted boundary layer. However, when $ku_\tau/\nu \geq 100$, the boundary layer is 'fully rough', that is, the effects of roughness saturate (the skin-friction coefficient becomes independent of Reynolds number) and the boundary layer becomes undistorted once again. At 'high' Reynolds numbers (the question of how high is 'high' is dealt with more fully below), the viscous length scale will become small and roughness becomes important. These issues alone constitute a vast topic and are currently a source of much debate, owing to the availability of new data at very high Reynolds numbers obtained under laboratory conditions, such as the 'superpipe' and HRTF facilities at Princeton University. See Zagarola and Smits (1998) and McKeon *et al.* (2004). As will be seen, these effects are important, and so need to be dealt with here, not least as a final caveat to those who regard the problem of turbulence as solvable by means of direct numerical simulations (DNS), inevitably at low Reynolds numbers. We will, however, refer to some DNS results that are used for a term-by-term assessment of the modelled equations.

At low Reynolds numbers, the prediction of transition is a particularly difficult issue and simple correlation techniques of long standing are still hard to beat in terms of their reliability. In many engineering flows, the background disturbances can be very large so that transition occurs quickly. Usually, they are ill posed too, certainly in engineering flows, but even in laboratory experiments too. In an experiment with very well-controlled boundary conditions, Gaster (2003) shows that periodic roughness acts as a set of 'receptivity' sites: with a roughness height as little as 63 μm in a boundary layer about 5 mm thick, amplification rates are changed significantly. Clearly then, any prediction method for, say, turbine blades or wings should consider the effects of erosion or deposition over time. For these reasons, Reynolds-averaged (RANS) models are unlikely to predict the growth of small disturbances accurately. In the case of large disturbances (e.g. 'bypass' transition), the effects are often treated by weighted functions of the diffusion term in a suitable transport equation. But it is unlikely that this is an adequate model of the physical processes (disturbances propagate and amplify at receptivity sites rather than diffusing) and so it is not clear that such a method will offer a prediction superior to that of a correlation.

Although boundary conditions can be modified abruptly with fetch, such as in a step change in roughness (or, more or less instantaneously in time with, say, a fast-response actuator), the turbulence is usually very slow to adapt. In Reynolds-averaged terminology, this means that the turbulence does not change immediately the mean strain rate changes. This is a principal cause for concern because it is not immediately obvious what might be the most appropriate time scale to represent these changes. In order to put sudden distortions into context, we define 'local equilibrium' to occur when the production, P, of turbulence kinetic energy, $k = \frac{1}{2}\overline{u_i^2}$, and its mean dissipation rate, ε, are equal. Therefore transport effects are negligible and energy is dissipated as soon as it is produced; this provides a good approximation to the energy budget in the log law region. More generally, one may define 'equilibrium' to be $P/\varepsilon = f(y/\delta)$ only. More often than not, turbulence models use a single time scale, k/ε. Simple models prescribing streamwise scale changes using ordinary differential equations are discussed below, together with, for even more complicated situations, 'split-spectrum' models, which permit the use of multiple time scales. For predictions of highly non-linear phenomena, we need to retain non-linearity of the governing equations for as long as possible. However, the results of (linear) rapid distortion theory (RDT) have been useful in providing insights into particular problems (see Savill 1987, Hunt and Carruthers 1990, Hunt and Morrison 2000). Savill discusses the rôle of RDT in helping to model the transport equations, with particular reference to the pressure-strain terms – it can give qualitatively useful results even outside the strict limits of its applicability. For the somewhat idealistic case of grid turbulence impinging on a moving wall, Magnaudet (2003) has shown that the RDT theory of Hunt and Graham (1978) offers a leading order approximation capable of describing both short- and long-time evolutions of shear-free boundary layers in the limit of high Reynolds number. Even so, it is useful to remember that the RDT approximation requires that the ratio $P/\varepsilon \sim (Re)^{\frac{1}{2}} \gg 1$, where, typically, the eddy Reynolds number $Re \approx 10^4$, at least. Thus we take '$\gg$' to mean $O(100)$. With the exception of shock-wave/boundary-layer interaction, this is a condition seldom attained in external engineering flows, but RDT is relevant to parts of the cycle in IC engines or in specific areas of gas/steam turbines in which the working fluid passes from a fixed blade row to a moving one (and vice versa).

Specifically, we examine both compressible and incompressible flows subject to large distortion(s) such as:

- 'extra' rates of mean strain,
- changes in surface condition – roughness,
- changes in streamwise pressure gradient,
- changes in flow species – separation and reattachment,
- free-stream turbulence,
- shock/boundary-layer interaction,
- boundary layers with embedded (usually streamwise) vorticity – this includes three-dimensional boundary layers in which the embedded vorticity is distributed.

These phenomena are all readily apparent in the global environment and are ubiquitous in both internal and external flows directly related to engineering practice. Usually they appear either singly or often together in complex interactions such as:

- shock/boundary-layer interaction,
- junction (secondary) flows (e.g. wing/body),

- vortical flows (e.g. tip-leakage flows, wing tips and delta wings),
- highly skewed three-dimensional boundary layers (e.g. wings),
- trailing-edge flows,
- transition,
- unsteady flows,
- atmospheric surface layer over an urban landscape, and crop or forest canopies.

The first six items above constitute the six 'application challenges' outlined by Atkins *et al.* (1999) for aerospace applications. Many of the above (vortical flows, shock-wave/boundary-layer interaction, unsteady flows, transition) are also key flow interactions in turbomachinery identified by Coupland (1999). Unsteady flows are taken to include buffeting, a specific fluid/structure interaction for which a linear treatment suffices and for which ad hoc treatments can be used with some reliability. This is not the case for flutter, which is non-linear and the problems of it are too complicated to be included here. The atmospheric surface layer offers particular difficulties over a change in surface roughness. Other effects such as stratification are dealt with elsewhere in this volume by Launder; intense sub-kilometre-scale roll vortices apparent in the surface layer (Wurman and Winslow 1998) have many of the characteristics we associate with embedded vortices in engineering flows.

The objectives of this chapter are an attempt at providing a framework for the modelling of the complex interactions above. This will, hopefully, provide a means of 'bridging the disconnect' between the intellectual rigour of the academic and the pragmatism required by workers in industry. Hence the objectives are to:

- summarise the current state of the art of the topics outlined above, providing, where appropriate, sources for more authoritative views;
- provide recommendations concerning the most appropriate turbulence models for the application challenges;
- identify gaps in the current state of the art and thus areas in which future research should be directed.

It is often said that modelling can be taken as a 'pacing item' for progress in turbulence. This has certainly been true as far as the end-user is concerned, but every so often, technology offers us a 'paradigm shift'. In this context, large-eddy simulation (LES) is offering just this, although it is probably too early to ascertain whether the community is merely jettisoning one set of problems for another. Similarly, the modelling requirements for LES are also likely to be a pacing item. However, in LES, there is the possibility of a greater degree of universality in modelling since, physically, the small scales exhibit a degree of universality not found in the large, energy-containing scales. However, this is most definitely not the case in near-wall turbulence where all the eddies are 'small', and not much bigger than the Kolmogorov length scale, η, leading to complicated spectral energy transfers. The universality of the small scales in general is currently the source of much debate, a review of which is beyond the scope of this chapter. In any case, the question of whether or not turbulence exhibits 'universal equilibrium', even when not distorted in the large scales, is largely irrelevant to successful LES modelling: a sufficient condition for a subgrid model to be considered sound is that it estimates accurately the low-order statistics of the energy transfer (Meneveau 1994). No review of the problems concerning the prediction of strongly distorted flows would be complete without an assessment of the successes (so far) of LES in this area and an assessment of its potential.

Aerodynamic design extends from a piecemeal approach, with potential flow calculations providing external boundary conditions, to the boundary layer calculation with displacement effects incorporated by an iterative viscous/inviscid interaction procedure, to RANS codes with appropriate mesh refinement in the region of surfaces or complex interactions. The former approach may not necessarily include the effects of free-stream turbulence and simple boundary-layer codes do not cater for surface-normal pressure gradients so clearly evident in shock/boundary-layer interactions. Similarly, the latter approach may lead to real problems with say, roughness effects in HP turbines, which are so crucial to performance losses, and which can be overlooked simply because of the difficulty in prescribing them with micron-size surface changes being beyond the resolution capabilities of most computational meshes. Only some of these difficulties are likely to be obviated by the use of LES for which there is a great need for 'off-the-surface' boundary conditions to make LES at engineering/meteorological Reynolds numbers viable. These problems are not new, the earliest consideration of the problems being by Robinson (1982). Indeed, who would have thought that much of the conditional-sampling and coherent-structure work of the 1970s and 1980s might now find application to what is one of the most important and difficult problems in LES?

Where temperature differences or contaminant concentrations are small, the fluid may be treated as one of constant property, with the major simplification for calculation methods that the energy equation can be solved consecutively, rather than simultaneously, with the momentum equations. For even smaller temperature differences, the velocity field drives the scalar field without the latter exerting any undue influence on the former. In this case, the scalar field is said to be 'passive'. Eddy-viscosity methods may then be extended to the prediction of heat transfer by using an empirical distribution (with some help from theory) for the turbulent Prandtl number, Pr_t (the ratio of turbulent diffusivity of momentum to that of heat) in the energy equation. The usual analysis for the law of the wall (see Section 2.1), together with empirical values for the von Kármán constant, κ, and its thermal analogue in the temperature log law, κ_θ, show that $Pr_t = \kappa/\kappa_\theta$, and that it is close to unity implying that, at least for undistorted flows where the molecular Prandtl number, $Pr \approx 1$, estimates for heat transfer can be obtained from Reynolds' analogy (the Stanton number, $St = C_f/2$, where C_f is the skin-friction coefficient). However, even in undistorted boundary layers in which the local-equilibrium approximation (and therefore the log law) is expected to be valid, $Pr_t = \kappa/\kappa_\theta$ decreases in the outer layer and increases in the viscous sublayer so that Reynolds' analogy will only ever be approximate. Still-relevant summaries of this difficult area are given by Launder (1978) and Kays and Crawford (1993). Buoyancy and stratification effects are dealt with elsewhere in this volume by Launder.

The detailed effects of strong distortion on scalar transport appear to be largely undocumented in the engineering literature, although the literature concerning the dispersion of passive contaminants in the environment is burgeoning. In practice, Reynolds-analogy models for heat transport do not work well for distorted boundary layers. The reasons are fundamental but clear: heat is, in fact, a poor marker of the instantaneous vorticity field. Following Batchelor and Townsend (1956), Morrison and Bradshaw (1991) suggest that high-wavenumber pressure fluctuations, and not viscous forces, drive the small-scale motion. Moreover, Guezennec *et al.* (1990) confirm that, except very near the wall, pressure-gradient forces dominate viscous forces, but play *no* rôle in the exchange of heat. Physically, this is because velocity and pressure fluctuations extend over much shorter distances than temperature fluctuations, the latter stages of mixing of the scalar field occurring by molecular diffusion even though large eddies are brought together by mean flow advection. Thus momentum exchange is more efficient and fluid is likely to retain its heat marking for significantly longer than its momentum. This shows that even the more successful Lagrangian dispersion models that rely on a complete neglect of molecular diffusion are at variance with a key physical process. Therefore it is hoped that DNS will be used more extensively than hitherto to highlight ways in which simple Reynolds-analogy-type models are likely to be of limited value and, more positively, suggest improvements to current practice.

Many industrial applications require predictions in compressible flow. It is usually assumed that for Mach numbers up to about five, Morkovin's (1962) hypothesis applies and that the time scales for the turbulence follow that of the mean flow. Thus, in consideration of such topics as shock-wave/boundary-layer interaction, it is assumed that Morkovin's hypothesis applies, and that density change effects need not be considered explicitly. Similarly, the transformed mean velocity exhibits a log law with approximately the same constants as those used in incompressible flow. See Bradshaw (1974) and Smits and Dussauge (1996) for further details. Huang *et al.* (1992) provide details of changes to model constants caused by the effects of compressibility.

The next section is devoted to a description of the physical effects we wish to model within the framework of multiple imposed length or velocity scales. Here it should be noted that only length scales are independent of the frame of reference, while mean velocity scales are not. Practically, this means we need to take particular care in defining velocity scales for the turbulence as distinct from those for the mean flow. We begin with undistorted flows (i.e. boundary layers with two imposed length scales only) that include Reynolds number effects and surface roughness. We then look at distorted boundary layers in terms of extra rates of strain and interacting shear layers. Section 4 offers general views on the major limitations of current RANS and LES modelling. Implicit in the present approach is the now-accepted wisdom (and a major outcome of the INI Programme) that no turbulence model is likely to be reliable in flow regions subject to different imposed length scales. This leads to the idea, first suggested by Kline (1981), of 'zonal' modelling. Section 5 attempts to address each of the above application challenges by providing a 'route-map' for the choice of a suitable turbulence model. No such recommendation would be complete without a clearly defined operation zone, outside of which a given model is unlikely to be reliable. We conclude with a brief summary of the areas in which there is a demonstrable need for future work.

2 Undistorted boundary layers

These are the simplest types of boundary layer that we consider and they help us to clarify what is meant by a distorted boundary layer. We should be reasonably confident that these types of boundary layers could be accurately predicted by algebraic models.

2.1 Reynolds number effects

Reynolds number similarity is an essential concept in describing the fundamental properties of wall turbulence. Unlike the drag coefficient for bluff bodies, that for a turbulent boundary layer continues to decrease with increasing Reynolds number because the small-scale motion near the surface is directly affected by viscosity at any Reynolds number. Table 5.1 indicates the scale of the problem. Turbulence models for design calculations, but validated in detail using laboratory data obtained in facilities at atmospheric pressure and temperature, are likely to involve extrapolation of quantities like drag coefficients over changes in Reynolds number of four orders of magnitude. Worse still, use of the log law,

$$\frac{U}{u_\tau} = \frac{1}{\kappa} \ln\left(\frac{y u_\tau}{\nu}\right) + C \qquad (5.1)$$

Table 5.1 *Typical Reynolds numbers*

	Re
Sun	10^{13}
Ocean/lower atmosphere	10^9
Boeing 747/submarine	10^8
Superpipe/HRTF	10^7
Natural gas/water main	10^7
Laboratory experiments at STP	10^4
DNS	10^3

as the surface boundary condition places much faith in the value of the von Kármán constant: a change in κ from 0.41 (an accepted value) to 0.436 (as recommended by Zagarola and Smits 1998) corresponds to a change in skin-friction coefficient of 2% and a change in drag coefficient of 1% at $Re \approx 10^7$. It should be remembered that classical derivations of the log law (Millikan 1938) use the method of matched asymptotics: that is the log law appears in the limit of infinite Reynolds number using both inner and outer scaling, and that at the lower Reynolds numbers met in real situations, higher-order Reynolds-number dependent terms could be required. In the limit $y^+ = yu_\tau/\nu \gg 1$ and $y/\delta \ll 1$, y becomes an important length scale and its neglect in calculations therefore requires a very good reason. The superpipe data (Zagarola and Smits 1998, McKeon *et al.* 2004) suggest that, for a log law to appear (that these higher-order terms are negligibly small), the von Kármán number, $R^+ = Ru_\tau/\nu \geq 5000$, and that once established, a power law appears for $600 < y^+ < 0.12R^+$. κ is a universal constant, so that one may realistically suppose that its value for engineering estimates, at least, is valid for both internal and external flows. Yet there is no universal agreement. Using independent measurements of the surface friction in a boundary layer, Österlund *et al.* (2000) suggest a value of κ as low as 0.38. However, the highest Reynolds number of this dataset is only $Re_\theta = 27$ 000, equivalent roughly to $R^+ \approx 1500$, somewhat lower than that at which true self-similarity (the log law) appears in the superpipe data. One of the attractive features of fully developed pipe flow is that it offers a very well conditioned flow for study: κ is obtained from friction factor data and then cross-checked with mean velocity profiles on *both* inner and outer scaling. With additional data and a revised analysis, McKeon *et al.* (2004) suggest $\kappa = 0.421$. The significant difficulty in estimating the value of κ using boundary layer data is the reliable estimation of the surface skin friction. Whatever the outcome of this debate, it is clear that more accurate experimental techniques have shown that boundary layers are far more sensitive to the surface condition than was previously thought. As a rule of thumb, it is likely that the wake region of a boundary layer becomes independent of viscosity at

$Re_\theta \approx 10\ 000$ (i.e. at the Reynolds number at which the log law appears), and not at $Re_\theta \approx 5000$, as previously thought and as deduced from the constancy of Coles' wake parameter.

At any point, the turbulence consists of a range of scales so that simple similarity of the higher statistics may only exist under a restrictive range of conditions. The situation is complicated by the fact that wall turbulence is highly anisotropic, owing to the different effects of the viscous constraint (the 'no-slip' condition) and the impermeability constraint. The latter leads to a reduction in the wall-normal v-component at a streamwise wavenumber, k_1, that is roughly inversely proportional to the distance from the wall, y. The effect of this 'blocking' or 'splatting' is to increase the wall-parallel components down to the viscous sublayer. This leads to the expectation that the spectrum for the v-component at low wavenumbers depends only on y, while spectra for the wall-parallel components depend on both y and an outer length scale such as the boundary layer thickness, δ.

A particularly useful concept that explains much of the foregoing is Townsend's theory concerning 'active' and 'inactive' motion (Townsend 1961). The active motion comprises the shear-stress, $-\rho\overline{uv}$-bearing motion which therefore scales on y and u_τ only. On the other hand, the inactive motion is of large scale, does not bear any significant shear stress, to first order, resides only in the u- and w-components and is the result both of large-scale vorticity and of irrotational pressure fluctuations generated in the outer layer (Bradshaw 1967b). Townsend (1956, 1976) described the inactive motion as a 'meandering or swirling' motion that contributes 'to the Reynolds (shear) stress much further from the wall than the point of observation', but not, as noted by Bradshaw (1967b), 'at the point of observation'. The purpose of these details is that they lead to the question of whether or not the Reynolds stresses are likely to exhibit simple similarity, as the mean velocity does. Unfortunately, the answer is 'no': Morrison *et al.* (2002, 2004) show, using data from the Princeton superpipe, that the horizontal stress, $\overline{u^2}$, exhibits no simple scaling with u_τ. From the above, it is also clear that the other horizontal stress, $\overline{w^2}$, also with contributions from inactive motion, will not either. On the other hand, the wall-normal stress, $\overline{v^2}$, is more likely to do so, but given that the concept of inactive motion is a linear, first-order approximation, it would be unwise to assume that $\overline{v^2}$ will show simple similarity. Moreover, laboratory data for $Re > 10^7$ are still lacking, although cross hot-wire data from the Princeton superpipe are beginning to appear.

On balance, there is a recipe to be offered, even if, while most of the ingredients are known, their relative proportions are not. Marušić *et al.* (1997) have suggested:

$$\frac{\overline{u^2}}{u_\tau^2} = B_1 - A_1 \ln\left[\frac{y}{\delta}\right] - V_g[y^+] - W_g\left[\frac{y}{\delta}\right], \tag{5.2}$$

where A_1 and B_1 are constants, V_g and W_g are functions that represent, respectively, a viscous deviation term operative at small y^+, and a wake deviation term effective at large y/R. It should be made clear that, while the logarithmic function has some theoretical support, in the form of a supposed self-similar k_1^{-1} range in the u-component spectrum, Morrison *et al.* (2002, 2004) suggest that the similarity is incomplete. Therefore, the form of (5.2) has rather less theoretical support than the log law (5.1). This also makes clear that A_1 and B_1, as well as any constants in the functions V_g and W_g, will be empirical. Morrison *et al.* (2003) show further that (5.2) requires an additional function that quantifies the inactive contribution near the wall. The equivalent equation for $\overline{w^2}^+$ will be the same as (5.2), with different constants, and that for $\overline{v^2}^+$ will be of the form

$$\frac{\overline{v^2}}{u_\tau^2} = B_2 - V_g[y^+] - W_g\left[\frac{y}{\delta}\right], \tag{5.3}$$

where the absence of the log expression indicates that this component is the active component – a manifestation of the 'blocking' effect. It is possible that the form of (5.3) would not need to be modified to include a term that represents the effect of inactive motion near the wall. The equivalent expression for $\overline{uv}^+$ would be the same as (5.3), with different constants, of course.

The above is predicated on the assumption that a single appropriate velocity scale, u_τ, is sufficient. As the velocity-defect and log laws show, u_τ is probably the best choice, although there are possible alternatives (Zagarola and Smits 1998, George and Castillo 1997, DeGraaff and Eaton 2000, Morrison *et al.* 2004). Whatever the choice, it is unlikely to be the mean velocity, which is frame dependent, while the turbulence statistics are not. Put statistically, this merely means that the first moment of a probability distribution function says nothing about the shape of the distribution, and therefore the higher moments. Fernholz and Finley (1996) provide the most complete review of the existing data to date. In summary, it seems best to use (5.2) and (5.3) as a functional form for correlations using data at higher Reynolds number when they become available. Correlations should also be used for the skin friction, rather than placing too much confidence in the constants of the log law (5.1). At least this approach has the virtue of consistency.

2.2 Roughness effects

For an estimate of skin-friction drag at high Reynolds numbers, roughness will become important. An equivalent expression for the log law written for a rough surface is

$$\frac{U}{u_\tau} = \frac{1}{\kappa} \ln\left(\frac{y}{k}\right) + B. \tag{5.4}$$

The dimensional analysis leading to the log laws (5.1) and (5.4) proceeds via the integration of

$$\frac{\partial U}{\partial y} = \frac{u_\tau}{\kappa y}. \tag{5.5}$$

Therefore, the length scale used to normalise y in the log argument is merely a constant of integration and may be freely chosen, as self-similarity implies. It is usually taken to be the dominant imposed length scale so that its influence on the additive constant is removed, that is, $k^+ = k u_\tau / \nu > 100$. If, however, $k^+ \le 100$, that is $k \approx \nu/u_\tau$, then three length scales are imposed, Millikan's (1938) overlap analysis becomes unworkable, and complete (self-)similarity is therefore not possible, that is, a universal log law is not possible. This rather simple analysis that shows that a log law cannot be universal in transitionally rough boundary layers appears not to be always appreciated. Rather, given the appearance of a log law on a smooth surface at sufficiently high Reynolds number as well as on a rough surface at sufficiently high roughness Reynolds number, many workers accept a log formulation for a transitionally rough surface in the form:

$$\frac{U}{u_\tau} = \frac{1}{\kappa} \ln\left(\frac{y}{k}\right) + B(k^+), \tag{5.6}$$

where the additive constant is, in fact, an additional roughness function. On the grounds of pragmatism *only* and in the absence of well-developed laws for transitionally rough surfaces, it would seem sensible to retain (5.6) as the basis of a correlation, provided the difficulties over the choice of surface origin can be overcome. Given the numerical difficulties of calculating boundary layers on rough surfaces, estimates of skin friction from predictions using codes that attempt resolution of the near-wall region or use the log law as a wall function should be strongly resisted. There are even stronger reasons for advocating correlation techniques: it is clear that many other roughness parameters are required to parameterise roughness effects and that large roughness elements exert a disproportionately large effect (Bradshaw 2000).

3 Types of distortion

We define a distorted boundary layer by its opposite, that is, by one that is *un*distorted and therefore in equilibrium in the sense that it exhibits approximate self-preservation. This means that a single velocity scale and a single length scale are sufficient to represent the mean velocity and the higher moments adequately, that is, they will be accurately predicted by an eddy-viscosity model. The log law is a special form of self-preservation. One definition of self-preservation that derives from the Falkner–Skan family of exact solutions for laminar boundary layers is one in which the free-stream velocity, $U_é \propto x^a$, where x is the streamwise fetch and a is a constant. Clauser's (1956) definition has a more physical appeal, in which the ratio of pressure-gradient forces to wall shear forces,

$$\beta = \frac{\delta^* \, \mathrm{d}p}{\tau_\mathrm{w} \, \mathrm{d}x},$$ (5.7)

is constant. Thus a distorted boundary layer may also be taken to be one in which β is not constant. Clauser also put forward his 'black-box' analogy in which a self-preserving boundary layer is suddenly perturbed and its behaviour is observed as it asymptotes to a new self-preserving state. This idea has a similar degree of physical appeal, and is one to which we return. However, it should be remembered that truly self-preserving boundary layers are very rare, requiring $\delta u_\tau / v = \text{constant}$, a condition attainable only with very specialised (or very simple) boundary conditions. For boundary layers, this condition is prohibitive, while for internal flows such a condition results in very simple flows, such as fully-developed pipe flow. Thus a more pragmatic definition of a distorted boundary layer is simply that of a complex flow (Bradshaw 1971, 1973, 1975, 1976).

Prandtl's boundary layer approximation assumes (using 'shear-layer axes': x, streamwise; y, surface-normal):

(1) $\mathrm{d}\delta/\mathrm{d}x$ is small, i.e. $\delta/x \ll 1$;
(2) gradients of normal stresses and $\partial p/\partial y$ are negligible. Then the only significant Reynolds stress gradient is $\dfrac{\partial(-\overline{\rho u v})}{\partial y}$, even though $\overline{u^2} \approx \overline{v^2} \approx \overline{w^2} \approx -\overline{uv}$. The x-component momentum equation is then simple, with resulting simplifications in the form of calculation method.

This can be generalised to other types of shear layer (e.g. jets, wakes, mixing layers) to give the thin-shear-layer approximation. In doing so, one has to be careful in making order-of-magnitude estimates of $\mathrm{d}\delta/\mathrm{d}x$ and the normal-stress gradients, both of which are larger in free shear layers than in boundary layers. With thin-shear-layer approximations and using shear-layer axes, the momentum equation is

$$U\frac{\partial U}{\partial x} + V\frac{\partial U}{\partial y} = -\frac{1}{\rho}\frac{\mathrm{d}p}{\mathrm{d}x} + \frac{\partial}{\partial y}\left(v\frac{\partial U}{\partial y} - \overline{uv}\right),$$ (5.8)

Table 5.2 *Effect of extra rates of strain (reproduced from Bradshaw 1975)*

		Effect of e on turbulence structure
Simple shear layer	$\dfrac{10e}{\partial U/\partial y} \ll 1$	negligible
Thin shear layer	$\dfrac{e}{\partial U/\partial y} \ll 1$	possibly significant[†]
Fairly thin shear layer	$\dfrac{10e}{\partial U/\partial y} < 1$	probably large

[†] as long as order $\left(\dfrac{e}{\partial U/\partial y}\right) <$ order $(d\delta/dx)$.

and the equation for the turbulence kinetic energy, $2k = \overline{q^2} = \overline{u^2} + \overline{v^2} + \overline{w^2}$, becomes

$$U\frac{\partial k}{\partial x} + V\frac{\partial k}{\partial y} = -\overline{uv}\frac{\partial U}{\partial y} - \frac{\partial}{\partial y}\left(\overline{pv} + \frac{1}{2}\overline{q^2 v}\right) - \varepsilon. \tag{5.9}$$

These equations will require additional terms for types of shear layer with growth rates larger than $\delta/x \ll 1$.

3.1 Extra rates of strain

Complex shear layers involve extra rates of mean strain, e, which appear explicitly as production terms in any transport equation. Thus the effect of e may be summarised as in Table 5.2, where '$\ll$' means a factor of 100, and '$<$' means one of 10. The effect of extra rates of strain also appears explicitly in some of the other terms in transport equations (notably the turbulent transport terms) but *not in the destruction terms*. The most important effect of extra strain rates is that their effect on the turbulence is very roughly ten times that which would be expected from the size of these extra terms alone. Therefore, in addition to the appearance of yet further terms such as surface-normal pressure gradients, prediction methods require empirical corrections for the effects of these extra strain rates. Figure 5.1 (Bradshaw 1975) shows examples of extra rates of strain such as streamline curvature, streamline divergence, bulk dilatation, and system rotation.

The concept of extra strain rates sits well within the current framework of defining distorted boundary layers by the additional imposed length and/or velocity scales. A time scale for the turbulence is k/ε; using the local-equilibrium approximation and taking $k \approx -\overline{uv}$, $k/\varepsilon \approx (\partial U/\partial y)^{-1}$ (Bradshaw 1996). Thus the imposition of

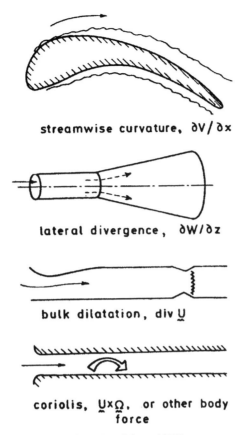

streamwise curvature, ∂V/∂x

lateral divergence, ∂W/∂z

bulk dilatation, div U

coriolis, U×Ω, or other body force

Fig. 5.1 Extra rates of strain (from Bradshaw 1975).

additional length and/or velocity scales is qualitatively equivalent to an additional rate of strain, e. Empirical corrections to, say, the mixing length, l, can be applied by

$$l = l_0 \left(1 + \sigma \frac{e}{\partial U / \partial y} \right), \tag{5.10}$$

where the empirical constant, σ, is of order ten, but of either sign. σ will depend on the extra strain rate considered. Near walls, when $\partial U / \partial y$ is large, the *local* effects of e are likely to be small, becoming larger further from the wall. An estimate of the combined *non-local* effects of e together with the *local* effects that depend directly on $\partial U / \partial y$ near the wall may be estimated by writing the dissipation length scale, L_ε:

$$L_\varepsilon^{-1} = \frac{\varepsilon}{u_\tau^3} = A_B \frac{1}{y} + A_S \frac{\partial U / \partial y}{u_\tau}, \tag{5.11}$$

where $A_B \approx 0.3$ and $A_S \approx 0.7$ are estimated empirically (Hunt *et al.* 1989). A physical justification for the first term on the right-hand side of (5.11 – in which the direct proportionality, $L_\varepsilon \propto y$, is the result of the assumption of a self-similar structure) is that blocking attenuates $\overline{v^2}$ at wavenumbers inversely proportional to the distance from the wall. This process is non-local in the sense that, near the wall (where $\partial U/\partial y$ is large), $\overline{v^2} \propto y^4$. Therefore the production of $-\overline{uv}$, $\overline{v^2}\partial U/\partial y \propto y^4$ also, and, in turn, this leads to a reduction in the production of $\overline{u^2}$, $-\overline{uv}\partial U/\partial y \propto y^3$. The second term on the right-hand side represents a local-equilibrium approximation of L_ε. These local effects are opposite to the increase of $\overline{u^2}$ by blocking or splatting, as in inactive motion. The relative importance of these two influences is determined by the values of the constants (Lee and Hunt 1989, Hunt 1984). Hunt *et al.* (1989) suggest that they are approximately the same for a wide range of boundary layers. A recent demonstration of the non-local blocking effects at low wavenumbers and the local shear effects at high wavenumbers is provided by Coates Ulrichsen (2003). Writing the additional energy production due to e as the (small) difference between the much larger production due to $\partial U/\partial y$, and $\varepsilon = u_\tau^3/L_\varepsilon$, the effect that e has on ε will be significant even if $\dfrac{e}{\partial U/\partial y}$ is small. Equation (5.5) suggests that this will lead to a significant change in $\partial U/\partial y$ also.

Additionally, distortions arise when two shear layers meet (see Fig. 5.2, taken from Bradshaw 1975), such as the interactions of corner flows in a duct (Prandtl's secondary flow of the 'second kind'), or when two boundary layers meet immediately downstream of an aerofoil or turbine blade. These interactions are inevitably non-linear, and immediate questions arise concerning the suitability of linear super-position techniques. Thus, would one simply take the average of the outer length scale in each boundary layer for an approximation of their combined equivalent? Such interactions are particularly relevant to turbomachinery (Bradshaw 1977, 1990, 1996, Purtell 1992, Lakshminarayana 1986). Specific interactions are dealt with below.

3.2 Changes in surface roughness

Smits and Wood (1985) provide a comprehensive review of the behaviour of distorted boundary layers under the influence of one or more sudden perturbations. When these are introduced at the wall (such as through changes in surface roughness) where the time scales are significantly shorter than the time scale of the energy-containing turbulence, then the boundary layer takes a very long time to adjust, the initial response being the development of an internal layer with scales determined by the local wall friction velocity. Such changes are often manifest by a jump in turbulent transport across the layer (usually expressed as a 'stress bore'),

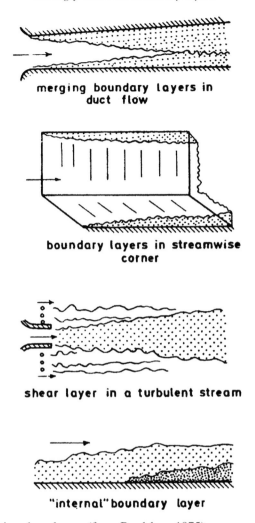

merging boundary layers in
duct flow

boundary layers in streamwise
corner

shear layer in a turbulent stream

"internal" boundary layer

Fig. 5.2 Interacting shear layers (from Bradshaw 1975).

leading to a failure of the local-equilibrium approximation and thus the modelling on which it is based (local equilibrium is also a special form of self-preservation). Intrinsic to the response is of course its non-linearity, and this becomes a significant problem in the case of multiple perturbations: the combined effect of two or more disturbances is seldom given by the summation of their separate effects.

Roughness changes are particularly important in meteorological applications, and largely for this reason, the effects are quite well documented. Nigim (1996) shows that the rate of recovery to equilibrium is generally very slow, but is much quicker in highly adverse pressure gradients: in such cases the gradient of Reynolds stress close to the wall is increased and it seems probable that this increases the rate at which the influence of the new wall condition is established. Lin *et al.* (1997) have used LES to examine the effects of roughness change in the surface layer: they suggest that the internal boundary layer following a rough-to-smooth or smooth-to-rough change in surface condition grows as the 4/5th power of the time. This result provides a useful calibration of numerical models.

3.3 Changes in streamwise pressure gradient

Smits and Wood (1985) also deal with changes in pressure gradient. The momentum equation (5.8) shows that, for a thin shear layer, the pressure gradient is experienced across the whole layer ($\partial p/\partial y = 0$). However, the transport equation for $\partial U/\partial y$ in a thin shear layer has a source term that consists of the gradients of the turbulent stress and the viscous stress, $\nu \partial U/\partial y$, only. In the outer layer, the latter is small and so $\partial U/\partial y$ cannot respond immediately the pressure gradient changes. Therefore the production term for k in (5.9) does not either, and so the turbulence in the outer layer responds only slowly. But close to the wall, $\nu \partial U/\partial y$ is large and balanced approximately by the pressure gradient. Thus $\partial U/\partial y$ is directly coupled to the pressure gradient and the turbulence responds more or less immediately.

More recently, Fernholz and Warnack (1998) and Warnack and Fernholz (1998) have completed an extensive study of boundary layers undergoing rapid streamwise strain followed by relaxation from it. Although complicated by the effects of both Reynolds number and axisymmetry, in all four cases of favourable pressure gradient, the log law breaks down and, moreover, the effects cannot be parameterised by a single variable. History (lag) effects are dominated by near-wall advection. Thus the comments concerning the failure of self-preservation (and thus the modelling assumptions based upon it) are just as relevant here also.

3.4 Changes in flow species – separation and reattachment

The best-documented example of such a change is flow over a backward-facing step (Driver 1991, Heenan and Morrison 1998a,b, Castro and Epik 1998, Spazzini *et al.* 2001, Furuichi and Kumada 2002). As such, it is one of a class of separated flows in which separation from a salient edge is followed by reattachment. We concentrate on the 'backstep' problem because it is a popular test case for RANS, LES and DNS alike. It represents a severe test for all turbulence models, as well as LES boundary conditions and models, comprising several zones with a distinctive feature in each. Thus the curved mixing layer at separation experiences streamline curvature and streamwise acceleration as well as carrying the history effects introduced by the change in species at separation – the turbulence hardly ever changes as soon as the mean strain rate does. As yet, there is no simple but accurate way of estimating the combined effects of multiple distortions that are inevitably non-linear (Smits and Wood 1985).

More subtly, the whole flow undergoes a global low-frequency oscillation, 'flapping' (Heenan and Morrison 1998a,b, Spazzini *et al.* 2001). Explanations of its cause vary, but a plausible two-dimensional explanation may be based on the imbalance between the entrainment rate of fluid into the bubble and the rate at which fluid is advected downstream, the balance being made up of a random shedding of vorticity. Fortunately, the time scale is so large (greater than ten times the time scale of the turbulence in the mixing layer) that conventional models can cope, at least as far as this unsteadiness is concerned. The flapping is, in fact, highly three-dimensional so that spanwise pressure gradients are set up. Thus there are significant mass fluxes across the span of the bubble during a flap. The use of a permeable surface by Heenan and Morrison (1998a,b) in the reattachment enables this spanwise (as well as streamwise) mass flux and inhibits the flapping motion.

Castro and Epik (1998) show just how slowly the reattaching layer relaxes to even a quasi-equilibrium condition. Moreover, the response is not monotonic, the turbulent stresses falling below those of the equilibrium condition before rising again. Physically, this is because of the splitting of the reattaching structures as opposed to a rapid distortion at impingement: large structures at reattachment are torn in two as opposed to their passing whole in opposite directions. What follows is a large transport of turbulent momentum towards the wall where the new internal boundary layer is growing, subject to a new surface condition.

3.5 Free-stream turbulence

Free-stream turbulence involves the imposition on the boundary layer of an additional single velocity, u_e, and a single length scale, L_e. While the wake component of the mean velocity profile is decreased (and can become negative), the skin-friction coefficient and the shape parameter are increased. Hancock and Bradshaw (1983) showed that the dependence of skin friction could be correlated with a single parameter

$$f = \frac{\left[\sqrt{\overline{u^2}}/U\right]_e}{L_e/\delta + 2}. \tag{5.12}$$

Roughly, a free-stream turbulence intensity of about 1% increases the surface skin friction by about 3%. Hancock and Bradshaw (1989) showed that the shear correlation coefficient is decreased owing to the large diffusion of energy into the boundary layer. Even so, the production and dissipation are still large by comparison, and therefore local-equilibrium results such as the log law remain valid. Thole and Bogard (1996) have investigated the use of very large free-stream turbulence levels ($10 \leq [\sqrt{\overline{u^2}}/U]_e \leq 20\%$) and show that large-scale turbulent eddies penetrate nearly to the wall. Unpublished work at Imperial College confirms that the turbulence intensities are increased, but only for scale sizes comparable to the boundary layer thickness and above. At smaller scales, the turbulence intensity is actually *decreased*. Generally therefore, it would appear that length scales for the boundary layer should be defined using the shear stress (e.g. $L_\tau = (-\overline{uv})^{\frac{3}{2}}/\varepsilon$) rather than the turbulence kinetic energy. The situation is analogous to the inactive motion of an undistorted layer, and free-stream turbulence may be viewed as the imposition of a large-scale disturbance on the near-wall motion, producing additional kinetic energy there by both large-scale vorticity and pressure fluctuations, which penetrate to the wall. When the interactions are very large, these ideas of linear superposition break down, of course.

Kalter and Fernholz (2001) have thoroughly documented the effect of free-stream turbulence on quite large bubbles in which the separating layer is already turbulent. They show that, with quite small free-stream intensities ($<6\%$) and with length scales of the same order as the boundary layer thickness, the bubble can be eliminated altogether. Like many other flows involving a change of species, the history effects are very long term. In a similar configuration, Na and Moin (1998) provide DNS data, albeit at low Reynolds number, which could be used for testing turbulence models. Other work concerning the effects of free-stream turbulence has tended to focus on quite large separations from salient edges (Cherry and Latour 1981, Saathoff and Melbourne 1997, Castro and Haque 1988), but the general effect is much the same: a shortening (but not the elimination) of the bubble.

Free-stream turbulence can have a particularly strong influence on boundary layers in turbomachines, and therefore on their heat transport properties. In this case, it is highly unlikely that Reynolds' analogy would hold even though this was the basis of much earlier work (McDonald and Fish 1973). In this context, free-stream turbulence is also very important in its effect on separated flows and on blade boundary layers, which may be of low Reynolds number and/or laminar and therefore susceptible to bypass transition. Castro (1984) showed that, at low Reynolds number, free-stream turbulence of length scale equal roughly to that of the boundary layer thickness causes an increase in skin friction and a reduction of the wake component that are less than those which would occur at higher Reynolds numbers. Blair (1983) found similar effects and Bandyopadhyay (1992) revised Blair's 'damping factor' to give

$$\beta' = [1 + 3\exp(-Re_\theta/425)]. \tag{5.13}$$

Moreover, there is a reversal in the Reynolds number effect when f (5.12) is small, so that he suggests

$$F = f\beta' \quad \text{for } f > 0.01,$$

and

$$F = f/\beta' \quad \text{for } f < 0.01. \tag{5.14}$$

The effect of free-stream turbulence is arguably never more important than in the context of bypass transition, simply because the increase in skin friction through transition can be as much as five to ten times that in a boundary layer that remains laminar at the same Reynolds number. Therefore heat transfer rates are similarly affected. Here, we are considering the process, in which the relatively slow, viscous transition by way of Tollmien–Schlichting (TS) waves is bypassed by an inertial effect with a much greater time scale, δ/U. Bypass mechanisms have been recently summarised by Jacobs and Durbin (2001). Using simulation data that offer a complete picture not easily matched in experiment, they show that streaks are, in fact, quite stable, only becoming receptive when lifted. This local breakdown then becomes recognisable as a spot that has its origin at the outer edge of the boundary layer, so earning the description of 'top-down' spots. These are not of the distinctive arrow-head shape (Emmons' spots), but rather with pointed regions facing not only downstream, but also upstream because they overtake the main turbulent motion. Only the low frequency modes with a convection velocity different to that of the free stream penetrate to the wall, a phenomenon known either as 'shear-sheltering' (Jacobs and Durbin 1998), or, more usually as 'receptivity', producing lower frequency modes that are amplified and elongated in the streamwise direction by the mean shear. This leads to the streaks (so-called Klebanoff modes) also seen in experiments (Matsubara and Alfredsson 2001). Intrinsic to the formation of spots are long 'backward' jets (i.e. they travel more slowly than the local mean velocity) that are the precursors to spot formation when they become locally unstable.

What constitutes the initial coupling between the free-stream turbulence and the appropriate parts of the boundary layer remains an open question. The physical description above that involves several different key processes is likely to undermine confidence in prediction methods based on simple modelling assumptions and make correlation a desirable route for progress, at least in the shorter term. Both Matsubara and Alfredsson (2001) and Reshotko (2001) identify transient growth processes as key. The point here is that it is non-modal, that is, it does not involve TS waves, and that, certainly in the initial stages, it is a linear process, even if subsequently non-linear effects must be included. Initial growth rates of the disturbance in the boundary layer often appear to be proportional to $x^{\frac{1}{2}}$, that is, with the same growth rate as that of the boundary layer itself. Transient growth also provides an explanation for roughness-induced transition (Reshotko 2001, Gaster 2003). Roach and Brierley (1990) provide one of the most established datasets (T3A and T3B), with data of apparently high quality. As such, they are often used for the validation of turbulence models (e.g. Savill 1993).

3.6 Shock/boundary-layer interaction

Smits and Dussauge (1996) provide a full and recent review of supersonic boundary layers undergoing a variety of perturbations. In the case of interaction with a shock, the boundary layer is subject to multiple perturbations (most of which are dealt with separately), and the resulting boundary layer is rapidly distorted by the jump in free-stream pressure. They also consider therefore the applicability of RDT. It is clear that large turbulent structures can distort the shock, leading to large oscillations in pressure gradient with local regions of separation. Moreover, shock curvature inevitably leads to the generation of vorticity and, as in the cases above, relaxation to equilibrium will be lengthy.

Given the complexity of the interactions, it is a slight comfort to find that the results of many different types of shock/boundary-layer interactions are similar. Even so, care is necessary in interpreting experimental data since the effects of a sudden pressure rise are complicated by those caused by, for example, a prolonged adverse pressure gradient, bulk compression/dilatation (Bradshaw 1974), streamline curvature and spanwise curvature. Thus experiments in which shocks are generated by a compression ramp (Adams 2000) will also exhibit the effects of concave streamline curvature. Similarly, the use of axisymmetric models to avoid three-dimensional effects will induce significant changes in flow direction that have large effects. In addition, such a model will exhibit the effects of spanwise curvature. Several turbulence amplification/generation mechanisms are present (Smits and Wood 1985). These include (a) 'direct' amplification as a result of the jump condition; (b) 'generation' of turbulence by incident acoustic and entropy fluctuations; (c) 'focusing' caused by distortions of the shock front; and (d) direct conversion of mean-flow energy into turbulence by shock oscillation. Of these, (a) and (d) are the most significant.

If the shock is strong enough, mere boundary layer thickening becomes a problem of local or even complete separation (Green 1970). Then the reflected waves and their interactions can be complicated and the interactions are not necessarily steady. Even with boundary layer thickening, it is not trivial to separate the dilatational effects of the extra rate of strain, $\nabla(\mathbf{U})$, from those of the concomitant streamline curvature, $\partial V / \partial x$, or even the effect of the sudden change in pressure gradient that appears directly in the equations of motion. Turbulence measurements in shock/boundary-layer interactions are relatively recent and were mostly performed by the Princeton group (Smits), or in France by Dussauge or by Délery and their colleagues. Immediately downstream of the shock, the Reynolds stresses increase sharply, and the amplification continues for some distance: this is partly because the pressure rise in the boundary layer is 'smeared' in the streamwise direction and partly because of the relatively slow non-linear response of the turbulence to sudden changes. However, details will vary from one experiment to another depending on how the shock is generated and the associated effects of, say, streamline curvature.

A complete review of test cases is provided by Fernholz *et al.* (1989), who also offer an assessment of the data. See also Fernholz and Finley (1981). In terms of a test case for turbulence models, this distortion represents something of a specialist interest, not only because of the complexity of the interaction, but also because of the special requirements of the numerical scheme (e.g. shock capturing, change in form of the equations below the sonic line).

3.7 Vortical flows

Bradshaw (1987) provides an extensive review of boundary layers with embedded streamwise vorticity, both discrete (secondary flows, junction flows) and distributed (three-dimensional boundary layers). Steady vortices embedded in turbulent boundary layers have been studied in detail, see Shabaka *et al.* (1985), Mehta and Bradshaw (1988), Cutler and Bradshaw (1993a,b). Given the significant anisotropy of vortex flows, a minimum requirement for a realistic prediction is that an eddy viscosity should have some directional sensitivity, or possibly better, that a full Reynolds-stress closure is used. However, Shabaka *et al.* (1985) and Mehta and Bradshaw (1988) show that eddy diffusivities for the Reynolds stress are very ill behaved, going to infinity where the stress gradients go to zero. The behaviour of the triple products is extremely complicated, so much so that bulk transport velocities for the transport terms (Section 4.1.2) are also ill behaved. Second-moment closure is similarly unsuccessful, especially if the vortices fill the boundary layer.

This area is likely to become increasingly important owing to the use of vortex generators for flow control. These can be stationary 'sub-boundary layer vortex generators', or, more likely, time-dependent 'on-demand' ('massless' or 'synthetic') jets with zero time-averaged mass flux, or local surface deformations that form periodic vortices, which lead to such effects as vorticity streaming. These are complicated processes, and control goals such as separation delay have to be modelled accurately. Information concerning time-dependent embedded vortices in turbulent boundary layers is virtually non-existent. Embedded vortices are also important in the context of laminar flow control on swept wings. The spatial evolution of both stationary and travelling cross-flow vortices arises from inflections in the velocity

profile during transition. Disturbances arise through receptivity to free-stream turbulence, noise and roughness (Joslin 1995).

Owing to the ubiquity of junction flows in both internal and external aerodynamics, the physical processes are quite well documented, although this does not
always mean that well-defined test data are available. Owing to the complexity of
the interactions, datasets are often partial, or the boundary conditions ill defined, or
not defined at all. Seal *et al.* (1997) and Khan *et al.* (1995) provide detailed analyses
of the vortex interactions for the purpose of improving understanding. Devenport
and Simpson (1992) provide comprehensive datasets against which established
turbulence models may be tested.

4 RANS and LES

4.1 RANS

Up to 1992, the most complete review of the capabilities and potential of RANS
models was given by Bradshaw (1992). For a more concise consideration, see
Bradshaw *et al.* (1996). Concentrating primarily on thin-shear layers, the principal conclusion was that second-moment closures are more accurate and adaptable,
especially in regard to complex flows, but that the greater cost involved was not
always taken into account by the participants in the collaboration. Significant weaknesses stem, not unexpectedly, from the weaknesses in the modelling assumptions
for the dissipation equation and from the treatment of the viscous wall region.
Spalart (2000) provides a more recent overview.

It is generally accepted that the 'constants' for turbulence models are, in fact,
not constant, and vary from one flow to another. For an exemplary use of DNS
data in helping to elucidate the behaviour of turbulence models, and particularly, a
term-by-term assessment of the 'constants' in $k \sim \varepsilon$ and standard Reynolds-stress
models, see Cazalbou and Bradshaw (1993). This therefore raises the issue of 'zonal
modelling' (Kline 1981, Ferziger *et al.* 1990), which we discuss below. We also
touch on some developments since 1992, with a view to suggesting ways in which
models might be improved and highlighting areas in which our understanding is
still incomplete. For a recent review of current models (not restricted to distorted
boundary layers) see Bardina *et al.* (1997).

4.1.1 Modelling of the dissipation equation

This is probably the most significant single intractable problem in RANS owing to
the lack of physical insight in the modelled form. Some progress has been made
by Jovanović *et al.* (1995) who suggest that the dissipation be decomposed into
homogeneous and inhomogeneous parts, as originally suggested by Chou (1945).
This suggestion does not appear to have been taken up. It is worth remembering

that one is really trying to measure the spectral drain from the energy-containing scales rather than the physical dissipation effected by the smallest scales. More empirical forms of a transport equation that offer the required length scale also exist, the most developed being the $k \sim \omega$ model (Wilcox 1998), where ω is the turbulence dissipation per unit turbulent kinetic energy. Its advantages over the $k \sim \varepsilon$ model are mixed and depend on the flow type. But it does have the advantage that its coefficients appear not to need Reynolds-number dependent adjustments near the wall.

4.1.2 Transport models

Gradient diffusion is not a realistic model for turbulent transport – turbulent eddies do not diffuse like molecules. Moreover, there are other more acceptable methods: Hong and Murthy (1986) use the concept of 'bulk convection'; Younis *et al.* (1999) use a rational procedure to account for the non-invariance of some diffusion models. Transport effects are usually large in distorted boundary layers (Section 5.4).

4.1.3 Zonal modelling

This technique, originally due to Kline (1981) (for a more easily available reference see Ferziger *et al.* 1990), derives from the recognition that no single model will cope adequately with a distorted boundary layer with established values for the 'constants'. A particularly difficult, well documented, example is that of turbulent flow over a backward-facing step, which can be characterised as a set of several interacting zones, all of which are distorted in one of the ways indicated above. The approach is to view a complex flow as a composite of several simpler ones in which the details of the model are tied to the local characteristics of the turbulence. Different zones have to be patched. Obviously, this is easier if adjacent zones have the same model but with different constants, i.e. a base model. The change from one zone to another should be as smooth as possible – a simple first-order, linear differential equation could be used to express the rate of change of the constants between zones. This example is especially difficult since some zones have multiple distortions: see above. The calculation of a complex flow by this method implies an iterative process in which the model constants are 'tuned' in a process of refinement by comparison with experimental data. The model can then be used in a truly predictive sense for *flows of the same type*.

The choice of base model is important: it should obviously be capable of providing reasonable accuracy at the desired level of detail. Therefore it would be unwise to use as a base model one that relies on the use of, say, an isotropic eddy viscosity for a flow in which anisotropy is a dominant feature (e.g. boundary layers with embedded vortices). The principal and, as yet, un-addressed difficulty of zonal modelling is the implementation of rational decisions for the type and extent of each zone that would make a prediction method using zonal modelling genuinely predictive, that is, it should function without user-defined input. Within the present framework of imposed length and/or velocity scales, a decision-making process can be invoked based on the length/velocity scales of the energy-containing eddies. This is dealt with in Section 5.5.

4.2 LES

A succinct review of the successes (up to 1990) and prospects of LES is given by Reynolds (1990). The area is developing rapidly – for more recent assessments, see Jiménez and Moser (2000) and Spalart (2000). The most intractable problem is the modelling of near-wall turbulence, in terms of both the modelling of the unresolved scales that are smaller than the filter or grid size ('subgrid' scales) and the implementation of an appropriate surface condition. This arises because near a surface, all the scales are 'small' so that the energy-containing scales may be smaller than the grid size. Voelkl *et al.* (2000), Nikitin *et al.* (2000) and Marušić *et al.* (2001) offer novel methods for near-wall modelling with a physical basis. Grid requirements make the use of approximate 'off-the-surface' boundary conditions very attractive (Piomelli and Balaras 2002). Subgrid-scale (SGS) modelling is particularly difficult near solid surfaces where the energy spectrum becomes complicated by the prevalence there of 'backscatter', the transfer of energy away from the small scales to the large ones (Piomelli *et al.* 1990). Meneveau (1994) has discussed which statistics a SGS model should reproduce in order to be considered sound. Such quantities include the low-order statistics of the energy transfer. Popular eddy-viscosity models present difficulties because they are absolutely dissipative: Mason and Thomson (1992), for example, show that only with the use of a stochastic backscatter model can the log law be obtained accurately even if it is used as a time-dependent surface boundary condition. The dynamic model of Germano *et al.* (1991) also presents problems concerning backscatter (Carati *et al.* 1995), although Porté-Agel *et al.* (2000) have generalised the model to improve predictions of a neutral atmospheric surface layer. Recent attempts by Redelsperger *et al.* (2001) to use SGS models based on the supposed self-similarity of surface-parallel velocity spectra as a function of streamwise wavenumber will probably lack generality: Morrison *et al.* (2002, 2004) have shown that spectra of the streamwise velocity component do not exhibit self-similarity even at very high Reynolds numbers. Further difficulties arise in the definition of a suitable SGS length scale, which is usually related simply to the grid size. The grid has to be highly anisotropic to capture the essential properties of the turbulence, but a model with a single length scale is unlikely to represent spectral transfer in wall turbulence accurately. Anisotropy of the turbulence near the wall extends to anisotropy in the SGS stresses. Further difficulties stem from the fact that, in order to obtain the required level of detail, some surface models are developed using linear stochastic estimation applied to simulation databases (Nicoud *et al.* 2001). However, Morrison *et al.* (2004) have recently demonstrated that the anisotropy of turbulence increases as the Reynolds number increases. Akhavan *et al.* (2000) have suggested that spectral energy transfers in a free jet comprise two effects: one involves forward scatter from non-local interactions that could be modelled by eddy viscosity; the other involves local interactions around the cut-off that generate bi-directional transfers. Dunn and Morrison (2003) show that this is also the case in wall turbulence. They also show that the pressure-gradient term effects negligible energy transfer, consistent with nearly all current LES models.

It is clear that some fresh ideas are required and it seems likely that modelling of the near-wall region is crucial. Robinson (1982) showed that the instantaneous velocity profile is likely to be far from the log law for the mean velocity, so much so that the log law is an unrealistic choice of surface boundary condition. Ideally, one would like the first mesh point to be in the log region (assuming that in a boundary layer far from equilibrium this is still a meaningful proposition). Marušić *et al.* (2001) have made measurements of the correlation between the streamwise velocity (at the first 'mesh' point) in a boundary layer and the local surface shear stress in order to improve the specification of a wall boundary condition. However, it seems that the problems of SGS modelling and implementation of a surface boundary condition (Piomelli *et al.* 1989) are inseparable. Yet models for near-wall turbulence in *equilibrium* flows are still in their infancy. These can be based on the extensive knowledge of near-wall structure and some ideas are beginning to emerge – Marušić *et al.* (2001). Dunn and Morrison (2003) have used a wavelet decomposition of the Navier–Stokes equations (see Meneveau 1991) to investigate the relationship between structure and energy flux. The basic question is whether near-wall 'ejections' and 'sweeps' may be used as the basis of a structure-based SGS model and surface boundary condition *simultaneously*. The work, in its early stages, is based on the assumption that ejections and sweeps in the local-equilibrium region constitute a 'first-order' inertial subrange, a sufficient criterion for which is that energy sources or sinks are a small fraction of the energy transfer. Formally, this does not require local isotropy in a wavenumber range in which the spectral shear correlation coefficient may be expected to be rapidly decreasing with increasing wavenumber (Saddoughi and Veeravalli 1994). Bradshaw (1967a) suggests a suitable criterion for a first-order subrange is that the Taylor microscale Reynolds number, $R_\lambda > 100$ only. This is a much less stringent condition than local isotropy, which is not justifiable even at high Reynolds numbers if the mean rate of strain is not small. Dunn and Morrison (2003) have shown that ejections and sweeps make up nearly all of the energy transfer between the resolved and subgrid motion, as well as being key determinants of the local wall shear stress.

It is likely that structure-based near-wall models will take some time to emerge. In the meantime hybrid RANS/LES techniques are likely to fulfil the rôle. In any case, the development of RANS models in practical situations may be necessary for the near-wall (under-resolved) region undergoing such distortions as roughness changes. This approach can be regarded as an extension of zonal modelling – see Peltier *et al.* (2000) and Speziale (1998).

5 The application challenges

Some general, technical comments that summarise the views expressed during the INI Programme are well worth making in order to emphasise their importance.

- There was a general consensus that there is no such thing as a 'universal' turbulence model. This has been generally accepted over the past twenty years or so and calls for pragmatic procedures such as 'zonal' modelling: see Ferziger *et al.* (1990). This pragmatic approach strongly influences the content of this review.
- Currently, we have a plethora of turbulence models, which have been largely developed for simple shear layers and there is a need to document existing models more fully so that, for a given flow, a 'best' model may be readily selected. Existing models should be made more robust, with validation including a specification of limits to accurate prediction.
- There is scope for development of turbulence models including the suggestions made in Section 4.1, especially in relation to modelling of the dissipation equation, wall functions for boundary layers far from equilibrium, and modelling of turbulent transport.
- Two advances in modelling require special mention, with the suggestion that these models be further developed. Durbin (1991, 1993) replaces local modelling of the pressure-strain term by an integral over a volume, so that near-wall inhomogeneities are represented more accurately. Similarly, the structure tensor of Kassinos *et al.* (2000) also represents a significant advance.
- Turbulence exhibits a universal state only at very high Reynolds numbers and then only at high wavenumbers. Thus Kolmogorov's hypotheses are essentially correct, although this is currently the source of some debate. The details of the Kolmogorov scales are irrelevant to RANS modelling, and probably to LES as well. However, the energy-containing scales (those that require modelling) *are* Reynolds-number dependent. This issue is addressed here only as a caveat in the interpretation of DNS data.
- DNS will not replace experiments, owing to the prohibitive rise in degrees of freedom with Reynolds number.
- There is a need for well-defined experiments involving parametric studies. Many existing experimental data are insufficient for assessing turbulence models, owing to the lack of attention paid to adequate specification of the boundary conditions or to the accuracy of the data themselves.
- The response times of the velocity spectra, $E(k)$, and the dissipation spectra $D(k) \sim k^2 E(k)$ are different, particularly for distorted or non-equilibrium flows. This calls for the further development of split-spectrum models (Schiestel 1987).

- There is a need for DNS at higher Reynolds numbers. It is conceivable that simulations at Reynolds numbers four times those of ten years ago will soon be possible (see Sandham's chapter in this volume).
- The use of LES in replacing DNS or experiments for fundamental studies and/or comparison with RANS solutions is of questionable value since numerical dissipation does not mimic physical dissipation correctly.
- The development of wall functions for calculations of distorted boundary layers is extremely important, both for use in RANS and LES. This might now be regarded as the 'pacing item' in determining our ability to predict this class of flows at high Reynolds numbers.

Here we address the application challenges by specifying a minimum level of closure required to achieve reliable results of integral parameters (e.g. the skin-friction coefficient, or displacement thickness). This is likely to mean that, in practice, we also require adequate modelling of the Reynolds stresses. Justification is provided for those cases in which full second-moment closure is recommended. *In principle*, closure at second-moment level will not require the ad hoc modifications required of eddy-viscosity based models. Such modifications for, say, effects of extra rates of strain do not offer significant improvements and, arguably, better modelling of the dissipation is more important. It is probably the case that two-equation modelling is the current standard in industry.

5.1 Vortical flows

This category deals with junction flows, embedded vortices existing either singly or in pairs, flow in curved and/or rectangular ducts, jets in a cross-flow and three-dimensional boundary layers with distributed streamwise vorticity. The exact transport equation for the streamwise vorticity, $\Omega_x = \dfrac{\partial W}{\partial y} - \dfrac{\partial V}{\partial z}$, is given by

$$\frac{D\Omega_x}{Dt} = U\frac{\partial \Omega_x}{\partial x} + V\frac{\partial \Omega_x}{\partial y} + W\frac{\partial \Omega_x}{\partial z} = \underbrace{\Omega_x\frac{\partial U}{\partial x}}_{1.} + \Omega_y\frac{\partial U}{\partial y} + \underbrace{\Omega_z\frac{\partial U}{\partial z}}_{2.}$$

$$+ \underbrace{\left(\frac{\partial^2}{\partial y^2} - \frac{\partial^2}{\partial z^2}\right)(-\overline{vw})}_{3.} + \underbrace{\frac{\partial^2}{\partial y \partial z}(\overline{v^2} - \overline{w^2})}_{4.} + \underbrace{\nu\nabla^2\Omega_x}_{5.} \qquad (5.15)$$

(e.g. Bradshaw 1987). Terms 1–5 represent:

1. Intensification of Ω_x by vortex stretching.
2. Augmentation of Ω_x by skewing of either of the other two vorticity components by mean strain. While vorticity is generated by the 'no-slip' condition, this augmentation is an inviscid process and is also known as Prandtl's 'first kind' of streamwise vorticity. This

is the exchange of vorticity between components, the most obvious example being a curved duct.

3. Generation of Ω_x by the anisotropy of the Reynolds stresses (Prandtl's 'second kind') which invariably occurs in the corners of ducts. In turbulent flows that are not exactly two-dimensional, mean strain rates can produce Reynolds stresses that make this term non-zero.
4. Very similar to 3, also involving anisotropy of the Reynolds stresses in the (y, z)-plane.
5. Viscous diffusion of mean vorticity.

By their very nature, these flows are highly anisotropic, and because vortices are stretched in the direction of the mean strain rate, induce significant transport in the (y, z)-plane. A three-dimensional boundary layer has distributed Ω_x. At the wall, the streamwise and spanwise momentum equations reduce to:

$$\frac{1}{\rho}\frac{\partial p}{\partial x}\bigg|_w = -\nu\frac{\partial \Omega_z}{\partial y}\bigg|_w, \tag{5.16}$$

$$\frac{1}{\rho}\frac{\partial p}{\partial z}\bigg|_w = -\nu\frac{\partial \Omega_z}{\partial y}\bigg|_w, \tag{5.17}$$

respectively. Equation (5.17) indicates that a spanwise pressure gradient is an inevitable concomitant of Ω_x, so that the thin-shear layer approximation is invalidated.

The dominant shear stresses are $-\rho\overline{uv}$ and $-\rho\overline{vw}$. The dimensional analysis that leads to the law of the wall and so to the log law leads also to the local-equilibrium approximation and the 'constant stress' layer. Dimensional analysis and the extension of the above to three-dimensional flows gives the mixing length formula:

$$\{(-\overline{uv})^2 + (-\overline{vw})^2\}^{\frac{1}{2}} = l^2\{(\partial U/\partial y)^2 + (\partial W/\partial y)^2\}. \tag{5.18}$$

The use of (5.18) in boundary layers with distributed Ω_x is well established (van den Berg 1975), but real problems will arise in the case of discrete vortices where transport effects are large, so invalidating the local-equilibrium approximation, and the shear stresses are not operating in the same direction as the strain rates. This creates particular difficulties in defining an accurate wall function, for which it is often assumed that a log law provides a suitable basis. Similar problems arise in the outer layer of any three-dimensional flow where the presence of Ω_x that is either distributed or appearing in more discrete vortices invalidates a simple relationship between shear stress and mean strain rate. Therefore (5.18) is of limited use.

In general it is unlikely that anything less than a full Reynolds stress closure will provide adequate predictions. Even then, special refinements will be required to produce adequate estimates of cross-stream intensities and shear stresses that appear in terms 3 and 4 of (5.15) and which control the generation of Ω_x. Devenport and Simpson (1992) suggest that the Cebeci–Smith (1974) and algebraic stress models provide satisfactory estimates of the shear stress in junction vortices. However, Liandrat *et al.* (1987) show that eddy-viscosity based models are satisfactory for single vortices, but partly fail with vortex pairs, such as a jet in a cross flow. It is therefore likely that the lowest level of modelling is likely to be a full Reynolds stress closure. While amenable to LES, vortex flows face difficulties in terms of the surface boundary conditions required. A particular advantage of LES of vortex flows lies in the calculation of pressure, so avoiding the inherent difficulties that arise in these flows in respect of the pressure–strain correlation that relies either on the anisotropy of the Reynolds stresses ('slow' term) or the anisotropy of production ('fast' term).

5.2 Transition

Given that this process depends on so many factors (surface finish, noise, external disturbances), it seems sensible to resort to simple correlation techniques which are quite old and could usefully be replaced by more modern ones using measurements specific to the application of the correlation. Such an approach has been used in the aerodynamic design of steam turbine blades (Morrison 1989, 1991).

In the case of natural transition, an eddy viscosity may be weighted by an 'intermittency' factor,

$$\gamma_{tr} = 1 - \exp\left(-G(x - x_{tr})\int_{x_{tr}}^{x}\frac{dx}{U_e}\right), \qquad (5.19)$$

where x_{tr} is the location of the start of transition, G is a spot-formation rate parameter (Dhawan and Narasimha 1958, Chen and Thyson 1971):

$$G = \frac{3}{C_1^2}\frac{U_e^3}{\nu^2}Re_{x_{tr}}^{-1.34}, \qquad (5.20)$$

where $Re_{x_{tr}} = \dfrac{x_{tr}U_e}{\nu}$ and $C_1 = 60.0 + 4.86M_e^{1.92}$ (Cebeci and Smith 1974).

The location of transition onset has been correlated by Abu-Ghannam and Shaw (1980), where, for $\lambda_\theta = \dfrac{\theta^2}{\nu}\dfrac{dU_e}{dx}$:

$$F(\lambda_\theta) = 6.91 + 12.75\lambda_\theta + 63.64\lambda_\theta^2; \quad \lambda_\theta < 0, \qquad (5.21)$$

$$F(\lambda_\theta) = 6.91 + 2.48\lambda_\theta - 12.27\lambda_\theta^2; \quad \lambda_\theta > 0. \qquad (5.22)$$

The value of Re_θ at transition onset is defined by:

$$Re_{\theta s} = 163 + \exp\left\{F(\lambda_\theta) - F(\lambda_\theta)\frac{Tu}{6.91}\right\},$$ (5.23)

where Tu is the free-stream turbulence intensity,

$$Tu = \frac{\sqrt{\frac{2}{3}k}}{U_e} \times 100.$$ (5.24)

Abu-Ghannam and Shaw (1980) use:

$$R_{\theta E} = 540 + 183.5(R_L \times 10^{-5} - 1.5)(1 - 1.4\lambda_\theta),$$ (5.25)

which gives a quadratic equation in θ_E. R_L is the transition-length Reynolds number.

For the case of transition occurring through a separation bubble, Horton (1969) has proposed the criterion:

$$\frac{\lambda_1}{\theta_s} = 4 \times 10^4 Re_{\theta s}^{-1},$$ (5.26)

where λ_1 is the streamwise distance between separation and transition (assumed, with some experimental support, to occur instantaneously) and θ_s is the momentum thickness at separation. Roberts (1980) subsequently reviewed the experimental data and suggested that Horton's correlation gives a delayed indication of bursting of the bubble and therefore proposed:

$$\frac{\lambda_1}{\theta_s} = 2.5 \times 10^4 \frac{\log[\coth(10 \times TF)]}{Re_{\theta s}}$$ (5.27)

to replace (5.26), where TF is a factor related to the turbulence intensity. Here we have left this undefined partly because these correlations are often specific to the application. For example, TF can be adjusted to account for straining effects in a turbine cascade. This really makes the point that workers in industry are probably advised to use their own correlations!

More recently, Savill (2002) has used parabolised Navier–Stokes (PSE) methods to predict the location of transition, followed by the more traditional approach of an intermittency-weighted eddy-viscosity model to complete the calculation. However, despite the appeal of pragmatism, such an approach lacks a physical description, as already made clear in Sections 1 and 3.5.

5.3 Shock/boundary-layer interaction

This application challenge is the most intractable one of all, owing primarily to the fact that this interaction is not one, but a varying mix of several perturbations. This leads to at least two problems. First, the perturbations often occur simultaneously, so that it is difficult if not impossible to dissociate the separate effects for the purposes of modelling. Even then, the combined effects will not be the linear sum of the separate ones. Second, the application of zonal modelling (Section 5.5) to shock/boundary-layer interaction will be limited because it is not yet clear what the important effects are in any given zone (Section 3.6). Therefore it is not yet possible to determine how individual terms in a model equation might best be adjusted from one zone to another. Below we consider the somewhat simpler case of flows dominated by a change of species (e.g. separation) which is amenable to zonal modelling.

The principal requirement of any model is that it should be able to cope with the sudden jump in Reynolds stresses. Not all of the stresses are amplified equally, implying that a stress transport model is the minimum level at which adequate prediction will be possible. However, this questions the adequacy of current datasets in calibrating existing Reynolds-stress models. This is therefore a key area in which a set of experiments with an incremental approach to shock/boundary-layer interaction (as a general term embodying all the effects mentioned in Section 3.6) is clearly necessary.

In the shorter term, prediction methods will continue to use a variety of approaches from algebraic modelling to standard two-equation models (Wilcox 1998) and (non-linear) RDT. Terms in the stress transport equations are almost bound to need term-by-term empirical adjustments to account for all the effects mentioned in Section 3.6. Yet, currently, we lack the knowledge for making such adjustments on a rational basis. Therefore attempts at implementing refinements to a full Reynolds-stress closure are not justified at present.

5.4 Change of species

We take this to be a general description of flows involving at least one region of either separation or reattachment (Section 3.4). In common with free-stream turbulence above a turbulent boundary layer (Section 3.5), these problems involve essentially the interaction of regions of turbulence governed by different length scales and velocity scales, as well as the relaxation from the effects of one (or more) imposed length scale(s) and/or velocity scale(s) to those of others.

The primary difficulty lies in determining how regions involving interactions should be treated. In short, we require a simple correlation of the form (5.12) that describes the response of key variables (e.g. skin friction, turbulent shear stress, turbulence kinetic energy). The relaxation of these variables from one species to another can be treated by simple ordinary differential equations such as:

$$C_\phi \frac{\mathrm{d}\phi}{\mathrm{d}x} = \phi - \phi_0,$$ (5.28)

where C_ϕ is a 'time' constant, or by

$$\frac{\partial \phi}{\partial t} = C'_\phi \frac{\partial^2 \phi}{\partial x^2},$$ (5.29)

where C'_ϕ is an exchange coefficient. Physically, (5.28) and (5.29) are justifiable based on what we know already: Wood and Bradshaw (1982) show that a plane mixing layer at reattachment experiences the effect of the wall before the point of mean reattachment via the pressure field. In fact this will be true for any sufficiently strong perturbation in subsonic flow. Similarly, the effects of reattachment persist for considerable streamwise distances and the behaviour is not monotonic (Heenan and Morrison 1998a,b, Castro and Epik 1998), as evidenced by such structural parameters as the shear correlation coefficient, R_{uv}, or $a_1 = -\overline{uv}/2k$. The purpose of (5.28) or (5.29) is to act as a 'governor' to changes in model constants going from one zone to another. While zones are determined by local variables, some recognition has to be taken of non-local effects: an obvious example is the location of zero strain rate in a wall jet where the shear stress does not go to zero.

Linear superposition is not necessarily a satisfactory approximation: how good is it and how might it be improved? Dean and Bradshaw (1976) show that linear superposition is accurate for the symmetrical interaction of shear layers in a duct. Physically this is justified by the observation that the large eddies 'time-share' and if their transport properties are similar, a time-average of the turbulence properties in the merging region is simply obtained as the fractional time contribution from each of the merging layers. This also implies that 'fine-grain' mixing is not important. For more asymmetric interactions (Andreopoulos and Bradshaw 1980, Weir *et al.* 1981) such as the trailing edge of a high-lift aerofoil when the differences between the scales of each layer are large, it is clear that such a simple summation will not be realistic because the cross-stream transport of key quantities will be asymmetric. Under these conditions it is necessary to decouple the Reynolds stresses from the mean velocity profile and this rules out eddy-viscosity models. This is a key assertion: the shearless mixing layer (see, for example, Heenan *et al.* 1994) is ample evidence that turbulence mixing proceeds by the imbalance of turbulence quantities and not by the mean velocity gradient. This also raises the question of the suitability of the gradient diffusion hypothesis (Section 4.1.2) and whether the hypothesis of bulk convection is better in the case of shear-layer interactions. Bulk convection velocities for the turbulence kinetic energy and shear stress are given by, respectively,

$$V_k = \frac{\overline{p'v}/\rho + \overline{kv}}{k}, \tag{5.30}$$

$$V_\tau = \frac{\overline{p'u}/\rho + \overline{uv^2}}{\overline{uv}}, \tag{5.31}$$

which are generally a better model of turbulent transport in the case of the shearless mixing layer (Heenan *et al.* 1994), and in flows dominated by discrete vortices (Shabaka *et al.* 1985, Mehta and Bradshaw 1988). However, for the latter case, these transport velocities are still too ill behaved to be of use.

5.5 Determination of zone dependencies

In an undistorted shear layer, large eddy 'lifetime' may be measured as the ratio of energy content to the rate of production of energy, $k/(-\overline{uv}\,\partial U/\partial y)$. Typically, in an undistorted boundary layer, $-\overline{uv} \approx k/3$ and $\partial U/\partial y \approx 0.3\,U_e/\delta$. Therefore a large eddy has a lifetime of $10\,\delta/U_e$, or equivalently, it travels ten boundary layer thicknesses before breaking up. Length scales are frame invariant; the mean velocity is hardly ever a scale for the turbulence. An appropriate velocity scale may be taken to be $(-\overline{uv})^{\frac{1}{2}} = u_\tau$. This is a better choice than $k^{\frac{1}{2}}$ which, close to a wall, receives non-local contributions from the large scales (so-called 'inactive' motion). This means that the non-local effects apparent in k make it an inappropriate choice of scale for determining the extent of a zone. Where distortions are local (even if their effects persist), it is important to use a local scale. Taking an undistorted flow to be one in which production of turbulence kinetic energy and its dissipation are in approximate balance, we have $L = (-\overline{uv})^{\frac{3}{2}}/\varepsilon$ as an appropriate length scale, and $T = (-\overline{uv})/\varepsilon$ as the corresponding time scale. A corresponding Reynolds number is $(-\overline{uv})^2/\varepsilon\nu$, rather than the more often-used $k^2/\varepsilon\nu$. In local equilibrium, both $L\,u_\tau/\nu$ and $T\,u_\tau^2/\nu$ are equal to $\kappa y^+ = \kappa y\,u_\tau/\nu$. Note that the use of ε rather than the production also allows us to define local scales that are not related to the mean motion, as in the case of the mixing length hypothesis.

In distorted boundary layers (additional imposed length or velocity scales), one can expect such quantities as L to change significantly over distances that are small compared with L. By way of example, a short length of surface roughness causes two new internal layers, one emanating from the smooth-to-roughness changes (S $\rightarrow$ R, thickness δ_{i1}), the next, from the rough-to-smooth change (R $\rightarrow$ S, thickness δ_{i2}). Andreopoulos and Wood (1982) show that, as the distance from the wall increases, L exhibits non-monotonic behaviour, decreasing below the equilibrium value for

$\delta_{i2} < y < \delta_{i1}$ and increasing above it for $y > \delta_{i1}$. Modelling of such a flow is not amenable to simple superposition techniques. Moreover, the relaxation distance to equilibrium is large: following an R → S step, the outward propagation of the internal layer increases the shear stress towards the fully developed rough-wall distribution, and this high level of shear stress will take a significant length to revert to the smooth-wall distribution.

It is proposed therefore that L and/or T may be used to form the basis of a predictive decision-making process for zonal modelling in which these parameters are used to delineate zones so that constants in a baseline model can be adjusted from one zone to the next. A suggested scheme is:

1. Obtain solution with baseline model using standard constants.
2. Identify deficiencies in veracity of solution.
3. Determine from the solution with baseline model fractional changes in L to form the basis of decisions in determining zone boundaries.
4. Make changes to model constants, either on ad hoc basis, or by making them functions of simple algebraic relationships such as those used to quantify the effects of extra strain rates, such as (5.10).
5. Use simple ordinary differential equations (5.30), (5.31) to take account of non-local effects.

Initially, this algorithm would be iterative, but experience and some code developments would obviate the need for this.

6 Conclusions

There is a clear need for the further development of existing turbulence models for the prediction of distorted boundary layers. Principally, this involves the acceptance that models and their constants need to be adjusted according to the specific nature of the distortion and that a single form of distortion can lead to different areas of a boundary layer in which the dominant features of the turbulence are not the same. In turn, this implies the need for zonal modelling which, unfortunately, lacks universal appeal. A substantial effort with international collaboration is required. Models require 'calibration' for specific types of distortion and constants recommended for each. Specific areas for development work include wall functions as well as more physically plausible models for dissipation, turbulent transport and the pressure-strain term. Multiple time scales should also be used. Suggestions for these areas have been put forward.

There is also a clear need for new experiments, specifically good ones that make up for the relative lack of funding for experimental work over the past 20 years or so. These should take advantage of new and improved experimental techniques such as PIV and need to be conceptually adventurous: a good experiment should have at its heart a hypothesis that requires testing. The onus is therefore on the experimenter to be able to, more than anything, conceive the 'right' experiment. This is not easy. An example might serve to illustrate the point: the law of the wall has for nearly 60 years been one of the few cornerstones in the whole subject of boundary layers, but its limitations (and those of the related assumptions of a constant-stress layer, and local equilibrium) in distorted boundary layers remain largely undocumented, with the result that it is called upon for use in situations for which it is ill suited. Thus there is a great need for well-defined experiments, with clear objectives. Given the importance of dissipation, new facilities should be geared around an appropriate degree of resolution to ensure that it can be measured. Often, this will require large-scale facilities. Owing to the increasing importance of DNS and LES, there is a very great need for facilities that enable measurements at full-size Reynolds numbers so that Reynolds number effects may be more fully documented.

Specifically:

- There is still a very significant requirement for high-profile, highly-accurate measurements in 'canonical' (undistorted) flows such as fully-developed turbulent pipe flow, turbulent boundary layers, jets and wakes over Reynolds-number ranges of at least three orders of magnitude. Of these, experiments of the first two are currently under way in the Princeton group, in respectively the superpipe and HRTF compressed-air facilities. The experiments are to test fundamental theories such as the log law.

- Measurements at high Reynolds number in compressed air have their limitations in terms of transducer resolution and streamwise fetch. The first difficulty may be improved by the use of improved technology (e.g. MEMS). The second is of fundamental importance – do shear layers 'remember' their origins? Is there, in practice, a limiting state attained by the turbulence, which may be described as self-similar? Answers to these questions really require the use of a large facility in which shear layers with a streamwise fetch of several thousand shear-layer thicknesses can be obtained. This second requirement also makes possible measurements with conventional transducers that have a resolution of the order of the Kolmogorov scales, and this makes possible investigations of high-wavenumber similarity. Such measurements should also be used for assessing wall functions in both RANS codes and 'off-the-surface' boundary conditions for LES. For economic reasons, such a tunnel would have also to be commercially relevant to near-market vehicle and environmental technologies.

- Measurements with modern instrumentation of shear-layer interactions are required of up to third-order (transport) statistics. An incremental approach is required, starting with symmetric interactions and leading toward strongly asymmetric interactions. These are to replace older experiments in which the source of the fluid in the interaction was identified using heat. The arguments of Section 1 (supported by measurements) indicate that the use of heat as a passive contaminant is not a reliable marker of momentum (and energy) transport. Such measurements should also be used to test the gradient diffusion hypothesis in these flows. The use of heat markers should be replaced by more sophisticated conditional-sampling techniques.

- The response of turbulent shear layers to a change of species is quite well documented. However, their response to time-dependent perturbations either local or global in nature is barely documented, if at all. The response is important in terms of flow control technology. At a more fundamental level, time-dependent forcing of turbulence is an important technique in elucidating the fundamental properties of turbulence, as suggested by Clauser in his 'black-box analogy'.
- A systematic study of shock/boundary-layer interaction is required, including contemporary heat transfer measurements and at both supersonic and hypersonic speeds. Attention should be paid to how the shock is generated so that its effects, as distinct from those produced by extra rates of strain, can be measured. A new programme is due to begin at Imperial College (R. Hillier, private communication).
- 'Experiments' involving the post-processing of DNS databases should continue to test fundamental concepts and ideas. They should **not** be used as the ultimate arbiter in the validation of existing turbulence models. In particular, we need a better understanding of wall turbulence in both equilibrium and non-equilibrium conditions: our understanding of the importance to energy transfer of near-wall dynamics is still poor and the level of detail is beyond current experimental techniques. Thus DNS databases at higher Reynolds numbers are required to ensure that the spectral transfer is representative of that at much higher Reynolds numbers. The development of models for LES may be regarded as the pacing item for the development of LES codes for practical situations.

Acknowledgements

I am indebted to the organisers of the Isaac Newton Institute Programme on Turbulence for the opportunity to have taken part. I am grateful to Professors Geoff Hewitt and Christos Vassilicos for their encouragement and patience in the writing of this chapter and to the referees for their very helpful comments.

References

Abu-Ghannam, B. J. & Shaw, R. 1980 Natural transition of boundary layers – the effects of turbulence, pressure gradient and flow history. *J. Mech. Eng. Sci.* **22**, 213–228.

Adams, N. A. 2000 Direct simulation of the boundary layer along a compression ramp at $M = 3$ and $Re_\theta = 1685$. *J. Fluid Mech.* **420**, 47–83.

Akhavan, R., Ansari, A., Kang, S. & Mangiavacchi, N. 2000 Subgrid-scale interactions in a numerically simulated planar jet and implications for modelling. *J. Fluid Mech.* **408**, 83–120.

Andreopoulos, J. & Bradshaw, P. 1980 Measurements of interacting turbulent shear layers in the near wake of a flat plate. *J. Fluid Mech.* **100**, 639–668.

Andreopoulos, J. & Wood, D. H. 1982 The response of a turbulent boundary layer to a short length of surface roughness. *J. Fluid Mech.* **118**, 143–164.

Atkins, C., Gould, A. R. B., Hills, D. P. & Hutton, A. 1999 The application challenges: an aerospace view. INI Programme note.

Bandyopadhyay, P. R. 1992 Reynolds number dependence of the freestream turbulence effects on turbulent boundary layers. *AIAA J.* **30**, 1910–1912.

Bardina, J. E., Huang, P. G. & Coakley, T. J. 1997 Turbulence modelling validation, testing, and development. NASA 110446.

Batchelor, G. K. & Townsend, A. A. 1956 Turbulent diffusion. In *Surveys in Mechanics* (eds. G. K. Batchelor & R. M. Davies). Cambridge University Press.

Blair, M. F. 1983 Influence of free-stream turbulent boundary layer heat transfer and mean profile development. Part II – Analysis of results. *ASME J. Heat Transfer* **105**, 41–47.

Bradshaw, P. 1967a Conditions for the existence of an inertial subrange in turbulent flow. Natl. Phys. Lab. Aero. Rep. No. 1220.

1967b 'Inactive' motion and pressure fluctuations in turbulent boundary layers. *J. Fluid Mech.* **30**, 241–258.

1971 Variations on a theme of Prandtl. *AGARD CP* 93. NATO.

1973 Effects of streamline curvature on turbulent flow. *AGARDograph* 169. NATO.

1974 The effect of mean compression or dilatation on the turbulence structure of supersonic boundary layers. *J. Fluid Mech.* **63**, 449–464.

1975 Complex turbulent flows. *ASME J. Fluids Engng.* **97**, 146–154.

1976 Complex turbulent flows. *Theoretical and Applied Mechanics* (ed. W. T. Koiter) p. 105. North-Holland Publishing Company.

1977 Interacting shear layers in turbomachines and diffusers. In *Turbulence in Internal Flows* (ed. S. N. B. Murthy). Proceedings of 1976 Project Squid Workshop. Hemisphere, Washington.

1987 Turbulent secondary flows. *Ann. Rev. Fluid Mech.* **19**, 53–74.

1990 Effects of extra rates of strain – review. In *Near-Wall Turbulence – 1988 Zaric Memorial Conference* (eds. S. J. Kline & N. H. Afgan) p. 106. Hemisphere.

1992 Collaborative testing of turbulence models. Final Report on AFOSR 90-0154.

1996 Turbulence modeling with application to turbomachinery. *Prog. Aerospace Sci.* **32**, 575–624.

2000 A note on "critical roughness height" and "transitional roughness". *Phys. Fluids* **12**, 1611–1614.

Bradshaw, P., Launder, B. E. & Lumley, J. L. 1996 Collaborative testing of turbulence models. *ASME J. Fluids Engng.* **118**, 243–247.

Carati, D., Ghosal, S. & Moin, P. 1995 On the representation of backscatter in dynamic localization models. *Phys. Fluids* **7**, 606–616.

Castro, I. P. 1984 Effects of free stream turbulence on low Reynolds number boundary layers. *J. Fluids Engng.* **106**, 298–306.

Castro, I. P. & Epik, E. 1998 Boundary layer development after a separated region. Part 2. Effects of free-stream turbulence. *J. Fluid Mech.* **192**, 577–595.

Castro, I. P. & Haque, A. 1988 The structure of a shear layer bounding a separation region. *J. Fluid Mech.* **374**, 91–116.

Cazalbou, J. B. & Bradshaw, P. 1993 Turbulent transport in wall-bounded flows. Evaluation of model coefficients using direct numerical simulation. *Phys. Fluids* A **5**, 3233–3239.

Cebeci, T. & Smith, A. M. O. 1974 *Analysis of Turbulent Boundary Layers*. Academic Press.

Chen, K. K. & Thyson, N. A. 1971 Extension of Emmons' spot theory to flows on blunt bodies. *AIAA J.* **5**, 821.

Chou, P. Y. 1945 On the velocity correlations and the solution of the equations of turbulent fluctuation. *Quart. Appl. Math.* **3**, 38–54.

Clauser, F. H. 1956 The turbulent boundary layer. *Adv. Appl. Mech.* **4**, 1–51.

Coates Ulrichsen, T. 2003 Turbulent boundary layers with varying shear. Final-year undergraduate project, Department of Aeronautics, Imperial College.

Coupland, J. 1999 Turbulence and transition for gas turbine aerodynamic flows. INI Programme note.

Cutler, A. D. & Bradshaw, P. 1993a Strong vortex/boundary layer interactions. Part I. Vortices high. *Expts. Fluids* **14**, 321–332.

 1993b Strong vortex/boundary layer interactions. Part II. Vortices low. *Expts. Fluids* **14**, 393–401.

Dean, R. B. & Bradshaw, P. 1976 Measurements of interacting turbulent shear layers in a duct. *J. Fluid Mech.* **78**, 641–676.

De Graaff, D. B. & Eaton, J. K. 2000 Reynolds-number scaling of the flat-plane turbulent boundary layer. *J. Fluid Mech.* **422**, 319–346.

Devenport, W. J. & Simpson. R. L. 1992 Flow past a wing-body junction – experimental evaluation of turbulence models. *AIAA J.* **30**, 873–881.

Dhawan, S. & Narasimha, R. 1958 Some properties of boundary-layer flow during the transition from laminar to turbulent motion. *J. Fluid Mech.* **3**, 418.

Driver, D. M. 1991 Reynolds shear stress measurements in a separated boundary layer flow. AIAA-91-1787.

Dunn, D. & Morrison, J. F. 2003 Anisotropy and energy flux in wall turbulence. *J. Fluid Mech.* **491**, 353–378.

Durbin, P. A. 1991 Near wall turbulence closure modeling without damping functions. *Theor. & Comp. Fluid Dyn.* **3**, 1–13.

 1993 A Reynolds stress model for near-wall turbulence. *J. Fluid Mech.* **249**, 465–498.

Fernholz, H. H. & Finley, P. J. 1981 A further compilation of compressible boundary layer data with a survey of turbulence data. *AGARD-AG-263*. NATO.

 1996 The incompressible zero-pressure-gradient turbulent boundary layer: an assessment of the data. *Prog. Aerospace Sci.* **32**, 245–311.

Fernholz, H. H. & Warnack, D. 1998 The effects of a favourable pressure gradient and of the Reynolds number on an incompressible axisymmetric turbulent boundary layer. Part 1. The turbulent boundary layer. *J. Fluid Mech.* **359**, 329–356.

Fernholz, H. H., Smits, A. J., Dussauge, J. P. & Finley, P. J. 1989 A survey of measurements and measuring techniques in rapidly distorted compressible turbulent boundary layers. *AGARDograph* 315. NATO.

Ferziger, J. H., Kline, S. J., Avva, R. K., Bordalo, S. N. & Tzuoo, K.-L. 1990 Zonal modeling of turbulent flows – philosophy and accomplishments. *Near-Wall Turbulence: 1988 Zoran Zaric Memorial Conference* (eds. S. J. Kline and H. H. Afgan). Hemisphere.

Furuichi, N. & Kumada, M. 2002 An experimental study of a spanwise structure around a reattachment region of a two-dimensional backward-facing step. *Expts. Fluids* **32**, 179–187.

Gaster, M. 2003 The influence of surface roughness on boundary layer transition. *Symposium on Advances in Fluid Mechanics.*

George, W. K. & Castillo, L. 1997 Zero-pressure-gradient turbulent boundary layer. *Appl. Mech. Rev.* **50**, 689–729.

Germano, M., Piomelli, U., Moin, P. & Cabot, W. H. 1991 A dynamic subgrid-scale eddy viscosity model. *Phys. Fluids* A **3**, 1760–1765.

Green, J. E. 1970 Interactions between shock waves and turbulent boundary layers. *Prog. Aerospace Sci.* **11**, 235–341.

Guezennec, Y., Stretch, D. & Kim, J. 1990 The structure of turbulent channel flow with passive scalar transport. In *Studying Turbulence Using Numerical Simulation Databases III*. Proc. 1990 Summer Program, CTR Stanford/NASA Ames, p. 127.

Hancock, P. E. & Bradshaw, P. 1983 The effect of free-stream turbulence on turbulent boundary layers. *J. Fluids Engng.* **105**, 284–289.

1989 Turbulence structure of a boundary layer beneath a turbulent free stream. *J. Fluid Mech.* **205**, 45.

Heenan, A. F. & Morrison, J. F. 1998a Passive control of pressure fluctuations generated by separated flow. *AIAA J.* **36**, 1014–1022.

1998b Passive control of backstep flow. *Exptl. Thermal Fluid Science* **16**, 122–132.

Heenan, A. F., Morrison, J. F., Iuso, G. & Onorato, M. 1994 Evolution of two-scale, shearless grid turbulence. *Proc. 2nd Int. Conf. on Experimental Fluid Mechanics*. Levrotto e Bella, Turin, pp. 321–330.

Hillier, R. & Cherry, N. J. 1981 The effects of stream turbulence on separation bubbles. *J. Wind Eng. Ind. Appl.* **8**, 49–58.

Hong, S. K. & Murthy, S. N. B. 1986 Effective velocity of transport in curved wall boundary layers. *AIAA J.* **24**, 361–369.

Horton, H. P. 1969 A semi-empirical theory for the growth and bursting of laminar separation bubbles. ARC CP No. 1073.

Huang, P. G., Bradshaw, P. & Coakley, T. J. 1992 Assessment of closure coefficients for compressible flow turbulence models. *NASA TM* 103882.

Hunt, J. C. R. 1984 Turbulence structure in thermal convection and shear-free boundary layers. *J. Fluid Mech.* **138**, 161–184.

Hunt, J. C. R. & Carruthers, D. J. 1990 Rapid distortion theory and the 'problems' of turbulence, *J. Fluid Mech.* **212**, 497–532.

Hunt, J. C. R. & Graham, J. M. R. 1978 Turbulence near plane boundaries. *J. Fluid Mech.* **84**, 209–235.

Hunt, J. C. R. & Morrison, J. F. 2000 Eddy structure in turbulent boundary layers. *Eur. J. Mechs. B – Fluids* **19**, 673–694.

Hunt, J. C. R., Moin, P., Lee, M., Moser, R. D., Spalart, P., Mansour, N. N., Kaimal, J. C. & Gaynor, E. 1989 Cross correlation and length scales in turbulent flows near surfaces. *Advances in Turbulence* (eds. H.-H. Fernholz & H. E. Fiedler). Springer-Verlag.

Jacobs, R. G. & Durbin, P. A. 1998 Shear sheltering and the continuous Orr–Sommerfeld equation. *Phys. Fluids.* **10**, 2006–2011.

2001 Simulations of bypass transition. *J. Fluid Mech.* **428**, 185–212.

Jiménez, J. & Moser, R. D. 2000 Large-eddy simulations: where are we and what can we expect. *AIAA J.* **38**, 605–612.

Joslin, R. D. 1995 Evolution of stationary crossflow vortices in boundary layers on swept wings. *AIAA J.* **33**, 1279–1285.

Jovanović, J., Ye, Q.-Y. & Durst, F. 1995 Statistical interpretation of the turbulent dissipation rate in wall-bounded flows. *J. Fluid Mech.* **293**, 321–347.

Kalter, M. & Fernholz, H. H. 2001 The reduction and elimination of a closed separation region by free-stream turbulence. *J. Fluid Mech.* **446**, 271–308.

Kassinos, S. C., Reynolds, W. C. & Rogers, M. M. 2001 One-point turbulence structure tensors. *J. Fluid Mech.* **428**, 213–248.

Kays, W. M. & Crawford, M. E. 1993 *Convective Heat and Mass Transfer*, 3rd edn. McGraw-Hill.

Khan, M. J., Ahmed, A. & Trosper, J. R. 1995 Dynamics of the juncture vortex. *AIAA J.* **33**, 1273–1278.

Kline, S. J. B. 1981 Universal or zonal modelling – the road ahead. In *AFOSR-HTTM-Stanford Conference on Complex Turbulent Flows* (eds. S. J. B. Kline, J. Cantwell & G. M. Lilley). Thermosciences Divn., Dept. Mech. Eng., Stanford University.

Lakshminarayana, B. 1986 Turbulence modelling for complex shear flows. *AIAA J.* **24**, 1900–1917.

Launder, B. E. 1978 Heat and mass transport. In *Turbulence*. Topics in Applied Physics, Vol. 12 (ed. P. Bradshaw). Springer-Verlag.

Lee, M. J. & Hunt, J. C. R. 1989 The structure of sheared turbulence near a plane boundary layer. In *7th Symposium on Turbulent Shear Flows*. Stanford University.

Liandrat, J., Aupoix, B. & Cousteix, J. 1987 Calculation of longitudinal vortices imbedded in a turbulent boundary layer. In *Turbulent Shear Flows* 5 (eds. F. Durst, B. E. Launder, J. L. Lumley, F. W. Schmidt & J. H. Whitelaw) pp. 253–277. Springer-Verlag.

Lin, C.-L., Moeng, C.-H., Sullivan, P. P. & McWilliams, J. C. 1997 The effect of surface roughness on flow structures in a neutrally stratified planetary boundary layer flow. *Phys. Fluids* **9**, 3235–3249.

Magnaudet, J. 2003 High-Reynolds-number turbulence in a shear-free boundary layer: revisiting the Hunt–Graham theory. *J. Fluid Mech.* **484**, 167–196.

Marušić, I., Uddin, A. K. M. & Perry, A. E. 1997 Similarity law for the streamwise turbulence intensity in zero-pressure-gradient turbulent boundary layers. *Phys. Fluids* **9**, 3718–3726.

Marušić, I., Kunkel, G. J. & Porté-Agel, F. 2001 Experimental study of wall boundary conditions for large-eddy simulation. *J. Fluid Mech.* **446**, 309–320.

Mason, P. J. & Thomson, D. J. 1992 Stochastic backscatter in the large-eddy simulations of boundary layers. *J. Fluid Mech.* **242**, 51–78.

Matsubara, M. & Alfredsson, P. H. 2001 Disturbance growth in boundary layers subjected to free-stream turbulence. *J. Fluid Mech.* **430**, 149–168.

McDonald, H. & Kreskovsky, J. P. 1973 Effect of free stream turbulence on the turbulent boundary layer. United Aircraft Res. Labs. Report No. M110887-1.

McKeon, B. J., Li, J., Jiang, W., Morrison, J. F. & Smits, A. J. 2004 Further observations on the mean velocity distribution in fully-developed pipe flow. *J. Fluid Mech.* **501**, 135–147.

Mehta, R. D. & Bradshaw, P. 1988 Longitudinal vortices imbedded in turbulent boundary layers. Part 2. Vortex pair with 'common flow' upwards. *J. Fluid Mech.* **188**, 529–546.

Meneveau, C. 1991 Analysis of turbulence in the orthonormal wavelet representation. *J. Fluid Mech.* **232**, 469–520.

1994 Statistics of turbulence subgrid-scale stresses: Necessary conditions and experimental tests. *Phys. Fluids* **6**, 815–833.

Millikan, C. M. 1938 A critical discussion of turbulent flow in channels and circular tubes. *Proc. 5th Int. Congr. Appl. Mech.* pp. 386–392. Wiley.

Morkovin, M. V. 1962 Effects of compressibility on turbulent flows. In *Méchanique de la Turbulence* (ed. A. J. Favre) pp. 367–380. CNRS.

Morrison, J. F. 1989 Modelling of physical phenomena in the Cebeci–Smith boundary layer calculation method. GEC Turbine Generators Ltd. Aerodynamics Group, Tech. Note, AGN 578.

1991 Modelling of physical phenomena in 3D viscous time marching (3DVTM). GEC Turbine Generators Ltd. Aerodynamics Group, Tech. Note, AGN 624.

Morrison, J. F. & Bradshaw, P. 1991 Bursts and sources of pressure fluctuations in turbulent boundary layers. *Proc. 8th Symp. Turbulent Shear Flows*. Paper 2–1.

Morrison, J. F., Jiang, W., McKeon, B. J. & Smits, A. J. 2002 Reynolds number dependence of streamwise velocity spectra in turbulent pipe flow. *Phys. Rev. Lett.* **88**, 214501.

Morrison, J. F., McKeon, B. J., Jiang, W. & Smits, A. J. 2004 Scalings of the streamwise velocity component in turbulent pipe flow. *J. Fluid Mech.* **508**, 99–131.

Na, Y. & Moin, P. 1998 Direct numerical simulation of a separated turbulent boundary layer. *J. Fluid Mech.* **374**, 379–405.

Nicoud, F., Baggett, J. S., Moin, P. & Cabot, W. 2001 Large eddy simulation wall-modeling based on suboptimal theory and linear stochastic estimation. *Phys. Fluids* **13**, 2968–2984.

Nigim, H. H. 1996 Recovery of equilibrium turbulent boundary layers downstream of obstacles. *Phys. Fluids* **8**, 548–554.

Nikitin, N. V., Nicoud, F., Wasistho, B., Squires, K. D. & Spalart, P. R. 2000 An approach to wall modeling in large-eddy simulations. *Phys. Fluids* **12**, 1629–1632.

Österlund, J. M., Johansson, A. V., Nagib, H. M. & Hites, M. H. 2000 A note on the overlap region in turbulent boundary layers. *Phys. Fluids* **12**, 1–4.

Peltier, L. J., Zajaczkowski, F. J. & Wyngaard, J. C. 2000 A hybrid RANS/LES approach to large-eddy simulation of high-Reynolds-number wall-bounded turbulence. *Proc. ASME FEDSM'00*, FEDSM2000-11177.

Piomelli, U. & Balaras, E. 2002 Wall-layer models for large-eddy simulations. *Ann. Rev. Fluids Mech.* **34**, 349–374.

Piomelli, U., Ferziger, J. & Moin, P. 1989 New approximate boundary conditions for large eddy simulations of wall-bounded flows. *Phys. Fluids* A **1**, 1061–1068.

Piomelli, U., Cabot, W. H., Moin, P. & Lee, S. 1990 Subgrid-scale backscatter in transitional and turbulent flows. *Studying Turbulence Using Direct Numerical Simulation Databases III. Proceedings of the 1990 Summer Program*, NASA Ames/Center for Turbulence Research, Stanford University.

Porté-Agel, F., Meneveau, C. & Parlauge, M. B. 2000 A scale-dependent dynamic model for large-eddy simulation: application to a neutral atmospheric boundary layer. *J. Fluid Mech.* **415**, 261–284.

Purtell, L. P. 1992 *Turbulence in Complex Flows – a Selected Review*. AIAA-92-0435.

Redelsperger, J. L., Mahé, F. & Carlotti, P. 2001 A simple and general subgrid model suitable both for surface layer and free-stream turbulence. *Boundary-Layer Met.* **101**, 375–408.

Reshotko, E. 2001 Transient growth: a factor in bypass transition. *Phys. Fluids* **13**, 1067–1075.

Reynolds, W. C. 1990 The potential and limitations of direct and large eddy simulations. In *Whither Turbulence? Turbulence at the Crossroads* (ed. J. L. Lumley) pp. 313–343. Springer.

Roach, P. E. & Brierley, D. H. 1990 The influence of a turbulent free-stream on zero pressure gradient transitional boundary layer development. Part I: Test cases T3A and T3B. In *Numerical Simulation of Unsteady Flows and Transition to Turbulence* (eds. O. Pironneau *et al.*) pp. 229–256. Cambridge University Press.

Roberts, W. W. 1980 Calculation of laminar separation bubbles and their effect on aerofoil performance. *AIAA J.* **18**, 25–31.

Robinson, S. K. 1982 An experimental search for near-wall boundary conditions for LES. AIAA-82-0963.

Saathoff, P. J. & Melbourne, W. H. 1997 Effects of free-stream turbulence on surface pressure fluctuations in a separation bubble. *J. Fluid Mech.* **337**, 1–24.

Saddoughi, S. G. & Veeravalli, S. V. 1994 Local isotropy in turbulent boundary layers at high Reynolds numbers. *J. Fluid Mech.* **268**, 333–372.

Savill, M. A. 1987 Recent developments in rapid distortion theory. *Ann. Rev. Fluid Mech.* **19**, 531–575.

1993 Evaluating turbulence model predictions of transition. *Advances in Turbulence IV* (ed. F. T. M. Nieuwstadt) pp. 555–562. Also as *Appl. Scientific Res.* **51**. Kluwer Academic Publishers.

2002 By-pass transition using conventional closures and new strategies in modelling by-pass transition. *Closure Strategies for Turbulent and Transitional Flows* (eds. B. E. Launder and N. D. Sandham) pp. 464–519. Cambridge University Press.

Schiestel, R. 1987 Multiple-time-scale modelling of turbulent flows in one-point closures. *Phys. Fluids* **30**, 722–731.

Seal, C. V., Smith, C. R. & Rockwell, D. 1997 Dynamics of the vorticity distribution in end-wall junctions. *AIAA J.* **35**, 1041–1047.

Shabaka, I. M. M. A., Mehta, R. D. & Bradshaw, P. 1985 Longitudinal vortices imbedded in turbulent boundary layers. Part 1. Single vortex. *J. Fluid Mech.* **155**, 37–57.

Smits, A. J. & Dussauge, J.-P. 1996 *Turbulent Shear Layers in Supersonic Flows*. AIP, Woodbury, New York.

Smits, A. J. & Wood, D. H. 1985 The response of turbulent boundary layers to sudden perturbations. *Ann. Rev. Fluid Mech.* **17**, 321–358.

Spalart, P. R. 2000 Strategies for turbulence modelling and simulations. *Int. J. Heat & Fluid Flow* **21**, 252–263.

Spazzini, P. G., Iuso, G., Onorato, M., Zurlo, N. & Di Cicca, G. M. 2001 Unsteady behaviour of back-facing step flow. *Expts. Fluids* **30**, 551–561.

Speziale, C. G. 1998 Turbulence modeling for time-dependent RANS and VLES: a review. *AIAA J.* **36**, 173–184.

Thole, K. A. & Bogard, D. G. 1996 High freestream turbulence effects on turbulent boundary layers. *J. Fluids Engng.* **118**, 276–284.

Townsend, A. A. 1956 *The Structure of Turbulent Shear Flow*. 1st edn. Cambridge University Press.

1961 Equilibrium layers and wall turbulence. *J. Fluid Mech.* **11**, 97–120.

1976 *The Structure of Turbulent Shear Flow*. 2nd edn. Cambridge University Press.

Van den Berg, B. 1975 A three-dimensional law of the wall for turbulent shear flows. *J. Fluid Mech.* **70**, 149–160.

Voelkl, T., Pullin, D. I. & Chan, D. C. 2000 A physical-space version of the stretched-vortex subgrid-stress model for large-eddy simulation. *Phys. Fluids* **12**, 1810–1825.

Warnack, D. & Fernholz, H. H. 1998 The effects of a favourable pressure gradient and of the Reynolds number on an incompressible axisymmetric turbulent boundary layer. Part 2. The boundary layer with relaminarization. *J. Fluid Mech.* **359**, 357–381.

Weir, A. D., Wood, D. H. & Bradshaw, P. 1981 Interacting turbulent shear layers in a plane jet. *J. Fluid Mech.* **107**, 237–260.

Wilcox, D. C. 1998 *Turbulence Modeling for CFD* (2nd edn). DCW Industries, Inc.

Wood, D. H. & Bradshaw, P. 1982 A turbulent mixing layer constrained by a solid surface. Part 1. Measurements before reaching the surface. *J. Fluid Mech.* **122**, 57–89.

Wurman, J. & Winslow, J. 1998 Intense sub-kilometer-scale boundary layer rolls observed in hurricane Fran. *Science* **280**, 555–557.

Younis, B. A., Gatski, T. B. & Speziale, C. G. 1999 Towards a rational model for the triple velocity correlations of turbulence. NASA TM-209134.

Zagarola, M. V. & Smits, A. J. 1998 Mean-flow scaling of turbulent pipe flow. *J. Fluid Mech.* **373**, 37–79.

6

Turbulence simulation

N. D. Sandham

University of Southampton

6.1 Introduction

Computer simulation of turbulent flows is becoming increasingly attractive due to the greater physical realism relative to conventional modelling, at a cost that is reducing with continuing advances in computer hardware and algorithms. The first turbulence simulations appeared over 30 years ago, since when we have seen increases in computer performance of over four orders of magnitude such that many of the canonical turbulent flows first studied by laboratory experiments can now be reliably simulated by computer. Examples include turbulent channels (Kim *et al.*, 1987), turbulent boundary layers (Spalart, 1988), mixing layers (Rogers & Moser, 1994), subsonic and supersonic jets (Freund, 2001, Freund *et al.*, 2000) and backward-facing steps (Le *et al.*, 1997). Where simulations can be reliably made, they provide more data than are available from laboratory experiments, even with modern non-intrusive flow diagnostics. In these situations they provide insight into the basic fluid mechanics. This can be at a very simple flow visualisation level, where a conceptual picture of what is happening in a flow can be quickly obtained from computer animations of key features, or at more advanced levels where the simulations provide statistical data to assist Reynolds-averaged model development. Indeed, several important recent turbulence models have come out of groups who do both simulations and modelling, examples being the Spalart & Allmaras (1994) model and Durbin's $K-\varepsilon-v^2$ (Durbin, 1995), and it is rare to come across a turbulence modelling paper that has not used simulation data as a reference.

Despite the major advances in simulation capability it is still not feasible to simulate reliably a great many practical flows. For some application areas it is just a question of the level of computer resources required, and with projected

Prediction of Turbulent Flows, eds. G. F. Hewitt and J. C. Vassilicos.
Published by Cambridge University Press. © Cambridge University Press 2005.

improvements in hardware many flows will become amenable to simulation-based approaches. In these cases users need to decide the right time to invest in the computer hardware and software resources required for simulation, rather than using Reynolds-averaged Navier–Stokes (RANS) approaches. If simulations turn out to be reliable and cost effective they will assist the design process, to the benefit of the company that made the decision to invest. If a wrong decision is made then money spent on expensive computers may well have been wasted. By contrast some flows, such as high Reynolds number near-wall flows, will perhaps never be amenable to the simulation approach. In these cases, however, there may still be a role for simulation and the decision to be made is whether a set of simulations encompassing the most important physical phenomena can lead to improvements in modelling those phenomena. It is never a bad thing to understand flows better by doing simulations. However, the insight obtained often does not directly lead to model improvements. Modelling is a complex business and, except for routine calibration exercises, depends on flashes of inspiration that cannot be programmed; it can take years before physical insight into a fluid flow is followed by a suitable modelling strategy, and ultimately some flows may never be well predicted with single-point models.

In this chapter we provide a background to simulation methods and discuss the advantages and disadvantages of the various possible approaches, hopefully to a level such that the reader can begin to make rational decisions as to whether a simulation, rather than modelling or other approach is appropriate to their turbulent flow problem. We begin in Section 6.2 with a review of the direct numerical simulation (DNS) approach, where one minimises errors by simulating the full Navier–Stokes equations. This is clearly the most desirable solution if computer resources allow. In Section 6.3 we discuss the large-eddy simulation (LES) approach, whereby only the large turbulence eddies are computed, while the smallest are modelled. We include in the discussion some recent developments on LES, including deconvolution approaches to recovering the small scales, and the popular detached eddy simulation (DES) approach, which switches between LES and RANS according to local criteria. We end (Section 6.4) with a summary of test cases together with recent developments in compressible flow LES and the rapidly emerging area of computational aeroacoustics, which, for particular problems such as jet noise, relies heavily upon turbulence simulation.

6.2 Direct numerical simulation

Direct simulation of turbulence aims to resolve all the relevant scales of turbulent motion and is increasing in realism (complexity and Reynolds number) as computers increase in performance. Moin & Mahesh (1998) gave a review of the role of DNS in turbulence research and Sandham (2002) gave an introduction with examples. Here we give a brief introduction and review some recent extensions in capability.

We restrict the discussion initially to incompressible flow for which the governing equations for the three velocity components u_i and pressure p are

$$\frac{\partial u_i}{\partial x_i} = 0, \tag{6.1}$$

$$\frac{\partial u_i}{\partial t} + \frac{\partial (u_i u_j)}{\partial x_j} = -\frac{1}{\rho}\frac{\partial p}{\partial x_i} + \nu \frac{\partial^2 u_i}{\partial x_j \partial x_j}, \tag{6.2}$$

with ρ the fluid density and ν the kinematic viscosity. The equations are nonlinear and unpredictable, in the chaotic dynamics sense that small changes in initial conditions will ultimately lead to different realisations.

Space permits only a brief overview of numerical methods for direct numerical simulation and the reader is referred to the cited literature for more details of the schemes that have been used. Many studies have been made with spectral methods, which decompose the flow variables into, for example, Fourier or Chebyshev modes, while finite difference and finite volume methods have also been widely used. Important characteristics of numerical methods for application to turbulence simulation include:

- accuracy and consistency (behaviour of the truncation error);
- wave resolution (behaviour considered from the modified wavenumber point of view);
- treatment of aliasing errors (especially for spectral methods);
- stability;
- efficiency.

DNS has to be time accurate, and fully implicit methods are usually ruled out on cost grounds. For incompressible flow, however, the continuity equation is a constraint that has to be satisfied implicitly, and for boundary layer flows it is usual to treat the viscous terms implicitly as well. Thus one usually ends up with explicit treatment of the nonlinear terms and implicit treatment of pressure and viscous terms. The actual time step that can be used is usually determined from a stability (Courant number) condition, rather than on accuracy grounds.

6.2.1 Turbulence scales and resolution requirements

Turbulence is broadband, covering a wide range of spatial and temporal scales. Estimates for the smallest scales are available from the Kolmogorov microscales, obtained from dimensional analysis assuming dependence only upon the fluid kinematic viscosity ν and the rate of dissipation of turbulence energy to heat ε. The length and time microscales are respectively

$$\eta_K = \left(\frac{v^3}{\varepsilon}\right)^{1/4} \tag{6.3}$$

and

$$\tau_K = \left(\frac{v}{\varepsilon}\right)^{1/2}. \tag{6.4}$$

If the length and velocity macroscales of the problem are L and U and if we assume that the rate of dissipation is proportional to the rate of production of turbulence energy at the large scales, i.e. scaling as U^3/L, we have

$$\frac{L}{\eta_K} = Re^{3/4}, \tag{6.5}$$

where the Reynolds number is given by $Re = UL/v$. Thus the difference between the largest and smallest length scales in turbulence increases as the Reynolds number increases. Since there are three spatial dimensions the number of grid points required to resolve turbulence increases as the cube of Eq. (6.5), i.e. as $Re^{9/4}$. When the time step (CFL) restriction is factored in, one ends up roughly with computational cost increasing as the cube of the Reynolds number. This means that for every doubling in Reynolds number we need to wait eight times as long for a numerical result. To get from the current boundary layer simulations at displacement thickness $Re = 2000$, perhaps taking one week on a supercomputer, up to a full scale on a large transport aircraft of say $Re = 64\,000$ we will have to wait 2^{15} weeks, or 630 years! This kind of scaling makes high Reynolds numbers impossible to simulate. However, many phenomena in turbulence appear to have a high Reynolds number asymptote (free shear layer growth rates, near-wall mean profile etc.) and so numerical simulations at a 'high-enough' Reynolds number to capture these phenomena can already contribute greatly to understanding and model development.

We use the term 'direct' numerical simulation to refer to computations where all relevant spatial and temporal scales are adequately resolved for the given application. Some applications, such as those requiring statistics involving higher derivatives, will require more resolution than others. The precise number of grid points needed depends also upon the numerical method. The preference for flows in simple geometries has been for spectral or high-order (greater or equal to fourth order) finite difference or finite volume methods. Detailed grid refinement studies show that, away from walls, grids of the order of $5\eta_K$ (recall that η_K is based on dimensional arguments alone) are sufficient for most purposes, such as prediction of mean flow and second moments of turbulence as well as all the terms in the turbulence kinetic energy transport equation. A rule of thumb is that second order methods require as much as a factor of two more points in all directions. However, such methods may be quite efficient per grid point and more suitable for extension to complex geometry. Given that DNS on present-day parallel computers tend to be limited by run time rather than memory, there are some applications where such methods can be used. Upwind methods, unless they be of high order, are generally considered to be too dissipative for use in turbulence simulation (analysis in terms of the modified wavenumber is needed to check the wave-resolving powers of such methods). Unstructured grid calculations (Laurence, 2002) are expected to require more nodes in total to compute a given phenomenon, but estimates of the additional computational cost are not yet available.

The usefulness of direct numerical simulations has increased with the rise in power of computers. Simulations carried out in the early 1980s on Cray X-MP computers (for example isotropic turbulence at low Reynolds numbers, or the early stages of transition) can now be carried out on personal computers. Larger computers can now be used to simulate higher Reynolds numbers and a wider range of geometries. Plots of computer performance against time show improvements ranging from 10 times speed up every 5 years to 10 times speed up every 10 years. In the light of this, let's consider again our example of a full-scale boundary layer calculation which would take 630 years on a current computer. If we take processor performance as doubling every two years, it will take 30 years to get a computer 2^{15} times as powerful, that will let us do the $Re = 64\,000$ calculation in one week. This kind of extrapolation does assume that computer performance can continue to grow at the same rate. Shrinking chip sizes do eventually run into quantum effects, but we can also expect larger and larger parallel computers being developed to continue to push the boundaries of what is possible. Nevertheless, it is clear that turbulent flow at very high Reynolds numbers is not going to be solved any time soon by the purely numerical approach. The tendency in practice appears to be to use the increased performance to move to more complex flows, rather than just extend existing flows to higher Reynolds numbers.

Implementation of DNS numerical methods for massively parallel computing entails additional considerations. The implementation of a spectral method described in Sandham & Howard (1998) used a global transpose method which involved the parcelling of small 'cubes' of data on each processor, labelling these with the address of the processor that requires the information, transferring the data, reading it by the receiving processor, and reconstructing the flow field. Both the PVM (parallel virtual machine) and MPI (message passing interface) libraries have been used successfully for this purpose. With the demonstrated success of such simulations up to 512 processors, there appears little to prevent simulations in the near future on machines with several thousands of processors. In passing, we note that another application of parallel processing to turbulence DNS is in the use of ensemble averaging to obtain averages for unsteady turbulence computations. In computing the temporal development of turbulence (and assuming that the time taken to converge to the base flow is not the critical element), one is often limited by the need to obtain ensemble averages by running the same code several times with independent initial conditions. This is easily done on a parallel computer by running multiple realisations at the same time. For example, in one 256-processor job, four realisations each requiring 64 processors, or eight realisations each requiring 32 processors, can be run at the same time.

6.2.2 Validation procedures

All codes for simulation of turbulence need validation, and the lack of relevant exact solutions of the governing equations makes this more difficult. A list of suitable tests is as follows. Ideally all of these would be satisfied for a simulation to be accepted. In practice this is only true of one or two flows to date.

- Validate code against analytic solutions and asymptotic limits. This could include comparisons with exact solutions of the Navier–Stokes equations, boundary layer solutions, linear growth rates of small disturbances, etc.
- Carry out systematic grid resolution studies, varying resolution in one direction at a time and checking convergence for quantities of interest, e.g. turbulence statistics.
- Carry out systematic box size variations to check that the computational domain is large enough to contain the relevant flow physics. Decay of two-point correlations to zero is a test that can be applied after a simulation has completed.
- Compute budgets of statistical quantities such as Reynolds stresses and check for balance.
- Carry out tests with a reduced time step to check for time discretisation errors.
- Compare results for the same problem between two different numerical methods, ideally from two separate, independently programmed codes. This is perhaps a little extreme, but controversial new results may well require this step before they are accepted.

For some flows the above criteria have been distilled down to a few 'rules of thumb'. For example, in free shear layer calculations if one preserves, say, six decades of roll-off in the energy spectrum, one may reasonably expect good second moment turbulence statistics. For attached turbulent flow over walls computed with high-order or spectral methods, grid spacings of streamwise $\Delta x^+ = 12$, spanwise $\Delta z^+ = 6$ and ten points for $y^+ < 10$ are usually reckoned sufficient for good statistics related to budgets of the Reynolds stresses. (Note that $(\cdot)^+$ variables have been normalised by the viscous length-scale ν/u_τ, where the friction velocity $u_\tau = \sqrt{\tau_w/\rho}$ with τ_w the wall shear stress.)

6.2.3 Post-processing DNS data

Besides accumulation of statistical data, perhaps the first thing a user wants from a simulation is a snapshot, or even computer animation, of the three-dimensional structure of the turbulence. Velocity vectors and planar contour plots are quite limited since turbulent flow is highly three-dimensional. Identification of 3D vortex structures from simulations is a recurring problem, partly because there is no universally accepted definition of a vortex: measures such as vorticity are difficult to conceptualise since they register both shear layers and vortices, which we usually like to think of as different phenomena.

One rigorous definition is in terms of local critical point theory where foci or centres occur wherever complex eigenvalues of the local velocity gradient tensor exist. In DNS this can be plotted using a known discriminant, which separates foci/centres from saddle point/node structures. (See Chong *et al.*, 1990, for a complete discussion.) Rather than plotting surfaces of the zero of the discriminant, it is usually more revealing to look at surfaces of (arbitrary) positive values of

$$Q = -\frac{1}{2} \frac{\partial u_i}{\partial x_j} \frac{\partial u_j}{\partial x_i}. \tag{6.6}$$

In practice this measure emphasises small-scale structure and may not easily identify the largest-scale structures. Another measure for identifying vortices that is commonly used is low pressure, relative to some arbitrary reference level. This certainly picks out the strongest structures, but may be ineffective in flows where mean pressure is changing spatially. Jeong & Hussain (1995) introduced a new definition based on the second eigenvalue λ_2 of a combination of the symmetric strain rate and antisymmetric rotation rate tensors $S_{ik}S_{kj} + \Omega_{ik}\Omega_{kj}$, where by definition $S_{ij} = \frac{1}{2}(\partial u_i/\partial x_j + \partial u_j/\partial x_i)$ and $\Omega_{ij} = \frac{1}{2}(\partial u_i/\partial x_j - \partial u_j/\partial x_i)$. This was developed following a detailed critique of other measures and is based on an approximation (steady, inviscid) to an equation for the pressure Hessian $\partial^2 p/\partial x_i \partial x_j$. Our experience for near-wall flows studied in Southampton is that this method is very similar to plotting surfaces of constant Q. An alternative definition employed by Kida & Miura (1998) uses the pressure Hessian itself, together with a projection of the local flow on a plane perpendicular to the vortex axis, making use of a two-dimensional discriminant to ensure that low pressure cores correspond to swirling flow. The resulting definition of 'vortex skeletons' is advantageous in that it does not require any user-defined threshold to be set.

Visualisation is really a matter of data reduction. A DNS produces many GBytes of raw data. The end user will require this to be reduced to a few images or animations that show the key phenomena. In the process the simulator will want to look at many aspects of the flow to try to make sure nothing important has been omitted from the final set of images. Suggestions of things to look at include

- contour plots of raw velocity and pressure in instantaneous realisations in all three principal co-ordinate directions, with several slices in each;
- velocity vectors and instantaneous streamlines, which may be helpful in some situations along with contours of derived quantities such as vorticity;
- surface plots and animations of Q and/or λ_2 and a relative pressure;
- Side-by-side comparison of instantaneous flow with the time- or phase-averaged flowfield.

6.2.4 Recent progress

Carefully controlled DNS are assisting in understanding the mechanisms of near-wall turbulence. This is an area where the main features are already captured at moderate Reynolds numbers and the simulations provide complete information to evaluate theories. The most obvious structural features are low-speed streaks, which are rather passive streamwise-elongated regions of relatively low momentum fluid, and vortex rolls, which are much shorter regions of intense transient vortical activity. It has been demonstrated by Jiménez & Pinelli (1999) and Jiménez & Simens (2001) that there exists an autonomous near-wall region that is capable of self-sustaining turbulence without external forcing. This is not to say that there will not be interaction between inner and outer regions of a turbulent boundary layer, but does encourage the idea that a complete model for the inner region is at least possible, and which can later be coupled to the outer region, perhaps via a 'smart' LES boundary condition. A simplistic picture of what is happening is a 'regeneration cycle', where vortices produce streaks and streak instability leads to new vortices. Ongoing theoretical work includes streak instability analysis (Andersson et al., 2001), low dimensional modelling (reviewed by Rempfer, 2003), Reynolds-stress anisotropy modelling (Nikitin & Chernyshenko, 1997) and application of non-normal mode (optimal perturbation) theory (Butler & Farrel, 1993). It is known that if streaks grow beyond a critical amplitude they become locally unstable and a short-duration 'burst' will occur, leading to new vortices. However, the idea that the observed vortices produce streaks is challenged by recent work of Baig & Chernyshenko (2003), who use a combination of passive scalars to show that streak spacing is a function of mean profile shape, independent of the vorticity field. In their view, streaks form by optimal non-normal mode growth of any (random) background disturbances, unrelated to the vortices seen in the vicinity of the streaks.

Although this work on near-wall turbulence is far from complete, it does indicate a role for simulations that we haven't discussed yet, namely the isolation of flow phenomena by modification of the Navier–Stokes equations. In the case of Jiménez &

Pinelli (1999) this was achieved by inserting a buffer zone into a near-wall turbulent flow, preventing the inner region nearest the wall interacting with the outer region, which was filtered back to a laminar flow. Extreme care is called for when carrying out such simulations, lest the modified equations introduce spurious effects. Nevertheless, this kind of approach can be quite useful when looking at cause-and-effect relationships in turbulent flows, and are the kind of experiments that certainly cannot be contemplated in the laboratory.

The technology of DNS is currently being pushed into more complicated and practically relevant flows. Wu *et al.* (1999) have carried out DNS of bypass transition in which passing wakes triggered turbulent spots in a flat plate boundary layer. The simulations were modelled on an experimental/modelling study by Liu & Rodi (1991), which in turn modelled the environment in which transition on a gas turbine blade may occur. In the simulations travelling wakes, using data from a precursor simulation, were imposed upon a laminar boundary layer profile. Some disturbances were immediately ingested into the boundary layer as streaks, but did not cause transition. Instead turbulent spots developed further downstream, near the boundary layer neutral point, where disturbances from the free-stream entered the boundary layer. The terminology 'top-down' spots was used to describe these turbulent regions, which appear different to the classical 'bottom-up' spots, triggered by disturbances imposed at the wall. An extension of this to the more practically relevant geometry of a low-pressure turbine stator passage was considered by Wu & Durbin (2001), who showed how the wakes from turbine blades are severely distorted as they pass through the stator passage. Near-wall vortices on the concave side of the blades were seen to form from this wake deformation mechanism, rather than from a Görtler-type of instability. A clear progression of this work is certainly feasible, leading to simulations of complete turbine stages, including transient separation bubbles and wake-induced transition to turbulence. Another area where DNS is contributing is in flow control. Continuing the turbomachinery theme, Yao & Sandham (2002) looked at DNS of trailing-edge flow control, via base blowing and a secondary splitter plate. At a base Reynolds number of 1000, based on trailing edge thickness and local flow velocity (note that this is about a factor of five below full-scale turbomachinery applications), they were able to carry out parametric studies of the control methods. For base blowing it was found to be better to blow slowly out of a wide slot, rather than rapidly from a narrow slot. Significant increases in base pressure (corresponding to reductions in drag) were observed for the rectangular plate configuration used, for moderate base mass flow rates. Secondary splitter plates were found to have an even more marked effect on base pressure coefficient, with increases of 25% achieved with secondary splitter plates with a length equal to the trailing edge thickness, increasing to 50% with a length equal to five times the base thickness. Figure 6.1 shows several visualisations

N. D. Sandham

(a)

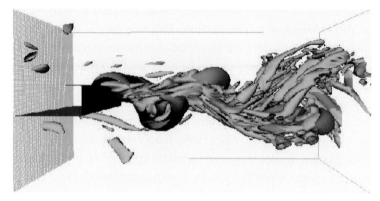

(b)

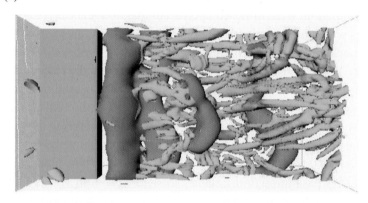

(c)

Fig. 6.1 Flow over a rectangular trailing edge (a) side view, (b) top view, (c) top view with base blowing. Dark surfaces are of constant pressure, enclosing a low pressure region, while light surfaces are of positive Q.

of the three-dimensional flow over the trailing edge, illustrating some of the flow visualisation points mentioned earlier. The dark surfaces are of constant pressure, enclosing low pressure regions, while light surfaces are of positive Q. It can be clearly seen how the base blowing, shown on Fig. 6.1(c), is effective in pushing turbulent structures further downstream and reducing the turbulence levels.

6.3 Large-eddy simulation techniques

The large-eddy simulation method has its roots in predictions of atmospheric flows in the 1960s, and like DNS has grown in importance as computers have increased in size and performance. Commercial computational fluid dynamics (CFD) codes increasingly offer options to carry out LES, and as a result the user community is expected to grow significantly in the next few years. For recent reviews of LES the reader can consult Fröhlich & Rodi (2002), Piomelli & Balaras (2002), Piomelli (1999), Lesieur & Métais (1996), Härtel (1996) and Ferziger (1996).

6.3.1 The Smagorinsky model

Large-eddy simulation works by low-pass filtering the equations of motion and modelling only the smallest scales of motion. In one spatial dimension a filtered variable $\tilde{u}$ is defined by

$$\tilde{u} = \int_{-\infty}^{\infty} G(x - x')u(x')\mathrm{d}x'. \tag{6.7}$$

Typical filter kernels $G(z)$ have a defined filter width Δ and include the Gaussian

$$G(z) = \frac{\exp(-z^2/\sigma^2)}{\sigma\sqrt{\pi}} \tag{6.8}$$

with $\sigma = \Delta/\sqrt{6}$, and the top hat filter

$$G(z) = \frac{1}{\Delta} \text{ for } z < \Delta \text{ (and zero otherwise).} \tag{6.9}$$

The filtering operation can be applied to the original Navier–Stokes equations (6.1) and (6.2) leading to

$$\frac{\partial \tilde{u}_i}{\partial x_i} = 0, \tag{6.10}$$

$$\frac{\partial \tilde{u}_i}{\partial t} + \frac{\partial(\tilde{u}_i \tilde{u}_j)}{\partial x_j} = -\frac{1}{\rho}\frac{\partial \tilde{p}}{\partial x_i} + \nu\frac{\partial^2 \tilde{u}_i}{\partial x_j \partial x_j} - \frac{\partial \tau_{ij}}{\partial x_j}, \tag{6.11}$$

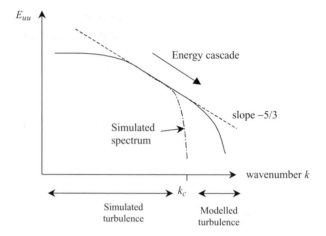

Fig. 6.2 Schematic representation of the one-dimensional energy spectrum E_{uu} as a function of wavenumber k, showing the effective spectrum simulated by the large-eddy simulation approach.

where we have assumed that the filtering operation commutes with the differentiation operators (see later discussion). The stress term

$$\tau_{ij} = \widetilde{u_i u_j} - \tilde{u}_i \tilde{u}_j \tag{6.12}$$

must be modelled in terms of the primary filtered variables to close the set of filtered equations. A simplified form of an LES spectrum is shown in Fig. 6.2. Below some cutoff wavenumber one should recover the usual $-5/3$ Kolmogorov spectrum, whereas above this cutoff the turbulence has to be modelled.

The first, and still very widely used, model for τ_{ij} is the Smagorinsky model, the form of which has its origins in the Richtmyer–von Neumann shock-capturing scheme. This can be written in the form of an eddy viscosity

$$\tau_{ij} - \frac{1}{3}\tau_{kk}\delta_{ij} = -2\nu_{\mathrm{T}}\tilde{S}_{ij}, \tag{6.13}$$

where $\tilde{S}_{ij} = \tfrac{1}{2}(\partial\tilde{u}_i/\partial x_j + \partial\tilde{u}_j/\partial x_i)$ is the strain rate tensor based on the filtered velocity field. The isotropic part of the stress term τ_{kk} is a scalar unknown and can be combined with $\tilde{p}$ to form a modified pressure variable. The eddy viscosity is expressed as

$$\nu_{\mathrm{T}} = (c_{\mathrm{S}}\Delta)^2 |\tilde{S}|, \tag{6.14}$$

where

$$|\tilde{S}| = \sqrt{2\tilde{S}_{ij}\tilde{S}_{ij}}. \tag{6.15}$$

Here Δ is the filter width and c_S is the Smagorinsky constant. With this basic Smagorinsky model, no explicit filtering is required – we work exclusively with filtered flow variables. In finite difference and finite volume schemes the filter width Δ is usually set equal to the grid spacing and the term 'sub-grid scale' (SGS) denotes the turbulence that has to be modelled. In three-dimensional calculations with grid spacing Δx_i in each direction it is usual to take $\Delta = (\Delta x_1 \Delta x_2 \Delta x_3)^{1/3}$ and the Smagorinsky constant as $c_S = 0.1$. Increasing c_S may be thought of as a method of increasing filter width and having more of the flow computed with the sub-grid model. One then needs more grid points to compute the same range of large scales, but one expects reduced influence of numerical issues (different numerical methods can have as big an influence as different sub-grid models, which leads to difficulty in making comparisons between different models implemented in different codes). We note in passing that, although derived from a very different perspective to the Smagorinsky model, the so-called 'structure function' modelling approach can be written (see Lesieur & Métais, 1996) in a form very similar to (6.14) and (6.15), with the addition of a term involving vorticity magnitude. This may lead to different local modelling of different structural features.

Near a wall this basic Smagorinsky model cannot be applied without further refinement. The eddy viscosity must be reduced to account for the wall proximity, for example by applying a van Driest multiplicative function f_{VD} to the eddy viscosity, with a simple form given by

$$f_{VD} = 1 - e^{-y^+/25}. \qquad (6.16)$$

It is possible to choose other functions that give more accurate near-wall asymptotic behaviour, however, the practical implications are small (there are worse approximations in LES than this). In meteorological applications it is common to introduce stochastic fluctuations in the near-wall region. These mimic the effect of backscatter – energy transfer from sub-grid scales to resolved scales. This additional realism has two benefits: firstly, improving the wall layer results (Mason & Thomson, 1992) and secondly, enabling the issue of predictability (sensitivity to random fluctuations) to be addressed (Leith, 1990). For engineering applications such explicit additional backscatter is not routinely used, but special treatment of the impermeable wall boundary condition is required. Rather than go all the way to the wall and apply no slip conditions, it is more efficient to apply a time-dependent wall shear stress condition that connects what is happening at the wall to events further away, possibly as far as the log law region. Schumann (1975) proposed

$$\tau_b = \langle \tau_b \rangle \frac{\tilde{u}_p}{\langle \tilde{u}_p \rangle}, \qquad (6.17)$$

where $\langle \cdot \rangle$ denotes some space-averaging procedure, $(\cdot)_b$ denotes the value on the boundary, and $(\cdot)_p$ denotes the control volume adjacent to the wall. An improvement to this is the procedure introduced by Werner & Wengle (1991), which integrates an approximation to the local mean flow to get a direct connection between the horizontal velocity components associated with the control volume next to the wall and the wall shear stress. To write the scheme, we introduce $R_{up} = u_p \Delta y / \nu$ as a Reynolds number based on the cell size Δy and velocity u_p (note that in the usual incompressible finite volume scheme the horizontal velocity components are located halfway up the faces of the control volume). Instantaneous wall shear stresses are assumed to be in phase with the velocities, with the magnitudes given by

$$
\frac{|\tau_{ub}|}{\rho u_p^2} = \frac{2}{R_{up}} \qquad \text{for } R_{up} \leq 70,
$$

$$
\frac{|\tau_{ub}|}{\rho u_p^2} = \frac{7.2 + a_p R_{up}}{R_{up}^{8/7}} \qquad \text{for } R_{up} > 70,
$$

(6.18)

with $a_p = (140^{4/7} - 7.2)/70$ for continuity of the function. (It should be noted that some of the constants have been rounded relative to the original to simplify the equations.) A similar procedure can be used for the other stress component τ_{wb}, using the local velocity w_p. As demonstrated by Werner & Wengle, the method is applicable to attached flows and to bluff body flows with large recirculation regions. Separation from a smooth surface is a much more difficult problem for LES, without using very high (quasi-DNS) resolutions.

6.3.2 Dynamic modelling

Large-eddy simulations of engineering flows are nowadays more likely to be based on dynamic formulations of the Smagorinsky model, originally developed by Germano *et al.* (1991). These rest upon the Germano (1992) identity

$$
L_{ij} \equiv \widehat{\bar{u}_i \bar{u}_j} - \hat{\bar{u}}_i \hat{\bar{u}}_j = \tau_{ij}^T - \hat{\tau}_{ij},
$$

(6.19)

where $\widehat{(\cdot)}$ represents a second filtering operation, called a 'test' filter, and

$$
\tau_{ij}^T = \widehat{\overline{u_i u_j}} - \hat{\bar{u}}_i \hat{\bar{u}}_j
$$

(6.20)

is the stress term arising from filtering the Navier–Stokes equations firstly with the original filter and then with the test filter. The important point to note is that, although we can't compute either τ_{ij} or τ_{ij}^T, we can compute the quantity L_{ij} just by combining filtered velocities.

The dynamic procedure can in principle be applied to any sub-grid model. For the Smagorinsky model given above we can equate L_{ij} to a term CM_{ij}, where M_{ij} is given by

$$M_{ij} = -2\left(\hat{\Delta}^2 \left|\hat{\tilde{S}}\right| \hat{\tilde{S}}_{ij} - \Delta^2 \widehat{\left|\tilde{S}\right| \tilde{S}_{ij}}\right), \qquad (6.21)$$

where $\hat{\Delta}$ is the test filter width. It is not possible to get a single value of the scalar C that works for all i, j combinations so some approximation is needed at this point. The procedure of Lilly (1992) selects C to minimise the square of the error $\varepsilon = L_{ij} - CM_{ij}$ by taking averages (denoted with $\langle \cdot \rangle$) over some sample of the flow:

$$C = \frac{\langle L_{ij} M_{ij} \rangle}{\langle M_{ij} M_{ij} \rangle}. \qquad (6.22)$$

If $\langle L_{ij} M_{ij} \rangle$ is found to be negative, we would get a negative eddy viscosity. Since this often leads to numerical instability it is usual to apply a limiter to $\langle L_{ij} M_{ij} \rangle$, for example by only allowing positive values. With C determined in this way and $\nu_T = C\Delta^2|\tilde{S}|$ there are no free parameters, save for the choice of filter type and ratio of filter widths $\hat{\Delta}/\Delta$. Operationally $\hat{\Delta}/\Delta = 2$ is usually used. The procedure defined by (6.22) is not unique and also introduces the problem of how to carry out the averaging operation. For flows that are homogeneous in one or more directions it is usual to average in those directions. For more general flows local averaging is required. Models that avoid averaging altogether (known as 'localisation' models, see for example Ghosal *et al.* (1995)) are attractive for complex geometry applications.

The dynamic modelling approach is especially well suited to transitional problems and relaminarisation, and can handle near-wall turbulence without additional corrections, so long as the near-wall structures are reasonably well resolved (within a factor of four of a DNS, for example, and perhaps even better in the wall-normal direction).

Both the basic Smagorinsky and dynamic Smagorinsky models are at heart rather basic eddy viscosity representations directly relating sub-filter stresses to the strain rates of the filtered flow variables. They actually work surprisingly well given that the normalised correlation between stress and strain rate is only of the order of 0.2 (Jiménez & Moser, 2000). Jiménez & Moser offer a partial explanation, pointing out that both the Smagorinsky and dynamic Smagorinsky approaches have an inbuilt self-adjustment property which makes them relatively insensitive to the precise value of the constant. If the constant is too low energy builds up in the high wavenumbers, leading to increased dissipation, and vice versa. Another key point from a practical point of view is that the models will always work well when the proportion of shear stress carried in the sub-grid model is significantly less than the shear stress of the large eddies.

6.3.3 Mixed models

Despite the successes of the dynamic Smagorinsky approach in comparisons of LES prediction with experiments (known as 'a posteriori' testing) it possesses a number of shortcomings. In practical implementations there is usually no permitted backscatter from sub-grid (strictly we should say 'sub-filter') to resolved terms. So-called 'a priori' tests, where sub-grid turbulence computed from filtered DNS is compared with predictions from eddy viscosity modelling, show that the models are in this respect extremely poor. Filter widths as small as possible are employed, but this introduces another potential problem in that the filter and test filter may be placed in a part of the spectrum close to the grid scale where numerical errors may distort results. Bearing these facts in mind, the hunt is still on for better formulations for the SGS terms, making optimum use of information already contained in the simulation variables.

Mixed models are SGS models that combine two different elementary models. The first example, proposed by Bardina *et al.* (1983), used the basic Smagorinsky model in combination with a scale similarity model,

$$\tau_{ij} = \widetilde{\tilde{u}_i \tilde{u}_j} - \tilde{\tilde{u}}_i \tilde{\tilde{u}}_j - 2\,(c_S \Delta)^2 \, |\tilde{S}| \tilde{S}_{ij}. \tag{6.23}$$

By itself the scale similarity part (the first two terms in (6.23)) performs well in a priori tests, but is considered insufficiently dissipative for practical applications. Another early model proposed for LES was the tensor eddy diffusivity model of Leonard (1974, 1997). For a Gaussian filter Leonard has shown that the shear stress terms can be represented exactly by

$$\tau_{ij} = \sum_{k=1}^{\infty} \left(\frac{\sigma^2}{2} \right)^k \frac{1}{k!} \frac{\partial^k \tilde{u}_i}{\partial x_l^k} \frac{\partial^k \tilde{u}_j}{\partial x_l^k}. \tag{6.24}$$

The full series would recover a 'DNS' result for τ_{ij} but is obviously not practical. Instead it was proposed to use only the first term as a sub-grid model:

$$\tau_{ij} = \frac{\sigma^2}{2} \frac{\partial \tilde{u}_i}{\partial x_l} \frac{\partial \tilde{u}_j}{\partial x_l}. \tag{6.25}$$

As with the scale similarity model, (6.25) by itself is insufficiently dissipative, and may become unstable. However, it has been combined with an eddy viscosity model to give good results (Vreman *et al.*, 1997, Winckelmans *et al.*, 2001, Leonard & Winckelmans, 1999).

Those that try mixed models usually come to the conclusion that they do better than the regular Smagorinsky or dynamic Smagorinsky models. A comprehensive study of models for the transitional compressible mixing layer is given by Vreman *et al.* (1997). In order of overall accuracy they had

1. dynamic mixed: Smagorinsky plus scale similarity;
2. dynamic mixed: Smagorinsky plus tensor eddy diffusivity;
3. dynamic Smagorinsky;
4. scale similarity;
5. tensor eddy diffusivity;
6. Smagorinsky.

The first three performed significantly better than the last three, while the last two were actually worse than no model at all, although this would depend on the numerical method used in the 'no model' test case. Further demonstration of the benefits of mixed models in other flows has been provided by Winckelmans *et al.* (2001) for decaying isotropic turbulence and channel flows, where a variant of model 2 above was used. Note however that Kobayashi & Shimomura (2003) report an inherent instability in near-wall turbulence with this approach. Obviously there is much work to do in classifying the performance of all the different possible combinations of models across a wide variety of flows and important numerical issues have to be addressed, so that results can be applied in different codes. However, a preliminary conclusion is that mixed models do well in both a priori and a posteriori calculations.

An explanation for the good performance of mixed models was offered by Shao, Sarkar & Pantano (1999). In this work a decomposition of the velocity field into a mean plus fluctuation $u = \langle u \rangle + u'$ is combined with the filtering operation. The shear stress is then written as $\tau_{ij} = \tau_{ij}^{\text{rapid}} + \tau_{ij}^{\text{slow}}$, where the 'rapid' part includes all terms with a mean velocity, and the 'slow' part includes terms only involving fluctuations. The slow part would always be present in LES, while the rapid part would only be active when mean velocity gradients were present, but would imply a rapid reaction of sub-grid stresses to changes in the mean velocity profile. Two models were analysed by Shao *et al.* The scale similarity model was found to predict energy transfer effects corresponding to the rapid part, while Smagorinsky could represent the slow terms. This offers a useful rationale for mixed models and suggests that such models will be essential to compute flows involving rapid changes of mean flow or flows that are far from a production equal to dissipation equilibrium.

6.3.4 Numerical errors in LES and guidelines for application

An often neglected aspect of LES is the influence of the numerical scheme on the results. Geurts (1999) presents the current state of LES as a delicate balancing act between competing sources of error, both numerical and modelling. To separate the many sources of possible error he uses explicit filtering, following Vreman *et al.* (1996a) and Ghosal (1996) and writes equations for general non-uniform filters, such as would be implicitly used in calculations on stretched grids. Additional terms appear due to the non-commutation of the filter operator with the differentiation operator. In Geurts & Leonard (2002) it is postulated that not all the errors may be removed by using optimised filters such as in Vasilyev *et al.* (1998). Demonstrations of typical magnitudes of errors are given in Geurts (1999) for various ratios of filter width to grid spacing Δ/h. Using data from Vreman's simulations, Geurts shows that for $\Delta/h = 1$ modelling errors are actually less significant than numerical errors for three different numerical schemes, while for $\Delta/h = 2$ numerical errors are reduced and modelling errors are larger. In fact there is cancellation of errors taking place, since the sign of the modelling error may be opposite to that of the discretisation error. This illustrates that for LES one can expect that research groups using different numerics may report different, even opposite, conclusions about the comparative performance of models and filters.

Geurts & Leonard (2002) concludes with guidelines for careful application of existing LES technology and addresses the issue of whether LES is actually ready for complex engineering calculations yet. The lack of a definitive 'yes' in their conclusions should be taken as a warning to those who propose that LES will replace RANS for practical calculations at high Reynolds numbers over the foreseeable future. The Geurts–Leonard guidelines are worth summarising again here to show how a new user may begin to gain trust in results obtained from a present-day LES code. (The validation procedures given for DNS in Section 6.2.2 are also relevant.)

- Validate the code against simpler theory and DNS databases.
- Use smoothly varying near-orthogonal grids and avoid dissipative numerical methods.
- Vary numerical and physical parameters (grid, box size, numerical method etc.).
- Use dynamic modelling (mixed models are especially recommended).
- Use explicit filtering and incorporate LES predictions at different Δ/h into the flow analysis.

6.3.5 Algorithmic approaches to LES

Several different but related techniques are based on the suggestion that an approximate inversion of the filtering operation can be used to infer sub-filter stresses, without the use of a turbulence model at all. The inverse modelling approach is laid out in Geurts (1997) and a related sub-grid estimation method of Domaradzki & Saiki (1997). Here we focus on a new deconvolution methodology as described by Stolz & Adams (1999) and Stolz *et al.* (2001a,b).

Given a filtering operator G such that $\tilde{u} = G * u$ an approximate inverse Q can be written as a truncated series

$$Q = \sum_{k=1}^{N} (I - G)^k, \qquad (6.26)$$

where I is the identity operator and N would typically be taken as 5. Stolz and Adams demonstrate the procedure using a family of non-negative filters based on Pade approximants. The cutoff wavenumber of the filter can be adjusted according to numerical method so that the numerical errors at high wavenumbers do not contaminate the solution. The model is completed by the use of a secondary filter, only active above the cutoff wavenumber of the primary filter. This secondary filter removes energy from the system, but acts only at the highest wavenumbers. It does entail the introduction of a new constant, but the authors claim the results are not very sensitive to the precise choice of this. A priori tests and full calculations for channel flow are given in Stolz *et al.* (2001a), showing good performance of the model for this standard test case.

A striking application of the approximate deconvolution procedure was given by Adams & Stolz (2002) for shocks in the one-dimensional Burgers and Euler equations. Here, model problems with discontinuities were solved without the use of explicit shock-capturing algorithms by reconstructing the steep gradients by deconvolution. This opens up the possibility of a unified sub-grid treatment that can handle sub-grid turbulence or near-discontinuities in an efficient manner. There are obvious applications to the problem of shock wave interaction with turbulent boundary layers, for which numerical methods are cumbersome and often inefficient.

A development from recent analysis is the Navier–Stokes alpha (NS-α) model, presented in Foias *et al.* (2001). A modified Kelvin's theorem is assumed to hold, using integration around a loop moving with a spatially filtered velocity. Equations comparable to the Navier–Stokes equations are obtained as

$$\frac{\partial \tilde{u}_i}{\partial x_i} = 0, \qquad (6.27)$$

$$\frac{\partial u_i}{\partial t} + \tilde{u}_j \frac{\partial u_i}{\partial x_j} + u_j \frac{\partial \tilde{u}_j}{\partial x_i} = -\frac{1}{\rho} \frac{\partial p}{\partial x_i} + \nu \frac{\partial^2 u_i}{\partial x_j \partial x_j}, \qquad (6.28)$$

where $\tilde{u}_i$ is a filtered form of u_i, or inversely

$$u_i = \tilde{u}_i - \alpha^2 \frac{\partial^2 \tilde{u}_i}{\partial x_j \partial x_j}. \qquad (6.29)$$

Numerical simulations of isotropic decaying turbulence with the NS-α equations are given by Chen *et al.* (1999), demonstrating some features of the model. The length scale α is fixed according to the smallest scale required to be accurately computed by the simulation. Compared to Navier–Stokes, the NS-α model shows a roll-off in energy spectrum of k^{-3} rather than $k^{-5/3}$ at high wavenumber k. Thus fewer modes are required to compute the same number of decades of the energy spectrum, with considerable savings in computer time compared to DNS of the Navier–Stokes equations. The model is not the same as conventional LES since an equation like (6.28) for $\tilde{u}_i$ would include time-dependent terms on the right hand side, which has not been tried in LES. A related approach to NS-α is to use a regularised (Leray) formulation of the Navier–Stokes equations. Geurts & Holm (2003) describe the method and present some sample calculations of LES at arbitrarily high Reynolds numbers, something that challenges conventional LES.

It is early days to make definitive conclusions about the various algorithmic approaches available. However, they do have several attractive features compared to the sub-grid modelling approaches. ADM is a procedure that extends in a much more straightforward manner to compressible flow than the modelling approach, while the NS-α and Leray-regularised methods separate the modelling from the numerical solution procedure.

6.3.6 LES/RANS hybrids

RANS and LES fix a length scale in different ways. In RANS it is determined by $K^{3/2}/\varepsilon$, or near a wall by κy, where κ is the Karman constant, while the smallest scale (comparable to a Kolmogorov scale) in a 'Smagorinsky fluid' (Muschinski, 1996) is $c_S \Delta$. Hybrid strategies are often employed near walls by switching length scales from filter width to wall distance as the wall is approached. Early examples are the methods of Schumann (1975) and Mason & Callen (1986).

A modern hybrid method that has gained considerable popularity is the detached eddy simulation (DES) approach, originally introduced by Spalart et al. (1997). The RANS model employed is the Spalart–Allmaras (1994) one-equation model, which has an eddy viscosity relation

$$\nu_T = \tilde{\nu} f(\tilde{\nu}/\nu) \tag{6.30}$$

and a transport equation for $\tilde{\nu}$ of the form

$$\frac{D\tilde{\nu}}{Dt} = g(u_i, \tilde{\nu}). \tag{6.31}$$

The full form of the functions $f(\tilde{\nu}/\nu)$ and $g(u_i, \tilde{\nu}, d)$ can be found in the original paper or in the text by Wilcox (2000). The important point from our perspective is that the right hand side contains an external length scale d. In the basic model this is the distance from the nearest wall, but this can be easily replaced by

$$\tilde{d} = \min(d, C_{\mathrm{DES}}\Delta), \tag{6.32}$$

where C_{DES} is a constant and $\Delta = \max(\Delta x_1, \Delta x_2, \Delta x_3)$. We can see that near the wall the model will reduce to RANS, but that away from the wall we will have a one-equation sub-grid model for an LES.

This basic DES model is now being applied to many flows. It appears especially useful for flows where separation occurs abruptly, for example from sharp corners or on massively stalled wings. Here, as noted by Spalart *et al.* (1997), the costs of doing reasonable LES are prohibitive, whereas DES can already be attempted. Separation from smooth surfaces is still a problem: if DES reduces to RANS at a wall we still have all the limitations of the underlying RANS model. Another problem that has emerged is in the blending region where the model changes from RANS to LES. Here Piomelli *et al.* (2002) see a step in the middle of the log law region and unrealistically long characteristic times. Similar issues were found by Temmerman *et al.* (2002) with a RANS/LES hybrid based on a one-equation model for the turbulence kinetic energy.

The practical issues concerned with application of DES often relate to the computational grid. Spalart (2001) identifies zones within a DES calculation that have different grid requirements. RANS regions have typical RANS requirements, down to Δy^+ less than 2 near walls with highly anisotropic grids (large spacings are allowed in wall-parallel directions). In the LES region the spatial extent of the fine grid, sufficient to resolve the turbulence, must be specified based on a guess as to where this region is. Uniform grid spacing $\Delta x = \Delta y = \Delta z$ is recommended in these regions. Overall, however, limits on grid stretching factors need to be applied, making the patched RANS/LES grid quite difficult to construct. It should be noted that the same issues apply in LES and even DNS of complex geometries, where the required grid can only be sensibly specified after a preliminary calculation has been carried out; these issues have been brought out in DES simply because it is being applied to some complex geometries already.

6.4 Applications and concluding remarks

6.4.1 LES test case studies

Rodi *et al.* (1997) and Rodi (2002) evaluate the status of LES using several bluff body test cases. These included the flow past a square-sectioned cylinder at $Re = 22\,000$, flow past a circular cylinder at $Re = 3900$ and $Re = 140\,000$, flow past a single wall-mounted cube, oriented face on to the oncoming flow at $Re = 40\,000$, and flow past a periodic array of wall-mounted cubes in a channel at

$Re = 3823$ based on the cube side. The flows are all characterised by large-scale vortex shedding and the test case results demonstrated clear superiority of LES over RANS, even though the LES simulations were significantly more expensive. No clear conclusions were reached regarding the relative performance of various sub-grid models, but comparisons did identify bad numerical practices to be avoided in LES, such as excessive grid stretching and the use of upwind schemes. Even at the modest Reynolds numbers considered, LES had problems where particular flow features, such as the thin reverse-flow region on the square cylinder test case, could not be resolved. For the circular cylinder case most practitioners avoided the higher Reynolds number subcritical test case. Those that tried it reported grid problems and a large influence of sub-grid model, which is perhaps not unexpected since affordable simulations could not fit enough grid points into the boundary layers without using extreme grid stretching. This test case is probably at the current limits of LES technology. The flows over wall-mounted cubes at low Reynolds numbers, on the other hand, were much better predicted by LES, and for these flows LES is already superior than RANS, albeit a factor of more than 25 more expensive.

A recent European collaborative project attempted LES of flow around an Aerospatiale-A airfoil at a chord Reynolds number $Re = 2 \times 10^6$. The results, described by Mellen *et al.* (2003), were generally disappointing, although perhaps not unsurprising given our previous discussions. Good results were only obtained when very high (and computationally expensive) numerical resolution was employed, suggesting that LES is currently not a cost-effective tool for aerofoil flows, even at these moderate Reynolds numbers. The largest calculations used 18 million grid points, but only computed a 1.2% of chord in the spanwise direction, which must be insufficient when separation occurs. Particular problems were caused by the process of transition to turbulence, which effectively had to be directly simulated. Turbulent trailing edge separation was also poorly predicted. An overall conclusion was that resolution was the key to good simulations. None of the various sub-grid treatments employed at low resolution were able to simulate adequately the physics of transition and separation.

6.4.2 Compressible flows

Although more expensive to compute than incompressible flows, high-speed compressible flows are increasingly being simulated, including the technologically important area of shock/boundary-layer interaction. This is an area where DNS has already proven its worth for theoretical developments, following simulations of homogeneous turbulence by Sarkar *et al.* (1991) and of mixing layers by Vreman *et al.* (1996b). The possible presence of shock waves, however, complicates the numerics. The principal issue in shock-wave/turbulence simulations is that good numerical methods for turbulence are generally inefficient for shock flows, while the best shock-capturing schemes are much too dissipative for accurate resolution of turbulence. Three main techniques are commonly used in shock-turbulence simulations: full shock resolution, essentially non-oscillatory (ENO) schemes, and hybrid methods, in which the method varies depending upon whether a shock wave is detected. The former two methods have proved too expensive for routine calculations and consequently hybrid methods have most commonly been used. The first DNS of a shock/turbulent boundary-layer interaction problem was a Mach 3 ramp flow case, simulated by Adams (2000). The simulation provided turbulence data in the interaction region where the shock oscillated due to the turbulence structures, and the turbulence in turn was changed by passing through the shock wave. In particular the Reynolds stresses were amplified by a factor as high as four, with normal stresses affected differently from shear stresses. The Reynolds analogy, commonly employed in modelling, was found to be invalid in the interaction region. As well as providing insight, this particular DNS provided a compressible flow test case for the approximate deconvolution model. Stolz *et al.* (2001b) demonstrated that this model served to capture sub-grid turbulent scales as well as the non-turbulent sub-grid scales associated with the shock wave. Another advantage of their approach was that the energy equation could be treated in exactly the same manner as the momentum equation. With conventional sub-grid modelling one would have to go term-by-term through the energy equation, decide whether a particular sub-grid term (there are six in all) needed modelling and if so, how to do the modelling.

LES is beginning to be applied to other shock/boundary-layer interaction problems. Garnier *et al.* (2002) have studied oblique shock impingement onto a Mach 2.3 turbulent boundary layer, while Sandham *et al.* (2003) simulated transonic flow over a circular arc bump. In the latter case the shock was relatively stationary, but there was strong streamwise variation of turbulence quantities in the separated shear layer immediately below the shock wave. Comparative studies of sub-grid models and numerical schemes are still in the early days for such flows.

6.4.3 Aeroacoustics

The field of computational aeroacoustics is growing rapidly and turbulence simulation is at the heart of recent advances. The framework for much of this work was set with Lighthill's famous acoustic analogy formulation, which rearranged the Navier–Stokes equations into a wave equation on the left hand side, representing wave propagation, and all the other terms (the 'sources') on the right hand side. In this way the acoustics part of the problem was essentially solved fifty years ago. However, further progress was limited by an inadequate representation of the sources in turbulent flow, so actually the problem of aeroacoustics is essentially a problem of turbulence, to which simulations are now contributing.

At a direct numerical simulation level (reinvented as 'direct noise computation' (DNC) by the acoustic community) we can simulate both the turbulence and the sound field. However, this is expensive even by DNS standards if one wants to compute anything more than the near field sound. Of the order of ten points per sound wavelength are needed to compute a sound wave accurately (depending on numerical method) so the grid requirements rapidly make this approach impractical. Additional difficulties relate to boundary conditions since acoustic calculations are highly sensitive to wave reflections from artificial computational boundaries. A whole range of 'buffer' zone boundaries have been developed. These are typically 'black box' regions around the edges of a computation where the Navier–Stokes equations are solved only in a modified form that attempts to damp any waves that reach the boundary. A healthy degree of scepticism is appropriate for these boundary conditions, since they are usually tuned specially for the problem under consideration and may not be appropriate to other flows where different wavelengths, wave amplitudes and wave orientations are present. Despite these additional difficulties several successful DNS of sound generation by turbulent flows have been published. Two of the most impressive have been the supersonic (Mach 1.92) jet sound calculation by Freund *et al.* (2000) and the high subsonic (Mach 0.9) jet calculation by Freund (2001). The supersonic jet noise problem is the easier proposition since the acoustic waves are stronger. If the sources of sound are also supersonic relative to the quiescent external flow then the sound will align predominantly with the Mach direction.

Extending the DNS beyond the immediate near field requires more effort. Freund *et al.*'s (2000) approach is to couple the DNS to a wave equation that is then solved by finite difference methods. This works if one is interested in computing out to a few tens of jet diameters, but to go further towards the acoustic far field alternative approaches are required. A method that is gaining in popularity involves the insertion of a surface into the computational domain. A generalisation of Lighthill's equations by Ffowcs Williams and Hawkings to include surfaces is then appropriate. The relevant terms can be accumulated on the surface during the calculation, followed by a post-processing exercise to extract the sound field. This approach has been demonstrated by Brentner & Farassat (1998) to be superior to the Kirchhoff surface approach used in acoustics.

Besides predicting the far field sound, the simulation approach can be used to study the physical nature of sound sources. Hu *et al.* (2003) extracted the acoustic sources from turbulent channel flow DNS and demonstrated the relative importance of fluctuating dissipation (monopole), wall shear stress (dipole) and Reynolds stress (quadrupole) sources of sound. In particular the influence of the wall shear stress dipole, a subject of considerable debate in the literature, has been clarified by these simulations. The zero wavenumber limit of both wall shear stress and wall pressure fluctuation spectra was found to be non-zero, implying a non-zero dipole source of sound. Predictions of the sound field suggest that this dipole source is dominant at low Mach numbers, but is overtaken by the quadrupole source as the Mach number increases.

6.4.4 Concluding remarks

DNS is a maturing technology, routinely used alongside laboratory experiments for scientific investigations of transitional and turbulent flow. As with any advanced technique it must be carefully applied, and there is scope for continued improvements in numerical algorithms, especially as more complex geometry flows are attempted. There is still much to be learned from simple geometry flows, however. We note that the highest Reynolds number boundary layer DNS is still that of Spalart (1988) at $R_\theta = 1440$. For its day this was an extremely large calculation, but hardware has already progressed to the point where one can conceive of further calculations at $R_\theta = 2880$ and $R_\theta = 5760$. Such simulations would complement experiments by providing complete datasets for Reynolds number trends to be studied and compared with theoretical predictions. Another interesting approach is the idea of the strained channel (Coleman *et al.*, 2000), where complex strain fields are imposed on turbulence in a simple geometry. This allows turbulence models to be checked for their ability to represent the turbulence response to (and relaxation from) imposed strain fields, clearly separated from issues of numerical resolution.

By contrast LES is not a mature technique. We have presented a variety of approaches, but it is not clear where these will line up in the final reckoning. More theoretical work is needed to place the actual techniques used on a sounder basis, and more careful studies need to be made of the comparative performance of models, both separately and in combination with particular numerical schemes, for more complicated flows than the isotropic turbulence and plane channel flow that have to date formed the main test beds for developing new models. Common practice at present in the engineering community is to use dynamic models and resolve near-wall structures such as low-speed streaks. A factor of four less resolution in all directions can be used relative to DNS, which represents a considerable saving. However, the scaling with Reynolds number is such that it is impossible to extend the method to applications such as flow over an aircraft wing. Such simulations are something of a 'poor man's DNS'. They are not sufficiently resolved to provide reliable data for investigations of flow physics and model validation (although they are undoubtedly a better representation of the flow physics than RANS models), while at the same time they are not demonstrating LES technology for applications, since the Reynolds numbers are still too low. Better wall treatments are needed to enable applications of LES to higher Reynolds numbers. Test cases involving DNS of flows with turbulent separation and reattachment should be of some assistance in developing models that have some useful range of applicability. Another hope for the future is that some of the fundamental knowledge being gained about near-wall turbulence mechanisms and control strategies can be applied to LES modelling.

References

Adams, N. A. 2000 Direct simulation of the turbulent boundary layer along a compression ramp at $M = 3$ and $Re\text{-}theta = 1685$. *J. Fluid Mech.* **420**, 47–83.

Adams, N. A. and Stolz, S. 2002 A subgrid-scale deconvolution approach for shock capturing. *J. Comp. Phys.* **178(2)**, 391–426.

Andersson, P., Brandt, L., Bottaro, A. and Henningson, D. S. 2001 On the breakdown of boundary layer streaks. *J. Fluid Mech.* **428**, 29–60.

Baig, M. F. and Chernyshenko, S. 2003 Regeneration mechanism of organized structures in near-wall turbulence. *Turbulence and Shear Flow Phenomena* **3**, Sendai, Japan, June 2003.

Bardina, J., Ferziger, J. and Reynolds, W. C. 1983 Improved turbulence models based on large eddy simulation of homogeneous, incompressible, turbulent flows. Report TF-19, Thermosciences Division, Mechanical Engineering Dept., Stanford University.

Brentner, K. S. and Farassat, F. 1998 Analytical comparison of the acoustic analogy and Kirchhoff formulation for moving surfaces. *AIAA J.* **36(8)**, 1379–1386.

Butler, K. M. and Farrell, B. F. 1993 Optimal perturbations and streak spacing in wall-bounded turbulent shear flow. *Phys. Fluids* **A5(3)**, 774–777.

Chen, S., Holm, D. D., Margolin, L. G. and Zhang, R. 1999 Direct numerical simulations of the Navier–Stokes alpha model. *Physica D* **133(1–4)**, 66–83.

Chong, M. S., Perry, A. E. and Cantwell, B. J. 1990 A general classification of three-dimensional flow fields. *Phys. Fluids* **2(5)**, 765–777.

Coleman, G. N., Kim, J. and Spalart, P. R. 2000 A numerical study of strained three-dimensional wall-bounded turbulence. *J. Fluid Mech.* **416**, 75–116.

Domaradzki, J. A. and Saiki, E. M. 1997 A subgrid-scale model based on the estimation of unresolved scales of motion. *Phys. Fluids* **9**, 2148–2164.

Durbin, P. A. 1995 Separated flow computations with the k–ε–v^2 model. *AIAA J.* **33(4)**, 659–664.

Ferziger, J. H. 1996 Large eddy simulation. Chapter 3 in *Simulation and Modelling of Turbulent Flows*, T. Gatski *et al.*, eds. Oxford University Press.

Foias, C., Holm, D. D. and Titi, E. S. 2001 The Navier–Stokes-alpha model of fluid turbulence. *Physica D* **152**, 505–519.

Freund, J. B. 2001 Noise sources in a low-Reynolds number turbulent jet at Mach 0.9. *J. Fluid Mech.* **438**, 277–305.

Freund, J. B., Lele, S. K. and Moin, P. 2000 Numerical simulation of a Mach 1.92 turbulent jet and its sound field. *AIAA J.* **38(11)**, 2023–2031.

Fröhlich, J. and Rodi, W. 2002 Introduction to large eddy simulation of turbulent flows. Chapter 8 in *Closure Strategies for Turbulent and Transitional Flows*, B. E. Launder and N. D. Sandham, eds. Cambridge University Press.

Garnier, E., Sagaut, P. and Deville, M. 2002 Large eddy simulation of shock/boundary-layer interaction. *AIAA J.* **40(10)**, 1935–1944.

Germano, M. 1992 Turbulence: the filtering approach. *J. Fluid Mech.* **238**, 325–336.

Germano, M., Piomelli, U., Moin, P. and Cabot, W. H. 1991 A dynamic subgrid-scale eddy viscosity model. *Phys. Fluids* **A3**, 1760–1765.

Geurts, B. J. 1997 Inverse modelling for large-eddy simulation. *Phys. Fluids* **9(12)**, 3585–3587.

 1999 Balancing errors in LES. In *Direct and Large-Eddy Simulation III*, P. R. Voke, N. D. Sandham and L. Kleiser, eds. Kluwer.

Geurts, B. J. and Holm, D. D. 2003 Regularization modelling for large-eddy simulation. *Phys. Fluids* **15(1)**, L13–L16.

Geurts, B. J. and Leonard, A. 2002 Is LES ready for complex flows? Chapter 25 in *Closure Strategies for Turbulent and Transitional Flows*, B. E. Launder and N. D. Sandham, eds. Cambridge University Press.

Ghosal, S. 1996 An analysis of numerical errors in large-eddy simulations of turbulence. *J. Comp. Phys.* **125**, 187–206.

Ghosal, S., Lund, T. S., Moin, P. and Akselvoll, K. 1995 A dynamic localization model for large-eddy simulation of turbulent flows. *J. Fluid Mech.* **286**, 229–255.

Härtel, C. 1996 In *Handbook of Computational Fluid Mechanics*, R. Peyret, ed. Academic Press, 284–338.

Hu, Z. H., Morfey, C. L. and Sandham, N. D. 2003 Sound radiation in turbulent channel flows. *J. Fluid Mech.* **475**, 269–302.

Jeong, J. and Hussain, F. 1995 On the identification of a vortex. *J. Fluid Mech.* **285**, 69–94.

Jiménez, J. and Moser, R. D. 2000 Large-eddy simulations: Where are we and what can we expect. *AIAA J.* **38(4)**, 605–612.

Jiménez, J. and Pinelli, A. 1999 The autonomous cycle of near-wall turbulence. *J. Fluid Mech.* **389**, 335–359.

Jiménez, J. and Simens, M. P. 2001 Low-dimensional dynamics of a turbulent wall flow. *J. Fluid Mech.* **435**, 81–91.

Kida, S. and Miura, H. 1998 Identification and analysis of vortical structures. *Eur. J. Mech. B/Fluids* **17(4)**, 471–488.

Kim, J., Moin, P. and Moser, R. 1987 Turbulence statistics in fully developed turbulent channel flow. *J. Fluid Mech.* **177**, 133–166.

Kobayashi, H. and Shimomura, Y. 2003 Inapplicability of the dynamic Clark model to the large eddy simulation of incompressible turbulent channel flows. *Phys. Fluids* **15(3)**, L29–L32.

Laurence, D. 2002 Large eddy simulation of industrial flows? Chapter 13 in *Closure Strategies for Turbulent and Transitional Flows*, B. E. Launder and N. D. Sandham, eds. Cambridge University Press.

Le, H., Moin, P. and Kim, J. 1997 Direct numerical simulation of turbulent flow over a backward-facing step. *J. Fluid Mech.* **330**, 349–374.

Leith, C. E. 1990 Stochastic backscatter in a subgrid-scale model: plane shear mixing layer. *Phys. Fluids* **A2**, 297–299.

Leonard, A. 1974 Energy cascade in large-eddy simulations of turbulent fluid flows. *Adv. in Geophysics* **18A**, 237–248.

 1997 Large-eddy simulation of chaotic advection and beyond. AIAA Paper 97-0204.

Leonard, A. and Winckelmans, G. S. 1999 A tensor diffusivity sub-grid model for large-eddy simulation. In *Direct and Large-Eddy Simulation III*, P. R. Voke, N. D. Sandham and L. Kleiser, eds. Kluwer.

Lesieur, M. and Métais, O. 1996 New trends in large-eddy simulation of turbulence. *Ann. Rev. Fluid Mech.* **28**, 45–82.

Lilly, D. K. 1992 A proposed modification of the Germano subgrid-scale closure. *Phys. Fluids* **4**, 633–635.

Liu, X. and Rodi, W. 1991 Experiments on transitional boundary layers with wake-induced unsteadiness. *J. Fluid Mech.* **231**, 229–256.

Mason, P. J. and Callen, N. S. 1986 On the magnitude of the subgrid scale eddy coefficient in large eddy simulations of turbulent channel flow. *J. Fluid Mech.* **162**, 439–462.

Mason, P. J. and Thomson, D. J. 1992 Stochastic backscatter in large-eddy simulations of boundary layers. *J. Fluid Mech.* **242**, 51–78.

Mellen, C. P., Fröhlich, J. and Rodi, W. 2003 Lessons from LESFOIL project on large-eddy simulation of flow around an airfoil. *AIAA J.* **41(4)**, 573–581.

Moin, P. and Mahesh, K. 1998 Direct numerical simulation: A tool in turbulence research. *Ann. Rev. Fluid Mech.* **30**, 539–578.

Muschinski, A. 1996 A similarity theory of locally homogeneous and isotropic turbulence generated by a Smagorinsky type LES. *J. Fluid Mech.* **325**, 239–260.

Nikitin, N. and Chernyshenko, S. 1997 On the nature of organised structures in turbulent near-wall flows. *Fluid Dynamics* **32(1)**, 18–23.

Piomelli, U. 1999 Large-eddy simulation: achievements and challenges. *Prog. Aero. Sci.* **35(4)**, 335–362.

Piomelli, U. and Balaras, E. 2002 Wall-layer models for large-eddy simulation. *Ann. Rev. Fluid Mech.* **34**, 349–374.

Piomelli, U., Balaras, E., Squires, K. D. and Spalart, P. R. 2002 Interaction of the inner and outer layers in large-eddy simulations with wall-layer models. *Engineering Turbulence Modelling and Experiments 5*, W. Rodi and N. Fueyo, eds. Elsevier.

Rempfer, D. 2003 Low-dimensional modelling and numerical simulation of transition in simple shear flows. *Ann. Rev. Fluid Mech.* **35**, 229–265.

Rodi, W. 2002 Large-eddy simulation of the flow past bluff bodies. Chapter 12 in *Closure Strategies for Turbulent and Transitional Flows*, B. E. Launder and N. D. Sandham, eds. Cambridge University Press.

Rodi, W., Ferziger, J. H., Breuer, M. and Pourquie, M. 1997 Status of large-eddy simulation: Results of a Workshop. *J. Fluids Engineering* **119(2)**, 248–262.

Rogers, M. M. and Moser, R. D. 1994 Direct simulation of a self-similar turbulent mixing layer. *Phys. Fluids* **6(2)**, 903–923.

Sandham, N. D. 2002 Introduction to direct numerical simulation. Chapter 7 in *Closure Strategies for Turbulent and Transitional Flows*, B. E. Launder and N. D. Sandham, eds. Cambridge University Press.

Sandham, N. D. and Howard, R. J. A. 1998 Direct simulation of turbulence using massively parallel computers In *Parallel Computational Fluid Dynamics*, D. R. Emerson *et al.*, eds. Elsevier.

Sandham, N. D., Yao, Y. F. and Lawal, A. A. 2003 Large-eddy simulation of transonic turbulent flow over a bump. *Int. J. Heat and Fluid Flow* **24(4)**, 584–595.

Sarkar, S., Erlebacher, G. and Hussaini, M. Y. 1991 Direct simulation of compressible turbulence in a shear flow. *Theor. Comp. Fluid Dyn.* **2**, 291–305.

Schumann, U. 1975 Subgrid scale model for finite difference simulations of turbulent flows in plane channels and annuli. *J. Comp. Phys.* **18**, 376–404.

Shao, L., Sarkar, S. and Pantano, C. 1999 On the relationship between the mean flow and subgrid stresses in LES of turbulent shear flows. *Phys. Fluids* **11(5)**, 1229–1248.

Spalart, P. R. 1988 Direct simulation of a turbulent boundary layer up to $R_\theta = 1410$. *J. Fluid Mech.* **187**, 61–98.

2001 Young person's guide to detached-eddy simulation grids. NASA/CR-2001-211032.

Spalart, P. R. and Allmaras, S. R. 1994 A one-equation turbulence model for aerodynamic flows. *La Recherché Aérospatiale* **1**, 5–21.

Spalart, P. R., Jou, W. H., Strelets, M. and Allmaras, S. R. 1997 Comments on the feasibility of LES for wings, and on a hybrid RANS/LES approach. *Proc. First AFOSR Conference on DNS/LES*, C. Liu and Z. Liu, eds. Greyden Press.

Stolz, S. and Adams, N. A. 1999 An approximate deconvolution procedure for large-eddy simulation. *Phys. Fluids* **11(7)**, 1699–1701.

Stolz, S., Adams, N. A. and Kleiser, L. 2001a An approximate deconvolution model for large eddy simulation with application to incompressible wall-bounded flow. *Phys. Fluids* **13(4)**, 997–1015.

2001b The approximate deconvolution model for large eddy simulations of compressible flows and its application to shock-turbulent-boundary-layer interaction with application to incompressible wall-bounded flow. *Phys. Fluids* **13(10)**, 2985–3001.

Temmerman, L., Leschziner, M. A. and Hanjalic, K. 2002 A-priori studies of a near-wall RANS model within a hybrid LES/RANS scheme. *Engineering Turbulence Modelling and Experiments 5*, W. Rodi and N. Fueyo, eds. Elsevier.

Vasilyev, O. V., Lund, T. S. and Moin, P. 1998 A general class of commutative filters for LES in complex geometries. *J. Comp. Phys.* **146**, 82–104.

Vreman, B., Geurts, B. and Kuerten, H. 1996a Comparison of numerical schemes in large eddy simulation of the temporal mixing layer. *Int. J. Num. Meth. Fluids* **22(4)**, 297–311.

Vreman, A. W., Sandham, N. D. and Luo, K. H. 1996b Compressible mixing layer growth rate and turbulence statistics. *J. Fluid Mech.* **320**, 235–258.

Vreman, B., Geurts, B. and Keurten, H. 1997 Large-eddy simulation of the turbulent mixing layer. *J. Fluid Mech.* **339**, 357–390.

Werner, H. and Wengle, H. 1991 Large-eddy simulation of turbulent flow over and around a cube in a plate channel. *Proceedings of Eighth Symposium on Turbulent Shear Flow*, Munich, Sept. 9–11, 1991.

Wilcox, D. C. 2000 *Turbulence Modeling for CFD*, DCW Industries, California.

Winckelmans, G. S., Wray, A. A., Vasilyev, O. V. and Jeanmart, H. 2001 Explicit-filtering large-eddy simulation using the tensor diffusivity model supplemented by a dynamic Smagorinsky term. *Phys. Fluids* **13(5)**, 1385–1403.

Wu, X. H. and Durbin, P. A. 2001 Evidence of longitudinal vortices evolved from distorted wakes in a turbine passage. *J. Fluid Mech.* **446**, 199–228.

Wu, X. H., Jacobs, R. G., Hunt, J. C. R. and Durbin, P. A. 1999 Simulation of boundary layer transition induced by periodically passing wakes. *J. Fluid Mech.* **398**, 109–153.

Yao, Y. F. and Sandham, N. D. 2002 Direct numerical simulation of turbulent trailing edge flow with base flow control. *AIAA J.* **40(9)**, 1708–1716.

7

Computational modelling of multi-phase flows

G. F. Hewitt

Imperial College of Science, Technology and Medicine, London

M. W. Reeks

University of Newcastle upon Tyne

1 Introduction

Computational modelling is assuming a greater and greater role in the study of multi-phase flows. Although it is not yet feasible to predict complex multi-phase flow fields over the full range of velocities and flow patterns, computational methods are helpful for a variety of reasons which include:

(1) They enable insights to be obtained on the nature and relative importance of phenomena and are a natural aid to experimental measurement. Indeed, it is often possible to compute quantities which cannot be readily measured.
(2) When coupled with experimental observations and empirical relationships, computational methods can give predictions which are reaching the stage of being useful in the design and operation of systems involving multi-phase flows, particularly for dispersed flow situations. This fact is reflected in the growing number of commercial computer codes which are available for application in this field.

In this chapter, we will deal first with the application of *single-phase* prediction methods in the interpretation of *two-phase* flows. Here, a brief description is given of the available turbulence models and examples cited of the application of this approach (flows in coiled tubes, horizontal annular flow and waves in annular flow).

   An important class of two-phase flows is that where one of the phases is *dispersed* in the other, for example dispersions of bubbles in a liquid (*bubble flow*), dispersions of solid particles in a gas or liquid (*gas–solids or liquid–solids dispersed flows*) and dispersions of droplets of one liquid in another liquid (*liquid–liquid dispersed flow*). Section 3 deals with these types of flow and begins with an overview (Sections 3.1

Prediction of Turbulent Flows, eds. G. F. Hewitt and J. C. Vassilicos.
Published by Cambridge University Press. © Cambridge University Press 2005.

and 3.2) of some of the key phenomena which affect the modelling (particles in turbulent structures, particle–wall interactions and particle–particle interactions in dense suspensions, etc.). The types of modelling methods (random walk models, multi-fluid models, PDF methods and more advanced models such as DNS, LES, etc.) are then reviewed in Sections 3.3 and 3.4. Section 3.5 then presents some examples of the application of *multi-fluid models* for dispersed flows. Increasingly, nowadays, the multi-fluid model is being applied to *multi-dimensional* dispersed flows. Basically, the phases are treated as *inter-penetrating media* and conservation equations are written for each phase, with *closure laws* written to describe their interaction. These closure laws are often empirically based. Examples are presented in which the model is implemented with various levels of sophistication in turbulence modelling. Applications with multiple bubble fields are also briefly reviewed.

In the multi-fluid model, the behaviour of the dispersed phase is represented only in terms of its phase fraction and velocity. The behaviour of individual dispersed phase elements (droplets, particles or bubbles) is lost in the modelling. There are a number of cases where it is advantageous to model ('track') the motion of the individual inclusions of the dispersed phase, and this type of modelling is introduced in Section 3.6; one form of such modelling ('bubble tracking') is described by way of example. Multi-fluid models for fluid–fluid dispersed flows (bubble and drop flows) do not usually take account of the deformability of the fluid–fluid interfaces (except implicitly through closure laws). Moreover, many two-phase flows occurring in practice are totally or partially *separated* (e.g. annular or stratified flows) and, again, interface deformation and structure is a crucial factor. Thus, in the classical multi-fluid models for such separated flows, special (and empirical) closure laws are required for interfacial friction, which is strongly affected by the interfacial structure. In Section 4.1, a brief discussion is given of the use of the multi-fluid model for separated flows.

For both dispersed and separated flows, there has been, in recent years, an increasing focus on the use of computational modelling to try to predict the detailed development of interfacial configuration. The ultimate objective of this work can be seen as to provide an alternative to the present empirical closure strategies. Though, for flows of engineering significance, this objective is often far from being achieved, some fascinating results and insights are already being generated. This work is reviewed in Sections 4.2 and 4.3 and begins (in Section 4.2) with a survey of available methods (boundary integral, VOF, level sets, front tracking, molecular dynamics, lattice gas, lattice Boltzmann). In Section 4.3, examples of calculations with the various methods are presented.

The chapter closes with a brief overview in Section 5.

2 Single-phase flow modelling applied to two-phase flow systems

Where the flow fields are separated, each field can be calculated by conventional single-phase computational fluid dynamics (CFD) methods, often using available commercial codes. In such calculations, the assumption is made of continuity of velocity and shear stress at the interface; this condition can be fulfilled by carrying out iterative calculations in which the shear stress and velocity of one phase at the boundary can be used as a boundary condition for the calculation of the other in a cyclical process. In many modern commercial computer codes, this process is executed automatically. A specific problem in this approach is the question of turbulence. It is normally assumed that turbulence is damped on either side of the interface, with continuity of shear stress being obtained by the existence of large velocity gradients in the vicinity of the interface and in both phases. To a first approximation, it may often be assumed that the gas phase flows over what approximates to a solid surface. However, in considering liquid phase calculations, it is important to take account of the velocity and shear stress distributions.

The single-phase flow field treatment can often be useful in dealing with separated flow fields such as those found in stratified flow, annular flow and (using computational domains which travel at the slug velocity) in slug flow. Inter-entrainment of the phases cannot, of course, be dealt with readily by this technique.

The calculation of single-phase turbulent flow fields is, in itself, extremely difficult, particularly at Reynolds numbers of engineering significance. In these cases, a *turbulence model* has to be used, and a brief review of some models is given in Section 2.1 below. Specific examples of the application of the single-phase flow methodology to gas–liquid flows are then given in Section 2.2, 2.3 and 2.4 for flows in coiled tubes, horizontal annular flow and flow in disturbance waves in vertical annular flow, respectively.

2.1 Turbulence modelling

For a turbulent single-phase flow, the momentum conservation equation takes the form:

$$U_j \frac{\partial U_i}{\partial x_j} = -\frac{1}{\rho}\frac{\partial p}{\partial x_i} + \frac{\partial}{\partial x_i}\left(\nu \frac{\partial U_i}{\partial X_j} - \overline{u_i' u_j'}\right), \tag{7.1}$$

where U_i is the time averaged velocity in the direction x_i, ρ the fluid density, p the pressure, ν the kinematic viscosity and u_i' the instantaneous velocity (relative to U_i) of the fluid in direction x_i. The term $\overline{u_i' u_j'}$ appearing on the right hand side is due to turbulent motion and the product $\rho \overline{u_i' u_j'}$ is called the Reynolds stress. It acts to promote the diffusion of momentum and cannot be obtained by further time-averaging alone and has to be modelled. Modelling of turbulent flows is a major topic in its own right and cannot be dealt with in detail here. A review by Gibson (1997) gives an entry into the subject. Here we give only a brief survey.

In the *standard k–ε model*, the turbulent motion is characterised by two quantities, namely k, the turbulent kinetic energy, and ε, the turbulent energy dissipation rate. These values are given by:

$$k = \frac{1}{2}\overline{u'_i u'_j}, \qquad (7.2)$$

$$\varepsilon = \nu \overline{\frac{\partial u'_i}{\partial x_j} \cdot \frac{\partial u'_j}{\partial x_i}}. \qquad (7.3)$$

The Reynolds stress tensor can be expressed in terms of mean velocity gradients and an effective turbulent kinematic viscosity ν_t (the Boussinesq approximation) as follows:

$$\overline{u'_i u'_j} = -\nu_t \left(\frac{\partial U_i}{\partial x_j} + \frac{\partial U_j}{\partial x_i} \right) \qquad (7.4)$$

with ν_t being given by:

$$\nu_t = c_\mu \frac{k^2}{\varepsilon}, \qquad (7.5)$$

where c_μ is an empirical constant whose value has been determined from measurement as 0.09. Thus, if the distribution of k and ε is known, then the momentum equation can be solved and the flow field calculated. In the k–ε model, transport equations are written for k and ε, which have the form:

$$\text{advection} = \text{production} - \text{diffusion} - \text{dissipation}, \qquad (7.6)$$

and these equations for k and ε, respectively, are as follows:

$$\frac{\partial k}{\partial t} + \frac{\partial (U_j k)}{\partial x_j} = \nu_t \left(\frac{\partial U_i}{\partial x_j} + \frac{\partial U_j}{\partial x_i} \right) \frac{\partial U_i}{\partial x_j} + \frac{\partial}{\partial x_j} \left(\frac{\nu}{\sigma_k} \frac{\partial k}{\partial x_j} \right) - \varepsilon, \qquad (7.7)$$

$$\frac{\partial \varepsilon}{\partial t} + \frac{\partial (U_j \varepsilon)}{\partial x_j} = c_1 \frac{\varepsilon}{k} \nu_t \left(\frac{\partial U_i}{\partial x_j} + \frac{\partial U_j}{\partial x_i} \right) \frac{\partial U_i}{\partial x_j} + \frac{\partial}{\partial x_j} \left(\frac{\varepsilon}{\sigma_\varepsilon} \frac{\partial \varepsilon}{\partial x_j} \right) - c_2 \frac{\varepsilon^2}{k}, \qquad (7.8)$$

where four new empirical constants have been introduced, namely $c_1 = 1.44$, $c_2 = 1.92$, $\sigma_k = 1.0$ and $\sigma_\varepsilon = 1.717$. The standard k–ε described above is valid only at high Reynolds numbers. Since the turbulent fluctuations are damped near a solid wall, there will be a region close to the wall where viscous effects are important. In this region, the local turbulence Reynolds number, defined by Jones and Launder

(1972) as:

$$Re_t = \rho k^2/\mu\varepsilon, \tag{7.9}$$

where μ is the dynamic viscosity of the fluid, will be small and the high Reynolds number models are no longer applicable. For this reason, the standard k–ε model cannot be used to calculate the flow right up to the wall through the viscous sub-layer. Use is made therefore of empirical laws of sufficient generality (such as the universal velocity profile) to connect the wall conditions such as shear stress and heat flux to the dependent variables just outside the viscous layer. This 'wall function' approach is not always very satisfactory, and a special low Reynolds number k–ε model can be employed, as proposed initially by Jones and Launder (1972). In this model, two of the empirical constants used in the original k–ε model (namely c_μ and c_2) are expressed (Launder and Sharma, 1974) as functions of the turbulence Reynolds number as follows:

$$c'_\mu = c_\mu \exp(-3.4/(1 + Re_t/50)^2), \tag{7.10}$$

$$c'_2 = c_2 \left[1 - 0.3 \exp\left(-Re_t^2\right)\right]. \tag{7.11}$$

The transport equations for k and ε are also modified to the forms:

$$\frac{\partial k}{\partial t} + \frac{\partial(U_j k)}{\partial x_j} = \nu_t \left(\frac{\partial U_i}{\partial x_j} + \frac{\partial U_j}{\partial x_i}\right)\frac{\partial U_i}{\partial x_j} + \frac{\partial}{\partial x_j}\left[\left(\nu + \frac{\nu_t}{\sigma_k}\right)\frac{\partial k}{\partial x_j}\right]$$
$$- \varepsilon - 2\nu\left(\frac{\partial k^{1/2}}{\partial x_i} \cdot \frac{\partial k^{1/2}}{\partial x_j}\right), \tag{7.12}$$

$$\frac{\partial \varepsilon}{\partial t} + \frac{\partial(U_j \varepsilon)}{\partial x_j} = c_1 \frac{\varepsilon}{k}\nu_t \left(\frac{\partial U_i}{\partial x_j} + \frac{\partial U_j}{\partial x_i}\right)\frac{\partial U_i}{\partial x_j} + \frac{\partial}{\partial x_j}\left[\left(\nu + \frac{\nu_t}{\sigma_\varepsilon}\right)\frac{\partial \varepsilon}{\partial x_j}\right]$$
$$- c'_2\frac{\varepsilon^2}{k} + 2\nu\nu_t\left(\frac{\partial^2 U_i}{\partial x_j \partial x_i}\right)^2. \tag{7.13}$$

The above equations are, of course, solved by finite element or finite volume methods; in the calculations described below, the CFX code was used, which offers a number of options for the differencing scheme. In most of the calculations described in this section, the quadratic upwind differencing option (QUICK) was employed.

Though the k–ε model and its variants are the most widely used, the development of better models is a major industry! Examples of alternative approaches are as follows.

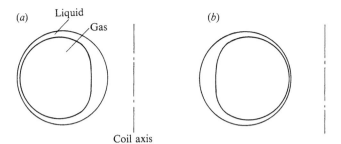

Fig. 7.1 (a) Inverted and (b) non-inverted film thickness distribution in annular two-phase flow in coiled tubes.

(1) **Reynolds stress transport models.** There are a number of turbulent fields where the Boussinesq approximation (Eq. 7.4) is not valid, the point of maximum velocity being different to the point of zero shear stress, for instance. As an alternative approach, transport equations can be written for the Reynolds stresses themselves. Thus, more accurate predictions can often be obtained, though at the expense of extra computer time. It should be pointed out that the Reynolds stress models have many fitted constants and usually use wall functions. Nevertheless, they are a useful option for more complex flows.

(2) **Direct numerical simulation (DNS).** For low Reynolds numbers (typically below 10 000), direct numerical simulation of turbulent flows is now feasible, though extra-ordinarily time consuming. Nevertheless, data obtained with DNS can often be useful in checking out many of the features of turbulence models.

(3) **Large eddy simulation (LES).** In this methodology, the larger eddies are directly simulated whilst the small eddies are represented by 'sub-grid models'. Wall functions are also required. LES is finding an increasing role in multi-phase flow systems and can be used up to Reynolds numbers of practical significance.

Even for single-phase flows, turbulence models often have a somewhat insecure basis. They should be used for two-phase flows with a certain degree of caution!

2.2 Flow in coiled tubes

In annular flow in coiled tubes, an interesting effect occurs which is illustrated in Fig. 7.1.

Intuitively, one would expect the liquid phase to be concentrated on the outside of the coil in such systems (Fig. 7.1 (b)) but, in fact, the reverse is often true with the liquid phase flowing on the *inside* of the bend (Fig. 7.1 (a)). The transition between the two conditions is often referred to as *inversion*. Clearly, such an effect has considerable significance in the many important applications of evaporation in coiled tubes.

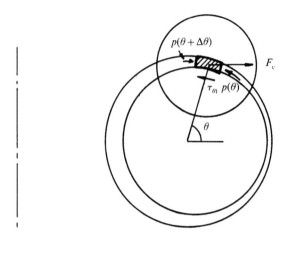

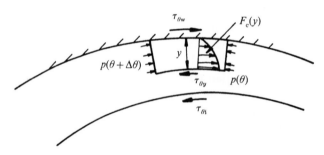

Fig. 7.2 Force balance on the liquid film in annular flow in coiled tubes. F_c is centrifugal force.

The application of CFD techniques to the interpretation of the inversion phenomenon in coiled tubes is described by Hewitt and Jayanti (1992) and this treatment is briefly summarised here. The approach was to treat the liquid film by a simple analytical model with boundary conditions given from a CFD treatment of the gas core. The basis of the film model is illustrated in Fig. 7.2. The circumferential shear stress $\tau_\theta(y)$ within the liquid film at a distance y from the wall is given by:

$$\tau_\theta(y) = \tau_\theta(0) - \frac{2}{D}\frac{\Delta p(\theta)}{\Delta\theta}y + \frac{\rho\tau_x^2\sin\theta}{\eta^2 R_c}\frac{y^3}{3} \qquad (7.14)$$

where $\tau_\theta(0)$ is the wall shear stress, D the diameter of the tube, $\Delta p(\theta)/\Delta\theta$ the circumferential pressure gradient, ρ the density of the liquid, η its viscosity and R_c the coil radius. τ_x is the axial shear stress and the third term on the right hand side of Eq. 7.14 represents the centrifugal force (assuming laminar flow in the film).

The wall shear stress is related to the interfacial shear stress $\tau_\theta(\delta)$ as follows:

$$\tau_\theta(0) = \tau_\theta(\delta) + \frac{2}{D}\frac{\Delta p}{\Delta \theta}\delta - \frac{\rho \tau_x^2 \sin \theta}{\eta^2 R_c}\frac{\delta^3}{3}. \tag{7.15}$$

The circumferential velocity distribution can be determined by integrating with the boundary condition that the velocity is zero at the channel wall:

$$w(y) = \int_0^y \frac{\tau_\theta(y)}{\eta}\,dy. \tag{7.16}$$

The circumferential mass flow rate in the liquid film is then determined by integrating the velocity profile across the film and multiplying by the density. The final result has the form:

$$\dot{m}_\theta(\theta) = c_\tau \delta + c_{\Delta p}\delta^2 - c_{cf}\delta^5, \tag{7.17}$$

where the three coefficients on the right hand side of Eq. 7.17 represent the contributions of circumferential shear stress at the interface, circumferential pressure gradient and centrifugal force on the liquid film to the circumferential flow.

It may be hypothesised that the point of inversion represents a condition where $\dot{m}_\theta(\theta)$ is zero at all points round the circumference. The critical film thickness ($\delta_{cr}(\theta)$) corresponding to this condition may be determined as a function of position via the above solution, provided that the interfacial circumferential shear stress and the circumferential pressure gradient are known. To obtain these latter two variables, CFD calculations for the gas core of the flow have to be invoked. Calculations of the flow field in the flow of a gas in a coiled tube were carried out using the CFX code. Calculations were done using a variety of turbulence models (see Jayanti *et al.*, 1990 and Jayanti, 1990). A modified form of the k–ε model gave improved results over the standard k–ε model for this case, but the closest agreement with experimental data for single-phase flow in coils was obtained using the Reynolds stress model. Figure 7.3 shows calculations of axial velocity obtained by this technique for a coil with coil diameter to tube diameter ratio of 25.9. The mean velocity in this case was 33 m/s. It was found that, over a range of Reynolds numbers between 50 000 and 200 000 the circumferential distribution of axial shear stress, circumferential shear stress and circumferential pressure gradient around the wall could be scaled by the mean axial shear stress $\bar{\tau}_x$, as illustrated in Fig. 7.4. $\bar{\tau}_x$ can be estimated from the CFD results or from a standard correlation for friction factor in coils. Thus, it is possible to estimate the components necessary to solve Eq. 7.4 to obtain $\delta_{cr}(\theta)$. Once the distribution of critical film thickness has been calculated, the axial liquid

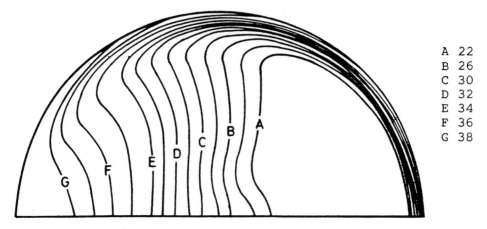

A 22
B 26
C 30
D 32
E 34
F 36
G 38

Fig. 7.3 Contours of axial velocity in a coil predicted by Reynolds stress model
($Re = 89\,000$, $D/d = 25.9$, mean flow velocity $= 33.0$ m/s) (Hewitt and Jayanti,
1992).

flow can be calculated at each circumferential position if the axial shear stress is
known. This can then be integrated around the circumference to give the total axial
liquid flow for the inversion condition. Thus, a complete solution of the problem
of inversion is possible.

Figure 7.5 shows the results for velocity profile as a function of circumferential
position for a typical inversion condition. As will be seen, though the circumferential
flow rate is zero at each point, the flow near the wall is passing inwards within the
coil and that near the interface is passing outwards. Thus, there is a secondary flow
within the liquid film (as, of course, there is also within the gas core).

The calculation procedure was compared with literature data for inversion and
agreed remarkably well. It is interesting to speculate what happens above and
below the inversion point. Above the inversion point, where the liquid tends to
move towards the inner surface of the tube, the liquid accumulates in that zone and
large waves are formed on the interface from which droplets are torn. The droplets
are accelerated in the gas phase and move towards the outer surface of the coil
where they deposit, forming a liquid film in which the net flow is, again, towards
the inside of the coil. Thus, a recirculation is set up which involves continuous
entrainment, acceleration, deposition and film flow back to the inner surface.

This is consistent with observations of flow in coils. If, on the other hand, the flow
is not inverted, then the liquid flows to the outer surface and would tend to remain
there as a stratified layer since any droplets entrained would fall back into this layer
as the flow proceeds round the coil. Again, this is consistent with experimental
observation.

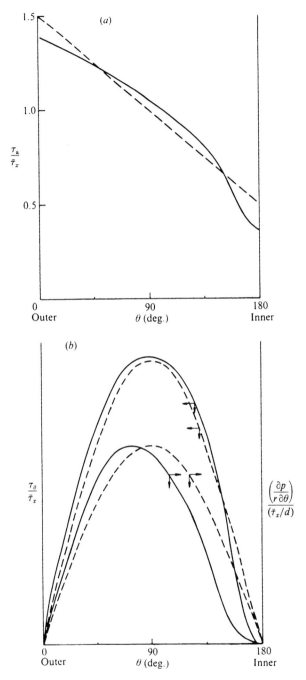

Fig. 7.4 Results from single-phase flow calculations in coiled tubes. (a) Axial wall shear stress. (b) Circumferential wall shear stress and pressure gradient. Dashed lines show approximate fits to curves (Hewitt and Jayanti, 1992).

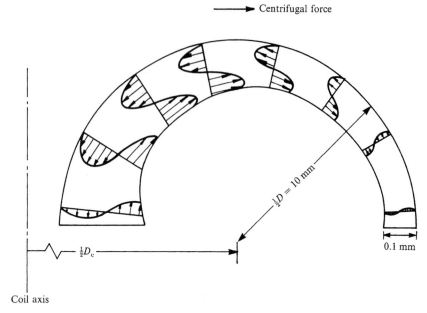

Fig. 7.5 Secondary flow in the liquid film in a coil calculated from thin film analysis with circumferential shear and pressure gradient from CFD calculations of gas core (Hewitt and Jayanti, 1992).

2.3 Horizontal annular flow

There has been curiosity for several decades about the precise mechanism of horizontal annular flow. In such a flow, gravity often causes a variation of film thickness around the periphery, with the film at the bottom of the tube being typically many times thicker than the film at the top of the tube. However, it is interesting to speculate why the liquid should flow at all at the top of the tube. What are the mechanisms which cause liquid transport against the direction induced by gravity? There have been three main mechanisms proposed for such transport:

(1) The thick film at the bottom of the tube has large waves on it and the interaction between the gas phase and these waves gives rise to droplets. Russell and Lamb (1965) proposed a mechanism of annular flow in which the droplets were entrained from the bottom film and diffused across the channel to the top, where they deposited, forming a new film which drained down to the bottom, the process being a continuous one.
(2) The large waves formed at the bottom of the tube may tend to be spread around the tube, and this mechanism was suggested by Butterworth (1971) as being responsible for the circumferential liquid transport.
(3) Since the interface of the film at the bottom of the tube is very much rougher than that at the sides and at the top, this variation of interfacial roughness may induce a secondary flow within the gas core, which gives rise to a circumferential shear stress. This

circumferential shear stress could transport the liquid film upwards at the interface, with gravity causing flow downwards in the inner zone of the film (this mechanism is somewhat similar qualitatively to the mechanism suggested above for coiled tubes except that gravity replaces centrifugal force). This mechanism was suggested by Laurinat *et al.* (1985) and received strong support from the work of Lin *et al.* (1985).

Though the third (secondary flow) mechanism appeared to fit the experimental data if an arbitrary distribution of circumferential shear was assumed, it was not possible to measure either the circumferential shear stress or the secondary flow distribution within the gas core. The only evidence which appeared to support the existence of such a secondary flow was that the maximum in the axial velocity profile was shifted towards the bottom of the tube, whereas, in the absence of a secondary flow, and with a rougher interface at the bottom than at the top, the maximum in velocity would tend to move towards the top. However, the question still remained as to whether the expected secondary flows would be strong enough to induce the predicted circulation. There was some evidence (based on known distributions of liquid film flow rate and film thickness) of the extent of the variation of roughness around the tube, and this evidence could be used as input to a CFD calculation of the flow field in the gas core. These calculations are described in detail by Jayanti *et al.* (1990) and will just be briefly summarised here.

To calculate secondary flows of the type postulated in horizontal annular flow, it is necessary to use models such as the Reynolds stress model, since simpler closures such as the k–ε model will not allow the prediction of such secondary flows in straight channels due to the assumption of isotropy of the turbulence. Using the CFDS-FLOW3D code, the distribution of velocities in the axial direction and in a direction normal to the tube axis were predicted assuming a variation of roughness around the tube periphery which was consistent with the experimental data for horizontal annular flow. The results are shown in Fig. 7.6. Secondary flows are, indeed, predicted and the location of the maximum velocity is displaced towards the bottom of the tube as expected. This seems to lend support to the idea of the secondary flow mechanism; however, the actual values of the circumferential stress are insufficient. Taking actual data for film thickness distribution in horizontal annular flow, the ratio of the circumferential shear stress τ_θ to the mean axial shear stress $\bar{\tau}_a$ needed to support the film may be calculated. The same ratio may be calculated from the CFD calculations and the values are compared in Table 7.1. As will be seen, the circumferential shear stresses are about an order of magnitude lower than those required to sustain this mechanism. Thus, the secondary flow explains the shift in the maximum in the gas velocity profile but does not explain the levitation of the liquid film. Further investigations of the mechanism indicated that the probable mechanism of transport was related to the formation of disturbance

Table 7.1

Angular position from bottom	Film thickness (mm)	$\tau_\theta/\bar\tau_a$ needed to support liquid film	$\tau_\theta/\bar\tau_a$ expected from secondary flow
45°	0.87	0.375	0.025
90°	0.31	0.157	0.020
135°	0.19	0.085	0.012

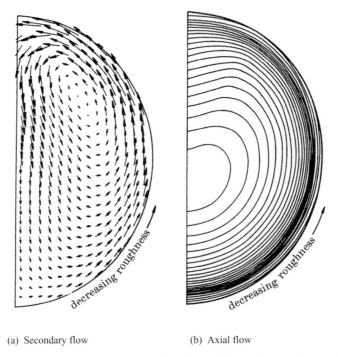

(a) Secondary flow   (b) Axial flow

Fig. 7.6 Velocity vectors and contours in tube with varying roughness (Jayanti *et al.*, 1990).

waves. Regarding the waves as a ring around the channel, their propagation was such that the ring sloped backwards in the direction of flow. The waves therefore act as a form of 'scoop' which scoops liquid from the bottom of the tube and transports it around the periphery of the wave to the top where it is shed and drains back down between the waves. Evidence for this particular form of the wave transport mechanism is given by Jayanti *et al.* (1990). The mechanism by which the waves spread around the tube has been recently clarified by Fukano and Inatomi (2003); we shall return to a description of this work in Section 4.3 below.

2.4 Disturbance waves in annular flow

The occurrence of large interfacial *disturbance waves* is a dominant feature in annular flow (see Hewitt and Hall Taylor, 1970). Some characteristics of the waves are as follows.

(1) Disturbance waves occur above a critical liquid film Reynolds number (around 250); below this Reynolds number, only small ripples occur on the interface.
(2) The frequency of the disturbance waves falls with distance and reaches an asymptotic value at large distances. There is some evidence that the initial frequency is close to that of turbulent bursts. The spacing between the waves is random.
(3) Wave velocities are typically 3–10 m/s and the waves are typically 1–2 tube diameters in extent.
(4) Shear stress measurements indicate that the wall shear stress is high underneath the waves.
(5) Visualisation and entrainment measurements confirm that the waves are the main source of entrained droplets, entrainment being a vital feature of annular flow.

Although there is some suggestion that the waves are associated with turbulence inception phenomena within the liquid film, their source and nature is still somewhat mysterious. In the work summarised here (and described in more detail in Jayanti and Hewitt, 1994), attention was focussed on the flow in the liquid film, the influence of the gas phase being made manifest by its influence on interfacial shear stress only. The influence of resolved normal stress (i.e. differences in net pressure force between the front and the back of the waves) was not taken into account. Further calculations by Wolf (1995), who calculated both the gas and liquid flow fields, indicated that this was a reasonable approximation.

A typical disturbance wave was chosen for study and had the following characteristics.

(1) The length of the disturbance wave region was taken as 20 mm followed by a substrate film region of 80 mm length. The disturbance wave was assumed to be of sinusoidal shape in its region of influence with a maximum amplitude of 0.75 mm, that is, five times the assumed substrate thickness of 0.15 mm.
(2) The film flow was assumed to be driven by air with a density of 2.15 kg/m^3 following at a mean velocity of 50 m/s. The tube diameter was assumed to be 30 mm.
(3) The wave was maintained in a stationary position in the computational domain with the wall moving at the wave velocity. The wave velocity may be estimated as that condition for which the forces on the film are in equilibrium. This velocity was calculated at around 2.6 m/s for the results summarised here (the velocity varied slightly with the turbulence model chosen), in good agreement with the values measured in experiments.

As was stated above, it is necessary to specify reasonable values for interfacial shear. For the substrate region, analysis of data by Shearer and Nedderman (1965) for annular flow in the sub-critical (ripples only) regime produced the following correlation for the substrate region interfacial friction factor f_i:

$$\frac{f_i}{f_G} = 0.856 + 0.00281 \, Re_G \frac{\delta_s}{D} \tag{7.18}$$

where f_G is the gas phase friction factor, Re_G the gas phase Reynolds number, δ_s the substrate film thickness and D the tube diameter. For the disturbance wave region, the equivalent sand roughness height was taken to be five times the local film thickness; this gives reasonable agreement with overall pressure drop measurements but is, of course, only an approximate estimate. The interfacial friction factor was then calculated using the standard Colebrook–White formula as a function of position. From the interfacial friction factor, the interfacial shear stress was calculated for both regions by the equation:

$$\tau_i = \tfrac{1}{2} f_i \rho_G U_G^2, \tag{7.19}$$

where ρ_G is the gas density and U_G the gas superficial velocity.

Results from the calculations on flows within the disturbance waves are illustrated in Figs. 7.7, 7.8 and 7.9, which show, respectively, the velocity vector distributions, the stream function contours and the turbulent viscosity distributions calculated in each case using the low Reynolds number k–ε model. It should be borne in mind that, for presentational purposes, the wave height has been exaggerated (i.e. the scales in the vertical and horizontal directions are different). It should be remembered that the wave height is 0.75 mm, whereas the zone occupied by the wave has an extent of 20 mm.

The first noticeable feature in the results is that of the recirculation pattern within the disturbance wave. This is evidenced by both the velocity vector plot (Fig. 7.7) and the stream function plot (Fig. 7.8).

Figure 7.9 shows distributions of turbulent viscosity. The low Reynolds number k–ε model predicts that the flow is turbulent only within the disturbance wave region.

Using CFDS-FLOW3D, it is possible to predict the temperature distribution in the computational domain for a situation with heat transfer. The simulation chosen was that with a fixed wall heat flux and a fixed interfacial temperature (the latter would be true for evaporation or condensation of a pure vapour). The results are illustrated in Fig. 7.10.

Perhaps the most interesting result from these calculations relates to average values for the heat transfer coefficient. Using the low Reynolds number k–ε model, a value of 6065 W/m^{2}K is calculated. The calculations were also carried out assuming

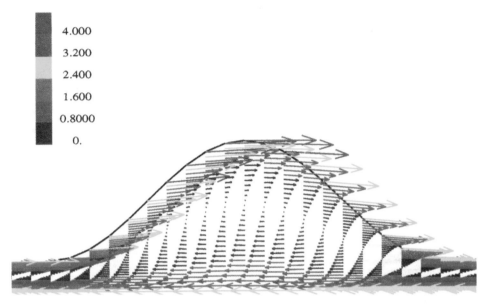

Fig. 7.7 Simulation of disturbance wave. Velocity vectors.

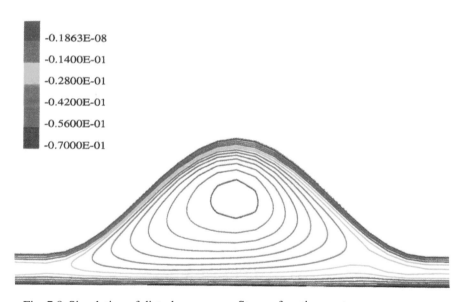

Fig. 7.8 Simulation of disturbance wave. Stream function contours.

laminar flow in the substrate and disturbance wave, and this gave a coefficient of 4505 W/m^{2}K. Classically, analysis of liquid film heat transfer has been carried out using one-dimensional analyses, assuming a flat interface and taking the average value of interfacial shear stress. A calculation of this kind (see Jayanti and Hewitt, 1994) produces a value of 15 650 W/m^{2}K. Such calculations are known to produce

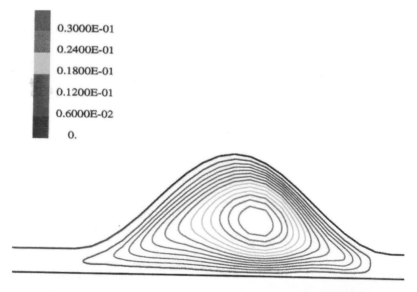

Fig. 7.9 Simulation of disturbance wave. Turbulent viscosity distribution.

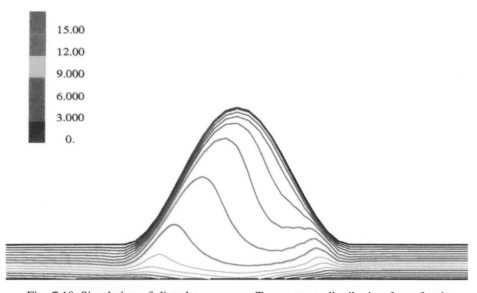

Fig. 7.10 Simulation of disturbance wave. Temperature distribution for a fixed
heat flux and fixed interface temperature.

coefficient values which are significantly higher than those obtained in experiments.
The present calculations using the low Reynolds number $k–\varepsilon$ model also give values
much lower than those based on averaged quantities, showing the limitations of the
averaging process. This has been confirmed in more recent calculations by Wolf
(1995).

3 Dispersed flows

3.1 Background

Dispersed multi-phase flows with droplets and particles abound in nature, from clouds, mist and fogs to the long-range transport of fine dust released in desert storms or in volcanic eruptions. They control the weather and influence the climate. They play key roles in many industrial energy processes – from spray drying, pneumatic and slurry transport, fluidised beds, to coal gasification and mixing and combustion processes. They can have a profound effect on our health and quality of life, e.g. the inhalation of very fine air-borne particulate (PM10s) damages respiratory functions, and has led to increased cardio-pulmonary mortality rates and allergic disorders.

Understanding the behaviour of these flows, through modelling and experiment, is therefore important in our control of the environment, improving our health, and in the design and improvement of industrial processes. It is crucial to the development of mitigation strategies to cope with climate change and to the spread of communicable diseases like Foot & Mouth. We need to know, for instance, how individual particles are dispersed by the carrier flow, and how they interact with one another and with the underlying flow. An important feature of all these processes is that they are statistical – they occur both randomly in space and time. So successful modelling of these processes means not only understanding the basic physics but how to average and in what way.

More precisely, dispersed multi-phase flows can be classified as either dilute or dense, the latter being dominated by inter-particle collisions and two-way coupling between the gas and dispersed (particle) phase. In dilute flows the mechanisms of importance are nucleation, agglomeration, condensation or evaporation, phase change, mixing and dispersion, near wall behaviour, particle–surface interactions (including impact, adhesion, deposition, erosion and re-entrainment/resuspension) and particle–field interactions. In dense flows the issues are fluidisation, two-way coupling, inter-particle collisions, clustering and bubble formation.

3.2 Important underlying features for simulation and modelling

3.2.1 Particles in turbulent structures

Whether in the carrier flow or generated in the wakes of particles themselves, turbulent structures play an important role in many particle/fluid processes. For instance, turbulent structures are influential in mixing and dispersion processes, in agglomeration and breakup, and in nucleation where their intermittency can give rise to bursts of nucleation. Coherent structures in turbulent boundary layers closely associated with the turbulent bursting and sweeping mechanisms (ejection and sweeps) have an important influence on particle deposition and resuspension,

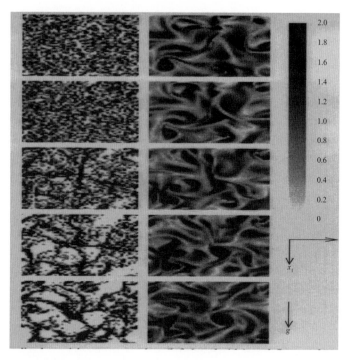

Fig. 7.11 De-mixing of particles in DNS homogeneous isotropic turbulence (Wang and Maxey, 1993).

on the phenomenon of roping, and on condensation and droplet formation where they have a marked influence on the local super-saturation. It is now well known that particles, depending on their response time to changes in flow (particle Stokes number), can be de-mixed by a turbulent flow (Crowe *et al.*, 1985). A turbulent flow field may be considered as a mixture of regions of vorticity and straining: in situations, for instance, when the flow consists of pairs of counter-rotating vortices, e.g. near a wall in a turbulent boundary layer, the region of strain occurs between the vortex pairs. Particles introduced initially, fully mixed with the flow, will be spun out of the regions of vorticity and segregate in the high straining regions between the vortices. So in the course of time particles de-mix/segregate – the phenomenon has been referred to as clustering or preferential concentration. Figure 7.11 shows an example of a DNS generated homogeneous isotropic turbulent flow containing particles which are fully mixed to begin with (Wang and Maxey, 1993). The left hand side set of pictures shows snapshots in time of the particle distribution, alongside of which are corresponding pictures of the scalar vorticity field of the underlying velocity flow field, where the dark regions correspond to regions of high vorticity. It is clear from the comparison between the concentration and vorticity field that particles are segregated in the regions between the vortices. Averaged over a sufficiently long period of time, the concentration is uniform, but it is clear that this de-mixing at the micro-scale level of the turbulence is important in processes like agglomeration (see e.g. Reade and Collins, 1998), and on two-way coupling between carrier and dispersed phases, where particles segregating in this way can have a profound effect on the dynamics of turbulence generation and dissipation (see e.g. Ahmed and Elghobashi, 2000). In other respects, understanding the way turbulent structures in jet flows break up droplets would lead to better data and understanding on droplet size distributions and their control. In order to develop better models for dispersed flow, new or existing methods (both experimental and theoretical) of analysing complex flows must be developed or refined, e.g. by proper orthogonal decomposition (POD) (Kunisch and Volkwein, 2002) to classify and identify these structures, to establish their statistics and elucidate the mechanisms by which they influence and control these fundamental processes.

3.2.2 Particle–wall interaction and near wall behaviour

A better understanding of particle/droplet wall interactions and the aerodynamic forces acting on a particle attached to solid/rough surfaces will lead to better models for particle sticking/resuspension and erosion and for droplet breakup and fragmentation. In particular, the influence of particle size, shape and surface roughness on surface adhesion and friction during or after impact needs to be investigated further, both experimentally and using the discrete element method (DEM), and the results implemented in more empirically based models.

Particle impact with a bed of particles has not been fully investigated either theoretically or experimentally, e.g. the mechanisms by which impacting particles lose energy to the bed, how their impacts can lead to ejections of particles from the bed, and how individual particles can be resuspended from the bed by aerodynamic forces, whether singly or as agglomerates. The dynamics of lift and roll and how that leads to particle resuspension require a more detailed study, both experimentally and by numerical simulation using DNS.

In the near wall region, a better understanding of the turbulent bursting mechanism in the presence of temperature gradients will lead to better models for turbulent/thermophoretic deposition of small particles. Here DNS and POD have important roles to play in identifying the turbulent structures responsible for the buildup of concentration near the wall and finally depositing particles on the wall. The effect of thermally generated forces can be so significant that devices called thermophoretic precipitators specifically exploit this mechanism to precipitate particles. These devices show great efficiency in collecting particles in the sub-micron range. In particle-laden gases, therefore, the influence of thermophoresis on deposition rates in a variety of flows and geometries is worthy of further study.

3.2.3 Particle–particle and particle–fluid interactions in dense suspensions

The focus of this work is on the understanding of the behaviour of fluidised beds where both sorts of interactions are important. Important considerations are the stability of the bed, bubble formation and coalescence, clustering and mixing of particles and how in particular these processes depend on the particle size and shape distributions. Consideration should be given to the form of the particle drag in these flows – how, for instance, it depends on the influence of other particles. To what degree is the motion of a particle influenced by the wakes of other particles and how does that affect dispersion and clustering? To what extent can these features be included in a rational way in the closure approximations of momentum and energy coupling between the two phases? An important statistical consideration when dealing with inter-particle collisions is the degree to which the assumption of 'molecular chaos' is justified, given the degree of correlation between the particle and the carrier gas. In cluster formation and breakup, the dynamics of impact and adhesion (contact mechanics) is an important area of study using simulation and separate effects' experiment.

3.3 Development of models, simulation and computational methods

There are traditionally two ways of modelling a dispersed flow: the Lagrangian tracking/random walk approach where an individual particle is tracked through a random flow field characteristic of the turbulence; and the Eulerian two-fluid approach in which the dispersed (particle) phase, like the continuous phase, is described by a set of continuum equations which represent the conservation of mass–momentum and energy of the dispersed phase within an elemental volume of the mixture. Both approaches are available in most commercial CFD codes and each has its advantages and disadvantages: in both cases there is a great potential for improvement as predictive tools for the design and control of particulate flows.

3.3.1 Application of random walk models

The attraction of using simple random walks is that in dilute flows (with no phase coupling and particle–particle interactions) they are relatively easy to implement and allow the basic physics (especially the behaviour at the wall) to be introduced in a straightforward manner. However, they are computationally intensive, which generally restricts their usage to modelling dispersion/deposition in relatively simple dilute flows and geometries. Their reliability usually depends on how well the random flow models the turbulent flow field of the gas. In simple random walk models, the underlying flow field has no structure, which leads to the problem of spurious drift in inhomogeneous flows. In many particle–fluid processes, the length scale and structure of the turbulence are important features, and considerable improvement can be made to random walk models if these features are incorporated at the appropriate level of the kinematics, whilst at the same time satisfying basic properties like continuity of flow. In this respect the application of POD or PIV can be a useful tool in identifying the relevant structures. Application of parallel processing can reduce computing times considerably, with the real possibility of using this method for analysing complex flows with both two-way coupling and inter-particle collisions.

3.3.2 Application of two-fluid models and other Eulerian models

Two-fluid modelling is a more efficient way of modelling a dispersed flow. In principle it can handle complex flows (with two-way coupling and inter-particle collisions) and geometries using the same CFD infra-structure as for single-phase flows (making it ideal for incorporation into commercial CFD software). However, even in the dilute case, there is an even greater uncertainty in the form of the constitutive relations for a dispersed flow than there is for a single-phase flow. Furthermore, the boundary conditions imposed at the wall necessary for solution are incompatible with the natural (physical) boundary conditions. What is needed is a rational approach upon which both the continuum equations and associated constitutive relations can be derived in a self-consistent manner, and at the same time provide a way of dealing directly with the particle–wall interactions and near wall behaviour without resorting to the application of artificial boundary conditions. With regard to both these requirements, the PDF approach is a promising development providing a rational framework for dispersed flows in the same way that kinetic theory provides a rational framework for the gas dynamics. Since this represents an important development in the modelling of dispersed flows, we discuss it briefly below.

Other approaches like ADE (advection diffusion equation) (see e.g. Young and Leeming, 1998) evaluate the particle velocity field separately from the particle concentration field and, in so doing, offer a much simpler and more attractive computational approach than the two-fluid approach, especially in calculating particle deposition in dilute flows. However, there are currently problems of closure that need to be overcome before it can be used as a general purpose model (Reeks, 2003).

Finally, an overriding weakness in commercial codes and models in general is an inability to deal with the influence of particle shape and size distribution (this is particularly true when bio-mass fuel is mixed with pulverised coal). Other weaknesses identified are the inability to deal with the change of phase and in droplet/liquid film interaction.

3.3.3 PDF approach to modelling turbulent dispersed flows

The probability density function (PDF) approach provides a general theoretical framework in which the behaviour of a reacting turbulent dispersed flow of particles can be formulated. This formulation refers not only to the particle transport by the carrier phase but also to the particle–wall interactions that constitute the natural boundary conditions of the flow. So the method can handle the physics of the particle–surface interactions, e.g. bouncing, deposition, resuspension, fragmentation etc. in a way that is part of the underlying formulation.

The PDF approach was first proposed by Buyevich (1971, 1972a, 1972b) and subsequently developed and refined by a number of researchers, principally Reeks (1980, 1991, 1993, 2003), Swailes and Darbyshire (1997), Hyland *et al.* (1999), Derevich and Zaichik (1988), Zaichik (1991), Minier and Pozorski (1997) and Simonin and his co-workers (1993). (See Simonin (2002) and Minier and Peirano (2001) for a comprehensive account of the PDF approach for gas–particle flows, and also Swailes *et al.* (1999) for how the PDF method deals with natural boundary conditions.) The approach is similar to the kinetic theory of gases in that the so-called continuum equations (the equations for the transport of average mass, momentum and energy) and their associated constitutive relations can be derived in a strictly formal way from a master equation or PDF equation analogous to the Maxwell–Boltzmann equation.

Two forms of the PDF approach are currently in use. The first approach, referred to as the kinetic method (KM), is based on the traditional form of PDF used in kinetic theory, involving the particle velocity **v** and position **x** at a given time *t*; and the second is based on a PDF which, in addition to **v** and **x**, includes the carrier flow velocity encountered by a particle with velocity **v** and position **x**. The second approach has been referred to as the generalised Langevin model (GLM) approach because the equations for the carrier flow are based on a generalised Langevin equation, similar to the Langevin equation used in Brownian motion, in which the carrier flow is driven by white noise, but in this case with response times of a more generalised nature. The GL equation of motion proposed by Simonin, Deutsch and Minier (SDM) (1993) is an extension of the GLM equations used by Pope (e.g. Haworth and Pope, 1986, Pope, 1991), as an analogue of the Navier–Stokes equation for turbulent flow. The use of such a GLM means that the process is realisable and the PDF equation is an exact equation for the model.

In the KM approach, the PDF equation requires a closure model for the 'phase space' diffusion current, the simplest such model (from both a practical and theoretical point of view) being a Boussinesq approximation involving velocity and spatial gradients of the PDF and fluid–particle diffusion coefficients that are functions of the particle displacements in velocity and position about a given point in particle phase space:

$$\langle u_i W \rangle = - \left(\langle u_i'(x,t) \Delta v_j \rangle \frac{\partial}{\partial v_j} + \langle u_i'(x,t) \Delta x_j \rangle \frac{\partial}{\partial x_j} \right) \langle W \rangle + v_{di} \langle W \rangle, \quad (7.20)$$

where $\Delta \mathbf{v}$ and $\Delta \mathbf{x}$ are displacements in velocity and position, respectively, for a particle starting out from some initial position and arriving at $\mathbf{v}, \mathbf{x}$ at time t. $\mathbf{u}(\mathbf{x}, t)$ is the carrier flow velocity at $\mathbf{x}, t$ and $\langle \rangle$ is an ensemble average. If these displacements constitute a Gaussian process about $\mathbf{x}, \mathbf{v}, t$, then the formula for the flux is exact. An important feature is the presence of the drift velocity $\mathbf{v}_d$ arising when the turbulence is inhomogeneous flow and is directly related to the flow structure and the de-mixing of the particle by the flow. It involves the compressibility of the particle velocity field along a particle trajectory that is associated with the displacements $\Delta \mathbf{v}$ and $\Delta \mathbf{x}$. In particular $\mathbf{v}_d$ is given by

$$\mathbf{v}_d = - \int_0^t \langle \mathbf{u}(\mathbf{x}, t) \nabla \cdot \mathbf{v}(\mathbf{x}, t|s) \rangle ds, \quad (7.21)$$

where $\nabla \cdot \mathbf{v}(\mathbf{x}, t|s)$ is the divergence of particle velocity field along a particle trajectory at time s that passes through $\mathbf{x}$ at time t. The formula is identical to the formula given by Maxey (1987) for the enhancement of gravitational settling of a particle under gravity, but it also contributes to the drift velocity in inhomogeneous shear flows, especially near a wall in a turbulent boundary layer (Reeks, 2003).

In the second approach the SDM GLM equation gives rise to a transport equation for the turbulent flux $\langle \mathbf{u}W \rangle$ which contracts to the same form as for the KM approach for Gaussian processes, but is more general since it is not based on any local equilibrium assumption but has no drift term (this is because the model for the turbulence uses a white noise function of time which induces no structure to the turbulence). However, as a PDF method, it provides a model for the dispersed flow and also for the underlying turbulence, as in the PDF approach developed by Pope which is receiving much attention as an alternative to the traditional RANS approach for turbulence. Both PDF approaches provide a set of mass–momentum and energy equations for the dispersed phase, together with the necessary constitutive equations which have been tested successfully on a variety of turbulent dispersed flows from separating, turbulent boundary layers and jets, and in some cases comparisons have been made with results from DNS and LES where the flows can be well simulated.

In most of these test cases, the particles have been solid inert particles (see Simonin, 2002 for many examples). But this has been chosen as a matter of simplicity and convenience – however, the PDF approach has recently been applied to evaporating droplets to find the distribution of sizes (Pandya and Mashayek,

2003) and recently extended to two-particle PDFs in modelling agglomeration and particle separation in turbulent flows (Zaichik and Alipchenkov, 2002).

Simonin (2002) has recently extended the PDF to account for particle collisions, in which case the KM PDF equation is precisely analogous to the Maxwell–Boltzmann equation. He has suitably modified the molecular chaos assumption to include the influence of the carrier flow before and after collision, and as a consequence has been able to model the particle kinetic stresses in the dispersed phase from dilute to dense to granular. The problem of dealing with natural boundary conditions has also been tackled using the solution PDF near the wall to generate suitable wall functions.

3.3.4 Application of DNS, LES, DEM, FCM and POD

All of these methods of numerical simulation are not suitable in general as predictive tools for complex flows and geometries. However, they can provide valuable information/data on the small-scale features of particle–fluid interactions that can be used to construct better closure models for use in more general purpose two-fluid models, etc. They have been widely used to validate two-fluid models in fairly simple flows, e.g. turbulent boundary layers but with complex interactions, including two-way coupling and inter-particle collisions. POD has already been mentioned as an important method for identifying turbulent structures from experimental and DNS data. The discrete element method (DEM) has been used successfully to simulate fluidised beds and to elucidate the particle–particle and particle–fluid interactions in fluidised beds. The force coupling method (FCM) is a very innovative approach recently developed for solving for the continuous phase with two-way coupling and a very accurate way of predicting drag reduction in turbulent boundary layers. LES probably has the greatest potential for application to complex flows/geometries but it is not yet clear how it can be adapted to deal with two-way coupling.

3.4 Computational methods

A variety of turbulence closure approximations can be incorporated into a number of computational schemes based on finite volume or finite element as in CFD for single-phase turbulence. Though both types of schemes are used in commercial CFD, in implementing changes for dispersed flows, the user usually does not have access to the source code and can only make modifications via user defined subroutines. Currently equation solvers for the particle kinetic stresses (as in single-phase turbulence) with closure approximations for the kinetic energy flux are the most reliable (because they include the most physics) but increase computational time. In general the equations for the dispersed and continuous phases are coupled, and more grid refinement is necessary than in single-phase problems to achieve the same accuracy/grid independence. So, for instance, in applying two-fluid Eulerian models to fluidised beds, 2D and 3D solutions are very sensitive to mesh definition and small-scale perturbations can grow to dominate the physical mechanisms. Whilst it is generally recognised that the use of particle–wall functions is a valuable way of handling natural boundary conditions, unlike their usage in single-phase flows, no method as yet has been developed for incorporating particle–wall functions into the solution scheme.

3.5 Example of multi-fluid modelling for dispersed flows

A wide range of applications of the *multi-fluid model* approach to industrial dispersed flow systems have been reported. It is beyond the scope of the present chapter to review all of these; rather, it has been decided to describe a few cases in more detail to illustrate the kind of calculation that is now achievable. In the *multi-fluid model* approach, separate equations are written for the respective phases, with interaction terms between the phases included. These sets of equations are then solved numerically to calculate the velocity and phase fraction fields. The equations are exemplified by those of Issa and Oliveira (1996) who write continuity equations for each phase k as follows:

$$\rho_k \left(\frac{\partial}{\partial t} \bar{\alpha}_k + \nabla \bar{\alpha}_k \bar{u}_k \right) = 0, \tag{7.22}$$

where $\bar{\alpha}_k$ is the time averaged void fraction and

$$\tilde{u}_k = \text{phase averaged velocity} = \overline{\alpha_k \mathbf{u}_k} / \bar{\alpha}_k. \tag{7.23}$$

The momentum equation for phase k is written as

$$\rho_k \left(\frac{\partial}{\partial t} \bar{\alpha}_k \tilde{\mathbf{u}}_k + \nabla . \bar{\alpha}_k \tilde{\mathbf{u}}_k \bar{\mathbf{u}}_k \right) = -\bar{\alpha}_k \nabla p + \bar{\alpha}_k \nabla . \tilde{\tau}_k$$

$$+ \nabla . \bar{\alpha}_k \tilde{\tau}_k^t + \rho_k \bar{\alpha}_k \tilde{\mathbf{g}} + \mathbf{F}_{\text{DK}}, \tag{7.24}$$

where

$$\mathbf{u}_k = \mathbf{u}_k + \mathbf{u}_k''. \tag{7.25}$$

The turbulent shear stress for phase k is given as:

$$\tilde{\tau}_k^t = -\frac{\rho_k \overline{(\alpha \mathbf{u}'' \mathbf{u}'')_k}}{\alpha_k}. \tag{7.26}$$

In typical applications of the multi-fluid modelling technique, the turbulent shear stress is calculated for the continuous phase using a single-phase model, and the turbulent shear stress of the dispersed phase is calculated from that of the continuous

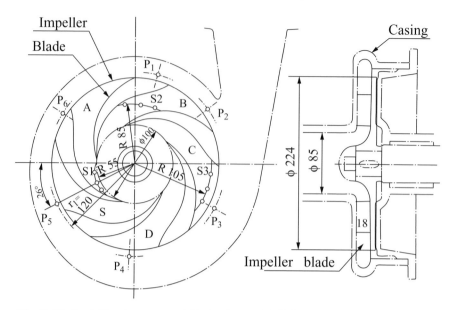

Fig. 7.12 Centrifugal pump modelled using multi-fluid model.

phase using various forms of *response function*. The closure relationships for inter-facial drag, etc. are always a specific difficulty in this kind of modelling and various semi-empirical relationships are used for the drag term F_{DK}.

An example of the application of this form of model using the k–ε model for turbulence is shown in Figs. 7.12–7.15 (Issa, private communication). The specific example considered is that of gas–liquid flow through the centrifugal pump illustrated in Fig. 7.12. With modern gridding techniques, it is possible to represent the extremely complex geometries (Fig. 7.13) and to obtain solutions for, say, continuous phase velocity (Fig. 7.14) and void fraction distribution (Fig. 7.15).

It is possible to use more advanced turbulence models in conjunction with the multi-fluid model, and Fig. 7.16 shows solid volume fraction contours calculated using the Reynolds stress turbulence model (Fluent Inc., private communication).

Most applications of the multi-fluid model to dispersed systems (bubble flow, dispersed droplet flow, dispersed solid particle flow, etc.) assume a *uniform* size for the discontinuous phase elements (bubbles, drops and particles). To capture many phenomena, however, it is desirable to take account of the fact that there is usually a distribution of dispersed phase element size. Studies taking account of this are reported by Tomiyama (1998), who wrote $(N + 1)$ equations to describe dispersed bubble flows in which the bubbles were represented by groups having N characteristic sizes. This gave improved predictions for bubble columns. Another factor which is important is that of bubble (or drop) breakup and coalescence, which lead to the development of bubble (or drop) populations as the flow proceeds through

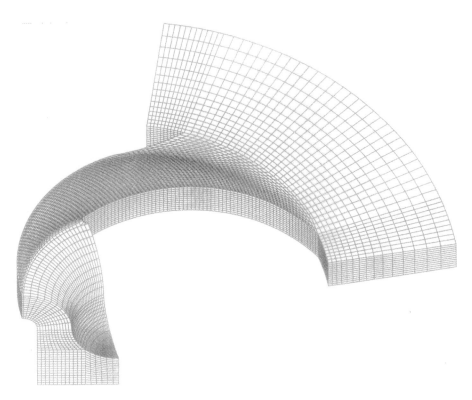

Fig. 7.13 Grid for modelling a centrifugal pump (Issa, private communication).

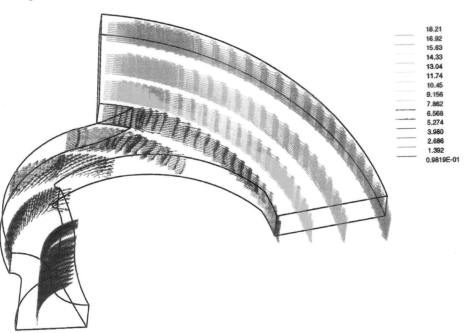

	18.21
	16.92
	15.63
	14.33
	13.04
	11.74
	10.45
	9.156
	7.862
	6.568
	5.274
	3.980
	2.686
	1.392
	0.9819E-01

Fig. 7.14 Continuous phase velocity contours in a centrifugal pump predicted by the multi-fluid model (Issa, private communication).

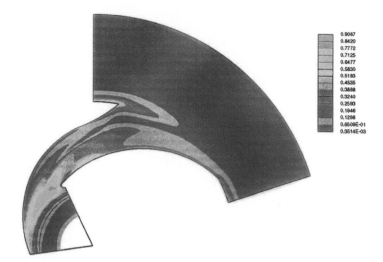

Fig. 7.15 Void fraction distribution at a given plane in a centrifugal pump predicted by the multi-fluid model (Issa, private communication).

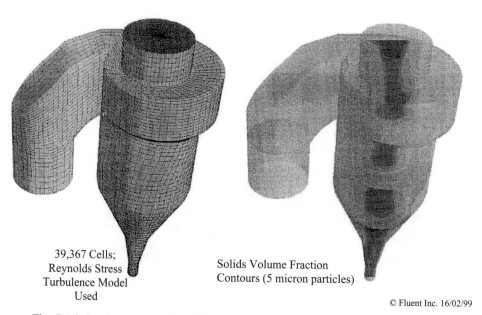

39,367 Cells;
Reynolds Stress
Turbulence Model
Used

Solids Volume Fraction
Contours (5 micron particles)

© Fluent Inc. 16/02/99

Fig. 7.16 Solids concentration distribution in a cyclone predicted using a multi-fluid model with Reynolds stress turbulence closure (Fluent Inc., private communication).

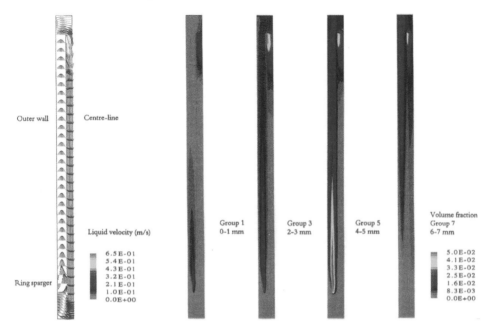

Fig. 7.17 Calculated distributions of concentrations of bubble groups of various sizes in a gas lift reactor surface using the MUSIG algorithm in CFX (Lo, 1999).

the equipment. Lo (1999) describes a new algorithm (MUSIG – multi-size group) which is implemented in the CFX code of AEA Technology. This model incorporates the model of Luo and Svendsen (1996) for bubble breakup and the model of Prince and Blanch (1990) for coalescence. Typical results calculated using the MUSIG algorithm are shown in Fig. 7.17 for the case of a gas lift reactor system in which circulation is induced by sparging bubbles into one side of the reactor.

3.6 Dispersed phase element tracking

In the multi-fluid model, the motion of specific dispersed phase elements (bubbles, drops) is not followed; rather, the dispersed phase is considered as a flow field which interacts with the continuous phase flow field. However, it is often instructive to track the position of individual phase elements (bubbles, drops, particles) as they move and interact with the turbulent flow field and with other dispersed phase elements.

An example of such phase element tracking methods is *bubble tracking*. Bubble tracking methods are reviewed by Tomiyama (1998) who distinguishes two forms of such models:

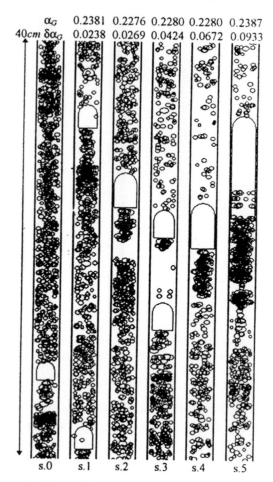

	α_G	0.2381	0.2276	0.2280	0.2280	0.2387
40cm	$\delta\alpha_G$	0.0238	0.0269	0.0424	0.0672	0.0933

s.0   s.1   s.2   s.3   s.4   s.5

Fig. 7.18 One-way bubble tracking simulation of bubble flow/slug flow transition (Tomiyama, 1998).

(1) *Two-way bubble tracking*. Here, the influence of the bubble field on the continuous flow field is considered. Calculations using this methodology tend to be rather complex and are limited in scope.

(2) *One-way bubble tracking*. Here, the bubble behaviour (wakes, coalescence, turbulence, etc.) is represented without modelling the liquid phase directly. Calculations by this means are much faster, and an illustration of results obtained is shown in Fig. 7.18. The bubble flow to slug flow transition mechanism illustrated in Fig. 7.18 corresponds closely to that observed experimentally, and this illustrates the power of the technique.

4 Modelling of systems with deformable interfaces

In fluid–fluid flows (gas–liquid, liquid–liquid, gas–liquid–liquid), the interfaces are usually deformed away from the idealised spherical shapes so often implicitly assumed in multi-fluid models. Deformations occur due to interactions between the fluids and due to the influence of body forces such as gravity. In the *multi-fluid model* for dispersed fluid–fluid flows discussed above, such interface deformations are not accounted for directly. They may, however, be included implicitly through closure relationships; for example, in bubble flows, such closures might include relationships between bubble rise velocity and diameter which account to some extent for departure from sphericity of the bubbles. However, the interface deformation can lead to the flows being *separated* rather than *dispersed*, extreme cases being stratified and annular flows. It is clear that the nature of the multi-fluid models for separated flows will be very different from those for dispersed flows, and we discuss such models briefly in Section 4.1.

The generality of closure relationships for multi-fluid modelling of both dispersed and separated flows is very doubtful, and there has been an increasing focus on direct prediction of the interfacial geometry including its evolution with time. A general review of methods for such predictions is given by Tryggvason *et al.* (2002). Here, only a brief description is given of some of the alternative methods (Section 4.2) with some illustrations of the results (Section 4.3).

4.1 Multi-fluid modelling of separated flows

Multi-fluid models are commonly used for annular and stratified flows. For these cases, the models are usually one-dimensional, a widely researched example being the *six-equation model* (including momentum energy and continuity equations for each of the two phases). This model is used in many nuclear reactor safety codes. The influence of the deformable interface between the phases is subsumed in closure laws (for instance, laws relating the interfacial friction to the liquid layer thickness in annular flows). Many such closure laws are available; the difficulty is that they are essentially empirical in nature and reflect limited data bases. Khor *et al.* (1997) give a taxonomy of interfacial friction relationships for stratified flows and evaluate these against a data base for three-phase liquid–liquid–gas stratified flows.

Slug flows have always been a challenge to the multi-fluid method. There are fundamental problems in applying the averaging procedures used in deriving the model for this case. However, rather than using averaging methods, such flows can be predicted as sequences of zones of stratified and single-phase liquid flow. The formation of slugs by Kelvin–Helmholtz wave growth in a stratified flow can actually be captured by the one-dimensional two-fluid model, and this leads to complete predictions of developed slug flows which are in encouraging agreement with experimental data (Issa and Kempf, 2003, Bonizzi and Issa, 2003a, 2003b).

Another problem in one-dimensional modelling is that of entrainment of discrete phase elements of one or more phases in the continuous regions of the separated phases. An example here is the entrainment of droplets into the gas core in annular flow. The multi-fluid model can sometimes be extended to such cases, for example by writing a further set of equations for the droplet 'field' in annular flow. However, this leads to a requirement for additional closure relationships (for instance, for the rates of entrainment and deposition of the drops).

Despite the above difficulties, multi-fluid modelling often represents the only viable methodology for many industrial multi-phase flow applications, as is reflected in its use in both nuclear and pipeline codes. New developments are more likely to be accepted if they are presented in the framework of such models.

4.2 Direct prediction of systems with deformable interfaces: survey of methods

The methods used for predicting flows with deformable interfaces have included:

(1) **Boundary integral methods.** Assuming that the fluid is inviscid and irrotational, the evolution of a flow with an interface can be reformulated as an integral equation along the boundary between the fluids. A classical paper on this method was that of Longuet-Higgins and Cocklett (1976), who used it to predict breaking water waves. The method has subsequently been used extensively, with a recent example being the work of Xue *et al.* (2001) on three-dimensional breaking waves. The method can be extended to laminar flows. For turbulent flows, however, the basic restrictions of the method render it generally unsuitable.

(2) **The VOF method.** In the VOF (volume of fluid) method (Harlow and Welch, 1965), the flow field in a two-phase flow is considered as a flow field of a single fluid which has physical properties which vary with a scalar which is transported by the flow. For an air–water flow, for instance, the density and viscosity may be considered to change between the extremes of air and water over a small range of variation of the scalar, as illustrated in Fig. 7.19, and this allows identification of the interface position within the computational domain. Earlier versions of the VOF method did not take specific account of surface tension, but, more recently, changes of pressure across the interface due to interface curvature and surface tension have been taken account of in the modelling. A detailed review of VOF methods is given by Scardovelli and Zaleski (1999). The nature of the VOF method means that it can be used (in principle, at least) for turbulent flows; it can be implemented in conjunction with the whole gamut of turbulence models ranging from Reynolds averaged Navier–Stokes (RANS) methods to direct numerical simulation (DNS). The VOF technique offers an extremely useful tool for investigating interfacial behaviour in multi-phase flows. One of the problems with the method is that the interface tends to diffuse. A wide variety of 'interface sharpening' techniques have been devised to offset this tendency.

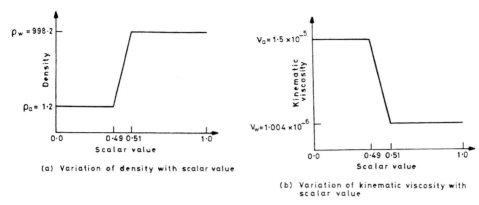

(a) Variation of density with scalar value

(b) Variation of kinematic viscosity with scalar value

Fig. 7.19 Variation of physical properties with scalar value in the VOF method.

(3) **The level set method.** In the level set method, a *level set function* ϕ is defined which is the distance from the interface (negative for gas and positive for liquid, say, and zero at the interface). The local density and viscosity are defined as a function of ϕ, and an advection equation is written for ϕ which allows the position of the interface to be followed. The level set method was originally proposed by Osher and Sethian (1988), and more recent generalisations of the method are reported by Suzzman *et al.* (1994) and Chang *et al.* (1996). The method is less diffusive than the VOF method.

(4) **Front tracking methods.** A methodology which avoids the interfacial numerical diffusion problems associated with VOF is the so-called *front tracking (embedded interface)* method. This has been developed, for instance, by Prof. Gretar Tryggvason and co-workers; earlier work is reported in the papers by Unverdi and Tryggvason (1992), Juric and Tryggvason (1996), Ervin and Tryggvason (1997) and Loth *et al.* (1997); more recent work is cited in Section 4.3 below. The interface is represented as an unstructured adaptive grid within a fixed Cartesian grid, as illustrated in Fig. 7.20. Grid points are added or subtracted from the interface grid as the interface changes shape. The density and viscosity change in a small thickness zone from the values for the gas phase to those for the liquid phase. In this sense, the method is similar to the VOF method, though numerical diffusion is avoided. Details of the computational bases of the method are given by Tryggvason *et al.* (2001).

(5) **Molecular dynamics modelling.** The above methods depend on the solution of equations (inviscid or Navier–Stokes) for the flow fields which are continuous in nature. Actually, of course, the fluids are not, in the limit, continuous and consist of individual molecules. The *molecular dynamics method* starts at a more fundamental level, calculating the detailed motions and interactions of a group of molecules and deducing the physical properties and bulk motions by suitable averaging. Typically, of the order of several hundred to several thousand molecules are used in the simulations, and the methods are computationally demanding even at this level. The methodology is clearly not well suited for directly investigating turbulent multi-phase flows; however, its use

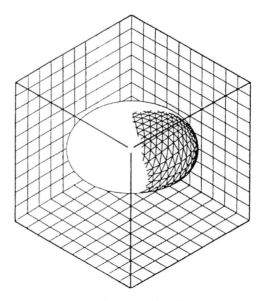

Fig. 7.20 Grid system used for the embedded interface method (Unverdi and Tryggvason, 1992).

can throw light on key phenomena occurring in such flows. An extensive review of molecular dynamics methods is given by Maruyama (2002).

(6) **Lattice gas automata (LGA) and lattice Boltzmann (LBM) methods.** Since molecular dynamics models are unlikely to be of interest in the foreseeable future for predicting turbulent flows, increasing attention has been focussed on lattice gas automata (LGA) and lattice Boltzmann (LBM) methods which model the motions of larger fluid elements. In the simplest form of LGA, particles (which may be regarded as 'clumps' of molecules) of unit mass and speed move from vertex to vertex of a discrete lattice in discrete timesteps, according to a set of rules. LGA methods are reviewed by Biggs and Humby (1998); a typical hexagonal lattice geometry is shown in Fig. 7.21. In more modern LGA models, Navier–Stokes hydrodynamics emerges from the simple LGA treatment and the models can thus predict both laminar and (in principle) turbulent flows. To represent a multi-phase system, particles of different 'colour' (red and black, say, for a two-phase system) are introduced into the lattice. Separation of the species into separate phases is induced by biasing collision outcomes in such a way that black particles head towards regions of highest black particle concentration, and vice versa for red particles. For flows involving solid surfaces, differences in wettability are simply imposed by colouring wall sites proportional to the wettability. If the 'red' fluid is perfectly non-wetting, for example, then all the boundary sites would be coloured black. LGA methods represent an alternative to combining DNS with VOF methods; however, they have been regarded as having hugely higher

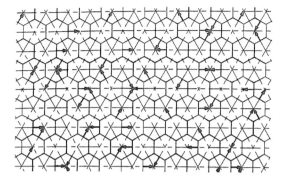

Fig. 7.21 Hexagonal lattice for use with the LGA method (Biggs and Humby, 1998).

storage and computational work requirements than the already-high requirements of DNS/VOF (Orszag and Yakhot, 1986). However, more recent estimates show that the differences, though still large, are now such as to allow these techniques to be used productively, taking advantage of their inherent simplicity and stability compared to DNS/VOF (Succi *et al.*, 1988). LGA is particularly well suited to parallel computing methods.

The lattice Boltzmann (LBM) method grew out of the LGA method. Essentially, it involves the replacement of the 'occupancy bits' of LGA with floating-point numbers that represent the average of the particle population. The need (in LGA) for averaging over a number of sites is removed in LBM and the lattice density is reduced. However, each site requires considerably more memory and it is not axiomatic that LBM is always more efficient. Also, the finite precision of digital computers mean that LBM is not guaranteed to march to equilibrium; stability problems may occur as they do in standard CFD.

Both LGA and LBM methods are restricted (in compressible flows) to relatively small Mach numbers (see Sankaranarayanan *et al.*, 2003).

4.3 Examples of direct predictions of flows with deformable interfaces

The objective of this section is to illustrate some of the results achieved using the methods described in Section 4.2. This subject area is a burgeoning one and new results are constantly appearing. Predictions are becoming possible for more and more complex systems.

An example of the use of ***boundary integral methods*** is reported by Ng (2002), who used the methods for laminar–laminar and laminar–turbulent stratified flows in circular tubes. Due to the influence of wetting and surface tension, the interface in such flows is not planar (as is often assumed). The interface shape can be predicted, however, if the fluid densities, surface tension and wetting angle are known. The

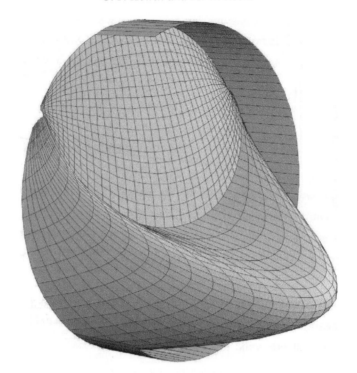

Fig. 7.22 Prediction of velocity profile in a stratified flow with a non-planar interface (Ng, 2002).

problem then arises of predicting the flow field for the very complex cross sections
of the respective fluids. This is an ideal application of boundary integral methods,
and results were obtained for a large range of conditions, typified by those shown
in Fig. 7.22 for a laminar–laminar flow with the less viscous fluid being the heavier.
The boundary element method cannot be used for turbulent flows; for the laminar–
turbulent case the effect of the turbulence was represented only in terms of the
influence of the turbulent zone on the interfacial shear stress of the laminar region.

The **VOF method** is perhaps the most widely used approach for predicting interfacial structures and development in multi-phase flows. Since it is incorporated now
into several of the commercial CFD codes, this facilitates its use by non-specialists
(though care needs to be exercised!). Quite complex interfacial developments can
be predicted, as are exemplified by the results shown in Figs. 7.23 and 7.24 for a
dam break in a horizontal tube and for a breaking wave, respectively.

Figures 7.25 and 7.26 illustrate calculations using the CFX code of the three-
dimensional tail of a slug in gas–liquid slug flow (Pan, 1996). Figure 7.25 shows the
change in slug tail shape as the mixture (total) velocity increases. At low velocities,
the shape is analogous to a dam break but, at higher velocities, the nose of the
'bubble' moves towards the centre of the tube with the shed of liquid draining
circumferentially around the tube. Figure 7.26 shows the effect of tube inclination

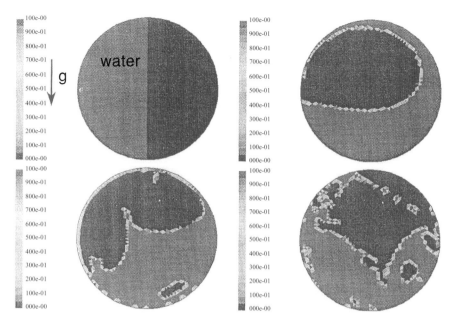

Fig. 7.23 VOF model for a dam break in a horizontal tube (Fluent Inc., private communication).

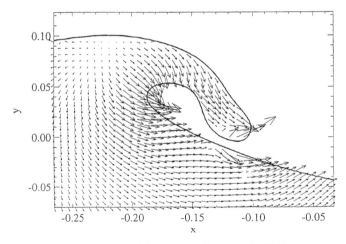

Fig. 7.24 VOF prediction of breaking wave (Chen *et al.*, 1999).

for a constant mixture velocity, with the bubble front becoming more and more asymmetric as the inclination angle decreases. The results obtained are in good agreement with experimental measurements.

A good illustration of the power of the VOF method in addressing phenomenological issues in two-phase flow is shown in Fig. 7.27, where the motions of bubbles of various sizes in linear shear fields have been calculated (Tomiyama, 1998). In

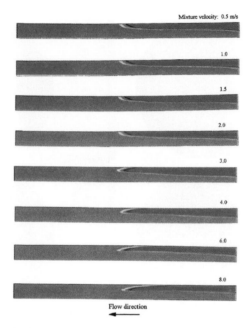

Fig. 7.25 Effect of mixture velocity on slug tail profiles in a 77.92 diameter tube.
Calculated by the VOF method using the CFX code (Pan, 1996).

Fig. 7.27, C is the non-dimensional gradient of the velocity, and the Eotvos number
(Eo) and the Morton number (M) are defined as follows:

$$Eo = \frac{g(\rho_L - \rho_G)d^2}{\sigma},$$ (7.27)

$$M = \frac{g\,\mu_L^4(\rho_L - \rho_G)}{\rho_L^2\,\sigma^3}.$$ (7.28)

Larger bubbles (i.e., with larger Eotvos numbers) migrate *up* the velocity gradient
whereas smaller bubbles (low Eotvos number) migrate *down* the velocity gradient.
This is consistent with experimental observations on bubble flows in tubes.

   Very detailed calculations can now be made of atomisation processes in, for
instance, diesel injectors. Leboissetier and Zaleski (2002) describe calculations
of this type and their results are exemplified in Fig. 7.28. The progressive wave
formation and breakup processes are clearly calculated. A very interesting finding
from this work was that the breakup process was strongly influenced by the presence
of turbulent vortices in the emerging liquid jet. In the practical case of a diesel
injector, such vortices arise in the boundary layer formed in the injector nozzle.
However, the VOF calculations can be carried out with (as shown in Fig. 7.28) or
without vortices. In the absence of the vortices, breakup of the jet was inhibited.

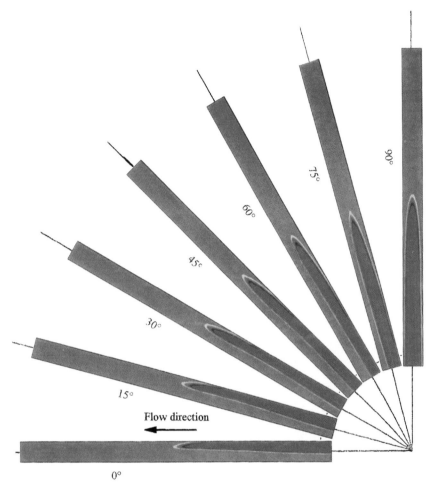

Fig. 7.26 Effect of tube inclination on slug tail profile at a mixture velocity of 4 m/s. Calculated using VOF method in CFX code (Pan, 1996).

As was stated above, the VOF method can be used in combination with the full spectrum of turbulence models. In the results shown in Figs. 7.25 and 7.26, the turbulent flow field was modelled using the standard $k–\varepsilon$ method (see Section 2 above). The VOF method can also be combined with other turbulence models, including direct numerical simulation (DNS). Hagiwara and co-workers (Hagiwara *et al.*, 2003, Iwasaki *et al.*, 2001) used DNS combined with VOF to study the behaviour of droplets in turbulent flows. Figure 7.29 shows some of their results for near wall droplet behaviour. It was found that the drops changed their orientation and shape as they approached the wall and that there was a complex effect on heat transfer arising from near wall droplet motion. This illustrates the enormous complexity of the hydrodynamic and heat transfer processes in such systems.

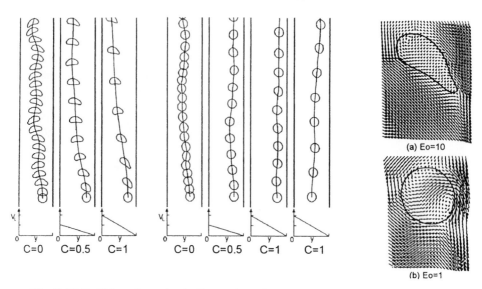

Fig. 7.27 Bubble trajectories in linear shear fields (Tomiyama, 1998).

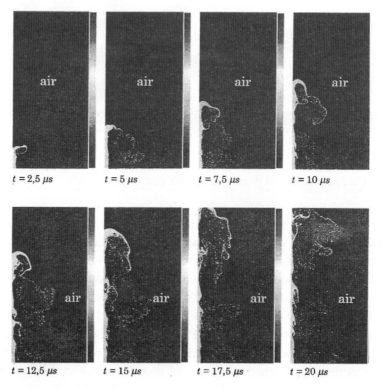

Fig. 7.28 Breakup of a diesel injector jet predicted by the VOF method (Lebois-setier and Zaleski, 2002).

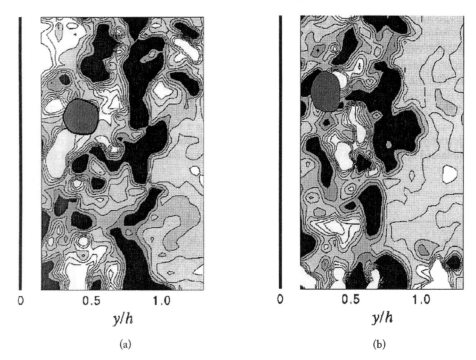

0   0.5   1.0   0   0.5   1.0

y/h y/h

(a)   (b)

Fig. 7.29 Droplet behaviour in the near wall region in a turbulent flow (Hagiwara *et al.*, 2003) (a) at the moment when the droplet changes direction from outward to wallward; (b) at the moment when the droplet changes direction from wallward to outward. Shading indicates vorticity contours.

A remarkable application of the *level sets* method is reported by Fukano and Inatomi (2003). This addressed the question of the mechanism of horizontal annular flow, as already introduced in Section 2.3 above. The objective was to investigate a wave spreading mechanism originally suggested by Fukano and Ousaka (1989), in which circumferential pressure gradients within the disturbance waves cause the transport of the liquid phase towards the top of the tube. The system was modelled as one which initially had a stratified flow, and the simulation followed the development of the waves and the consequential liquid transport. The results for the process of annular flow formation are shown in Fig. 7.30. As will be seen, the prediction demonstrates the formation of the large waves and their spreading around the circumference leading to annular flow. The simulation also allowed the prediction of the circumferential pressure gradients and confirmed the Fukano and Ousaka hypothesis.

An example of the application of the *front tracking technique* is shown in Fig. 7.31 for the case of film boiling on a horizontal surface. The successive pictures show the development of the interface and the shading indicates the vapour temperature; the hot vapour delays pinching off of the stem.

Fig. 7.30 Process of annular flow formation predicted by the level sets method (Fukano and Inatomi, 2003).

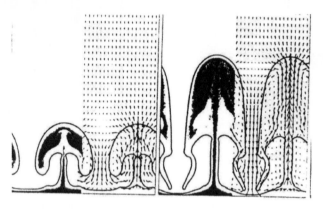

Fig. 7.31 Computation of film boiling using the front tracking method (Juric and Tryggvason, 1998).

The front tracking technique has been pursued actively by Professor Tryggvason and co-workers and has been applied not only to bubble flows (Esmaeeli and Tryggvason, 1998, 1999, Bunner and Tryggvason, 2002a, 2002b) but also to drop flows (Han and Tryggvason, 1999, 2001, Mortazavi and Tryggvason, 2000), instability in interfaces between stratified layers (Zhang *et al.*, 2002, Tauber *et al.*, 2002) and solidification (Al-Rawahi and Tryggvason, 2002). Examples taken from this later work are shown in Figs. 7.32–7.34 as follows.

(1) Figure 7.32 (from Esmaeeli and Tryggvason, 1999) shows the development of the flow field and bubble position and shape as a function of time after the release of a square array of bubbles in a liquid. The complexity of the bubble behaviour and flow field even in this highly idealised (two-dimensional) case is a striking demonstration of the formidable challenge in predicting the cases of interest in engineering applications.

(2) Figure 7.33 (from Bunner and Tryggvason, 2002a) shows a stage in the predicted development of a three-dimensional array of rising bubbles. This study was able to reproduce many of the observed features of real bubble flows, such as the rotation of bubbles around each other following a close encounter, and the formation of void waves.

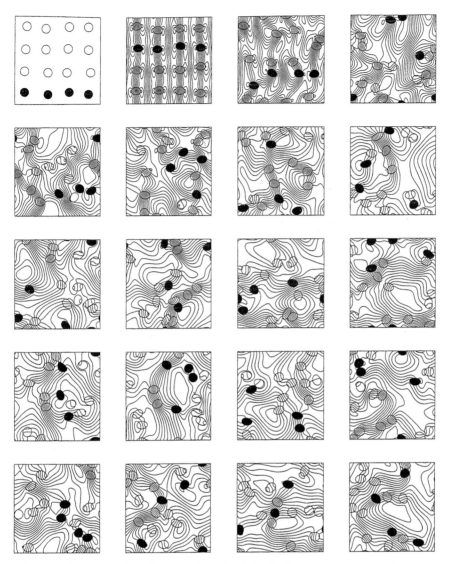

Fig. 7.32 Unsteady rise of sixteen bubbles initially in a two-dimensional square array (Esmaeeli and Tryggvason, 1999). Diagrams show streamlines and bubble positions and shapes. The initial positions of the bubbles are shown in the top left-hand corner frame, and time increases to the right and down. In each frame, four bubbles from one of the original horizontal rows (and not always the same row) are coloured black to show how the bubbles spread in the two-dimensional fluid.

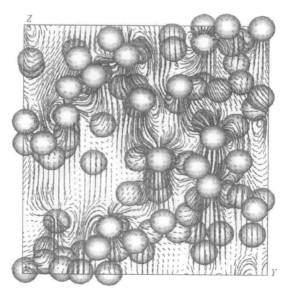

Fig. 7.33 Three-dimensional array of bubbles on a predicted bubble flow (Bunner and Tryggvason, 2002a).

(3) Figure 7.34 (from Tauber *et al.*, 2002) shows the development of instability at the sheared interface of two equal density fluids. As will be seen, even in this simple case, the evolution is highly non-linear.

As was stated above, the ***molecular dynamics method*** cannot really be regarded as a serious contender for the prediction of turbulent multi-phase flows. However, it is a useful technique in gaining an understanding of micro-scale phenomena. Figure 7.35 shows a molecular dynamics simulation of a liquid droplet forming on a solid surface (Maruyama, 2002). Here, the calculations were for 32 000 molecules. The distribution is time-varying, and the diagram shows a 'snapshot' at a particular time and also the distribution of average density.

The nucleation of a given phase (vapour or liquid) in a superheated or subcooled continuum of the other phase is a matter of great importance in many multi-phase systems. A key question is whether the surface tension of the (initially very tiny) nucleating bubbles or drops is equal to the macroscopic value. Molecular dynamics calculations of surface tension by Nijmeijer *et al.* (1988) and Alejandre *et al.* (1995) indicate that the macroscopic surface tension is achieved for surprisingly small clusters of molecules (of the order of several thousand). The molecular dynamics method can also be used directly in simulating nucleation and other processes; see Maruyama (2002) for a review.

Applications of the ***lattice gas automata*** method are reviewed by Biggs and Humby (1998); a typical application is the prediction of the evolution of

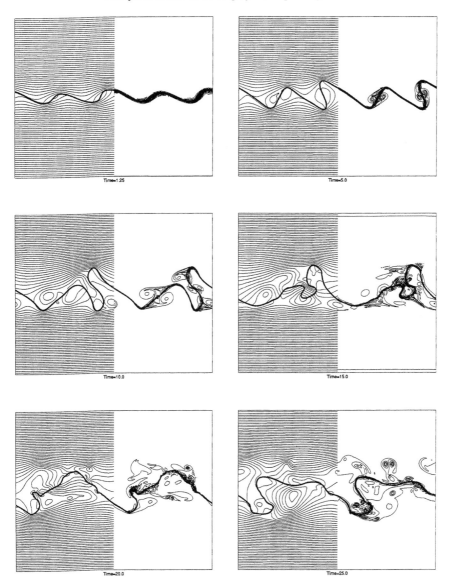

Fig. 7.34 Evolution of the instability of a sheared fluid interface (Tauber *et al.*, 2002).

micro-emulsions, as illustrated in Fig. 7.36 (Boghosian *et al.*, 1996). Other recent applications are to the prediction of the motion of suspended particulate materials in flow through porous media (Humby *et al.*, 2001, Biggs *et al.*, 2003).

There are relatively few reports of comparisons between alternative methods for the calculation of systems with deformable interfaces. Thus, the comparative study by Sankaranarayanan *et al.* (2003) of bubble rise in the fully developed oscillatory

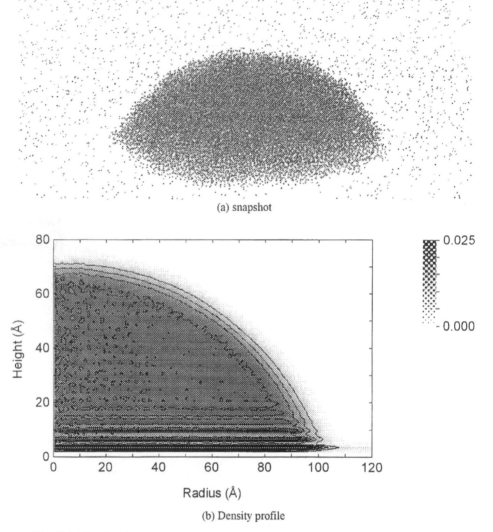

(a) snapshot

(b) Density profile

Fig. 7.35 Molecular dynamics simulation of a droplet forming on a solid surface (Maruyama, 2002).

regime using both the front tracking and **lattice Boltzmann** methods is of particular interest. The results are shown in Fig. 7.37; as will be seen, the predictions from the two methods (one based on the Navier–Stokes equations and the other based on a 'particulate' representation of the fluid and random motions of the 'particles') agree remarkably well.

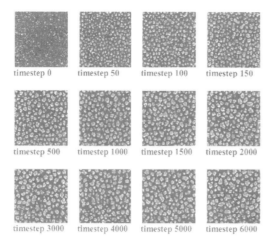

timestep 0   timestep 50   timestep 100   timestep 150

timestep 500   timestep 1000   timestep 1500   timestep 2000

timestep 3000   timestep 4000   timestep 5000   timestep 6000

Fig. 7.36 Time evolution of an oil-in-water micro-emulsion with surfactant present, illustrating the role of the surfactant in stabilising the emulsion (Boghosian *et al.*, 1996).

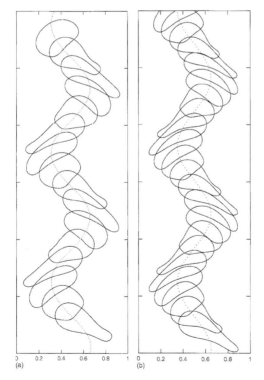

Fig. 7.37 Comparison of front tracking and lattice Boltzmann methods for bubble rise in the oscillatory regime (Sankaranarayanan *et al.*, 2003).

5 Overview

Computational methods are assuming an ever-increasing role in research, design and operation of systems with multi-phase flow. Commercial CFD codes are increasingly used for the prediction of a wide range of systems involving dispersions (bubble flows, fluid–solid flows, liquid–liquid flows and three-phase dispersed flows). For such dispersed flow systems, the commercial CFD vendors have competed with each other to adopt research results at an early stage (e.g. the two-way coupling between dispersed phase and continuous phase turbulent motions). Industrial applications have ranged from stirred tank reactors to sewage treatment plants, from two-phase pumps to combustion chambers, etc.

Research into dispersed flow systems has been driven by this background of industrial need and is ongoing. This research ranges from PDF methods (which treat the systems in a manner analogous to the kinetic theory of gases) to DNS of fluid and particle motions, from Lagrangian tracking methods to improved Eulerian methods. As in the case of single-phase flows, current (and foreseeable future) computational resources limit the detail at which industrial dispersed flow systems can be calculated. Thus, closure models are required. Often, these closure models are based on those developed for single-phase turbulent flows, and suffer not only from the problems for single-phase flows (identified in the other chapters of this book) but also from the uncertain effects of the presence of the other phase(s). Nevertheless, the use of CFD methods may present the only option for gaining a better, at least qualitative, understanding of the complex flow systems found in industrial dispersed flow applications. A detailed description of methods for dispersed flows is given in Section 3 above.

Many multi-phase *fluid* flows involve partial or complete *separation* of the phases. This separation may be mainly spatial (as in annular and stratified gas–liquid flows) or both spatial and temporal (as in gas–liquid slug flows). A common characteristic for all such flows is the *deformability* of the interfaces. The interfacial deformations may lead to the formation and growth of disturbances at the interface, resulting in a continuously changing interfacial configuration; examples here are the formation of waves in stratified flow leading (in a way not yet completely understood) to the formation of slugs, and the formation of the disturbance waves which dominate annular flow. Interfacial deformability also affects dispersed fluid–fluid flows (such as gas–liquid bubble flows), but its effects are often subsumed into the closure relationships with reasonable success; the development of generic closures for separated flows is proving an elusive goal.

Another characteristic of separated flows is that of *entrainment* of discrete elements of one phase into continuous regions of the other phase. Examples here are entrainment of water drops into the oil layer of a two-phase oil–water stratified flow, entrainment of droplets of liquid into the gas core of annular flow and entrainment of bubbles into the liquid slugs in slug flow. Flows with such entrainment may be regarded as being partly dispersed and partly separated and are thus more complicated than either fully separated or fully dispersed flows.

It is not surprising, in view of the complicating features described above, that computational methods for separated flows are much less advanced than those for dispersed flows. That is not to say that computational methods are not useful in studying such flows, but rather that prediction has to be closely allied to experimental observation. There are a number of ways in which computation is proving extremely helpful for separated flows.

(1) In evaluating phenomenological issues. Here, the use of single-phase CFD to study the respective separated regions can be crucial in deciding mechanisms. Several examples of this type of approach were given in Section 2.
(2) In providing a framework for the application of experimental results to system prediction. Here, the prime example is the use of the one-dimensional multi-fluid model in predicting steady state and transient flows in nuclear reactor and pipeline systems. Experimental data can be employed in developing closure laws for such systems and, despite the limitations in the range of such data, useful predictions can still be obtained. This topic is briefly reviewed in Section 4.1.
(3) In understanding the nature of interfacial phenomena. Here, techniques have been applied which range from those operating at the micro/nano scale (molecular dynamics, LGA, LBM) to those which solve the phasic conservation equations together with equations describing the properties of the interfaces (boundary element, level sets, VOF, front tracking). These methods are described in Section 4.2, with examples being given of their application in Section 4.3.

Computational methods can be expected to play an ever growing role in the study of multi-phase flows. They cannot and will not replace experimental measurement and observation; however, the combination of computation and experiment greatly enhances the effectiveness of each and is surely the way forward!

References

Ahmed, A. M. and Elghobashi, S. (2000) On the mechanisms of modifying the turbulent structure of turbulent homogeneous shear flows by dispersed particles. *Phys. Fluids* **12**(11), 2906–2929.

Alejandre, J., Tildesley, D. J. and Chapela, G. A. (1995) Molecular dynamics simulation of the orthobaric densities and surface tension of water. *J. Chem. Phys.* **81**(6), 1351–1361.

Al-Rawahi, N. and Tryggvason, G. (2002) Numerical simulation of dendritic solidification with convection: two-dimensional geometry. *J. Comput. Phys.* **180**, 471–496.

Biggs, M. J. and Humby, S. J. (1998) Lattice gas automata methods for engineering. *Trans. IChemE.* **76A**, 162–174.

Biggs, M. J., Humby, S. J., Buts, A. and Tuzin, U. (2003) Explicit numerical simulation of suspension flow with deposition in porous media: influence of local flow field variation on deposition processes predicted by trajectory methods. *Chem. Eng. Sci.* **58**, 1271–1288.

Boghosian, B. M., Coveney, P. V. and Emerton, A. N. (1996) *Proc. Roy. Soc.* **455**, 1221.

Bonizzi, M. and Issa, R. I. (2003a) On the simulation of three-phase slug flow in nearly horizontal pipes using the multi-fluid model. *Int. J. Multiphase Flow* **29**, 1719–1747.

(2003b) A model for simulating gas bubble entrainment in two-phase horizontal slug flow. *Int. J. Multiphase Flow* **29**, 1685–1717.

Bunner, B. and Tryggvason, G. (2002a) Dynamics of homogeneous bubbly flows. Part 1. Rise velocity and microstructure of the bubbles. *J. Fluid Mech.* **466**, 17–52.

(2002b) Dynamics of homogeneous bubbly flows. Part 2. Velocity fluctuations. *J. Fluid Mech.* **466**, 53–84.

Butterworth, D. (1971) Air–water annular flow in a horizontal tube. UKAEA Report No. AERE-R6687.

Buyevich, Y. (1971) Statistical hydromechanics of disperse systems. Part 1. Physical background and general equations. *J. Fluid Mech.* **49**, 489–507.

(1972a) Statistical hydromechanics of disperse systems. Part 2. Solution of the kinetic equation for suspended particles. *J. Fluid Mech.* **52**, 345–355.

(1972b) Statistical hydromechanics of disperse systems. Part 3. Pseudo-turbulent structure of homogeneous suspensions. *J. Fluid Mech.* **56**, 313–336.

Chang, Y. C., Hou, T. Y., Merriman, B. and Osher, S. (1996) A level set formulation of Eulerian interface capturing methods for incompressible flows. *J. Comput. Phys.* **124**, 449–464.

Chen, G., Kharif, C., Zaleski, S. and Li, J. (1999) Two-dimensional Navier–Stokes simulation of breaking waves. *Phys. Fluids* **11**, 121–133.

Crowe, C. T., Gore, R. and Troutt, T. R. (1985) Particle dispersion by coherent structure in free shear flows. *Particulate Sci. and Tech.* **3**, 149–158.

Derevich, I. V. and Zaichik, L. I. (1988) Precipitation of particles from a turbulent flow. *Izvestiyia Akademii Nauk SSR, Mekhanika Zhidkosti i Gaza* 96–104.

Ervin, E. A. and Tryggvason, G. (1997) The rise of bubbles in a vertical shear flow. *J. Fluids Engineering* **119**, 443–449.

Esmaeeli, A. and Tryggvason, G. (1998) Direct numerical simulation of bubbly flows. Part 1: Low Reynolds number arrays. *J. Fluid Mech.* **377**, 313–345.

(1999) Direct numerical simulation of bubbly flows. Part II: Moderate Reynolds number arrays. *J. Fluid Mech.* **385**, 325–358.

Fukano, T. and Inatomi, T. (2003) Analysis of liquid film formation in a horizontal annular flow by DNS. *Int. J. Multiphase Flow* **29**, 1413–1430.

Fukano, T. and Ousaka, A. (1989) Prediction of the circumferential distribution of film thickness in horizontal and near-horizontal gas–liquid annular flows. *Int. J. Multiphase Flow* **15**, 403–419.

Gibson, M. M. (1997) Turbulence modelling. *International Encyclopaedia of Heat and Mass Transfer*, pp. 1227–1234.

Hagiwara, Y., Takagaki, S. and Yuge, T. (2003) Effects of a droplet on near wall transport phenomena in turbulent downward liquid–liquid flow. *J. Enhanced Heat Transfer* **10(1)**, 103–115.

Han, J. and Tryggvason, G. (1999) Secondary breakup of axisymmetric liquid drops. I. Acceleration by a constant body force. *Physics of Fluids* **11(12)**, 3650–3667.

(2001) Secondary breakup of axisymetric liquid drops. II. Impulsive acceleration. *Physics of Fluids* **13(6)**, 1554–1565.

Harlow, F. M. and Welch, J. E. (1965) Numerical calculation of time-dependent viscous incompressible flow of fluid with free surfaces. *J. Phys. Fluids* **8**, 2182–2189.

Haworth, D. C. and Pope, S. B. (1986) A generalized Langevin model for turbulent flow. *Phys. Fluids* **29**, 387–405.

Hewitt, G. F. and Hall Taylor, N. S. (1970) *Annular Two Phase Flow*. Oxford, Pergamon Press.

Hewitt, G. F. and Jayanti, S. (1992) Prediction of film inversion in two-phase flow in coiled tubes. *J. Fluid Mech.* **236**, 497–511.

Humby, S. J., Biggs, M. J. and Tuzin, U. (2001) Explicit numerical simulation of fluids in reconstructed porous media. *Chem. Eng. Sci.* **57**, 1955–1968.

Hyland, K. E., Reeks, M. W. and McKee, S. (1999) Derivations of a pdf kinetic equation for the transport of particles in turbulent flow. *J. Phys. Math. Gen.* **32**, 6169–6190.

Issa, R. I. and Kempf, M. H. W. (2003) Simulation of slug flow in horizontal and nearly horizontal pipes with the two-fluid model. *Int. J. Multiphase Flow* **29**, 69–95.

Issa, R. I. and Oliveira, P. J. (1996) Validation of two-fluid model in shear-free mixing layers. ASME Fluids Engineering Division Conference Volume, FED-Vol. 236, pp. 113–120.

Iwasaki, T., Nisimura, K., Tanaka, M. and Hagiwara, Y. (2001) Direct numerical simulation of turbulent Couette flow with immiscible droplets. *Int. J. Heat and Fluid Flow* **22**, 332–342.

Jayanti, S. (1990) Contribution to the study of non-axisymmetric flows. Ph.D. Thesis, University of London (Imperial College).

Jayanti, S. and Hewitt, G. F. (1994) Hydrodynamics and heat transfer in wavy annular gas–liquid flow: a computational fluid dynamics study. Multiphase Systems Programme Report MPS/59 (Imperial College). (See also *Int. J. Multiphase Flow* **40**, 2445–2660, 1997.)

Jayanti, S., Wilkes, N. S., Clarke, D. S. and Hewitt, G. F. (1990) The prediction of turbulent flows over roughened surfaces and its application to interpretation of mechanisms of horizontal annular flow. *Proc. R. Soc. Lond.* **A431**, 71–88.

Jones, W. P. and Launder, B. E. (1972) The prediction of laminarisation with a two-equation model of turbulence. *Int. J. Heat and Mass Transfer* **15**, 301–314.

Juric, D. and Tryggvason, G. (1996) A front-tracking method for dendritic solidification. *J. Comp. Phys.* **123**, 127–148.

(1998) Computations of boiling flow. *Int. J. Multiphase Flow* **24**, 387–410.

Khor, S. H., Mendes-Tatsis, M. A. and Hewitt, G. F. (1997) One-dimensional modelling of phase holdups in three-phase stratified flow. *Int. J. Multiphase Flow* **23**, 885–897.

Kunisch, K. and Volkwein, S. (2002) Galerkin proper orthogonal decomposition methods for a general equation in fluid dynamics. *SIAM Journal on Numerical Analysis* **40(20)**, 492–515.

Launder, B. E. and Sharma, B. I. (1974) Application of the energy-dissipation model of turbulence to the calculation of flow near a spinning disk. *Letters in Heat Mass Transfer* **1**, 131–138.

Laurinat, J. E., Hanratty, T. J. and Jepson, W. P. (1985) Film thickness distribution for gas–liquid annular flow in a horizontal pipe. *Physico-Chemical Hydrodynamics* **6**, 179–195.

Leboissetier, A. and Zaleski, S. (2002) Influence des conditions amont turbulentes sur l'atomisation primaire. *Combustion* **2(1)**, 75–87.

Lin, T. F., Jones, O. C., Lahey, R. T., Block, R. C. and Murase, M. (1985) Film thickness measurements and modelling in horizontal annular flow. *Physico-Chemical Hydrodynamics* **6**, 197–206.

Lo, S. (1999) Application of population balance to CFD modelling of bubbly flows via the MUSIG model. Fourth International Conference on Gas–Liquid and Gas–Liquid–Solid Reactor Engineering, Delft, August 23rd–25th 1999.

Longuet-Higgins, M. S. and Cocklett, E. D. (1976) The deformation of steep surface waves on water. *Proc. Roy. Soc. London*, Ser. A **358**, 1.

Loth, E., Taeibi-Rahni, M. and Tryggvason, G. (1997) Deformable bubbles in free-shear layer. *Int. J. Multiphase Flow* **23**, 977–1001.

Luo, H. and Svendsen, H. F. (1996) Theoretical model for drop and bubble breakup in turbulent dispersions. *AIChE. J.* **42**, 1225–1233.

Maruyama, S. (2002) Molecular dynamics methods in microscale heat transfer. *Exchange Design Update* (Update journal for the *Heat Exchanger Design Handbook*, Begell House Inc., New York) **9(1/2)**, 2.13.8–1 to 2.13.8–26.

Maxey, M. R. (1987) The gravitational settling of aerosol particles in homogeneous turbulence and random flow fields. *J. Fluid Mech.* **174**, 441–465.

Minier, J. P. and Peirano, J.-P. (2001) The PDF approach to turbulent polydispersed two-phase flows, *Physics Reports* **352**, 1–3.

Minier, J. P. and Pozorski, J. (1997) Derivation of a pdf model for turbulent flows based on principles from statistical physics. *Phys. Fluids* **9(6)**, 1748–1753.

Mortazavi, S. and Tryggvason, G. (2000) A numerical study of the motion of drops in Poiseuille flow. *J. Fluid Mech.* **411**, 325–350.

Ng, T. Z. (2002) Interfacial structure in stratified pipe flow. Ph.D. Thesis, University of London, June 2002.

Nijmeijer, M. J. P., Bakker, A. F., Bruin, C. and Sikkenk, J. H. (1988) A molecular dynamics simulation of the Lennard-Jones liquid–vapor interface. *J. Chem. Phys.* **89(6)**, 3789–3792.

Orszag, S. A. and Yakhot, V. (1986) Reynolds number scaling of cellular-automaton hydrodynamics. *Phys. Rev. Lett.* **56**, 1691–1693.

Osher, S. and Sethian, J. A. (1988) Fronts propagating with curvature-dependent speed: algorithms based on Hamilton–Jacobi formulations. *J. Comp. Phys.* **79**, 12–49.

Pan, L. (1996) High pressure three-phase gas/liquid/liquid flow. Ph.D. Thesis, University of London.

Pandya, R. V. R. and Mashayek, F. (2003) Non-isothermal dispersed phase of particles in turbulent flow. *J. Fluid Mech.* **475**, 205–245.

Pope, S. B. (1991) Application of the velocity-dissipation probability density function model to inhomogeneous turbulent flows. *Phys. Fluids A* **3**, 1947–1957.

Prince, M. J. and Blanch, H. W. (1990) Bubble coalescence and break-up in air-sparged bubble columns. *AIChE. J.* **36**, 1485–1499.

Reade, W. C. and Collins L. R. (1998) Collisions and coagulation in the infinite Stokes number regime, *Aerosol Sci. Tech.* **29**, 493–509.

Reeks, M. W. (1980) Eulerian direct interaction applied to the statistical motion of particles in a turbulent fluid. *J. Fluid Mech.* **83**, 529–546.

(1991) On a kinetic equation for the transport of particles in turbulent flows. *Phys. Fluids A* **3**, 446–456.

(1993) On the continuum equations for dispersed particles in nonuniform flows. *Phys. Fluids A* **5(3)**, 750–761.

(2003) Comparison of recent model equations for particle deposition in a turbulent boundary layer with those based on the PDF approach, Paper # FEDSM2003-4573 Gas-particle Flow, Fourth ASME-JSME Joint Fluids Engineering Conference, Honolulu, Hawaii, USA, July 6–11, 2003.

Russell, T. W. F. and Lamb, D. E. (1965) Flow mechanisms of two-phase annular flow. *Can. J. Chem. Eng.* **43**, 237–245.

Sankaranarayanan, K., Kevrekidis, I. G., Sundarsan, S., Lu, J. and Tryggvason, G. (2003) A comparative study of lattice Boltzmann and front-tracking methods for bubble simulations. *Int. J. Multiphase Flow* **29**, 109–116.

Scardovelli, R. and Zaleski, S. (1999) Direct numerical simulation of free-surface and interfacial flow. *Ann. Rev. Fluid Mech.* **31**, 567–603.

Shearer, C. J. and Nedderman, R. M. (1965) Pressure gradient and liquid film thickness in co-current upwards flow of gas–liquid mixtures: application to film cooler design. *Chem. Eng. Sci.* **20**, 671–683.

Simonin, O. (2002) Statistical and continuum modelling of turbulent reactive particulate flows. In *Combustion and Turbulence in Two-phase Flows*, Von Karman Institute (VKI) Lecture Series, 1996–2002, VKI publication.

Simonin, O., Deutsch, E. and Minier, J.-P. (1993) Eulerian prediction of the fluid/particle correlated motion in turbulent two-phase flows. *Appl. Sci. Research* **51**, 275–283.

Suzzman, M., Smereka, P. and Osher, S. (1994) A level set approach for computing solutions to incompressible two-phase flow. *J. Comp. Phys.* **114**, 146–159.

Swailes, D. C. and Darbyshire, K. F. F. (1997) A generalised Fokker–Planck equation for particle transport in random media. *Physica A* **242**, 38–48.

Swailes, D. C., Devenish, D. A., Sergeev, Y. A. and Kurdyumov, V. N. (1999) A PDF model for dispersed particles with inelastic particle wall collisions. *Phys. Fluids* **11**, 1859–1868.

Tauber, W., Unverdi, S. O. and Tryggvason, G. (2002) The nonlinear behaviour of a sheared immiscible fluid interface. *Phys. Fluids* **14(8)**, 2871–2885.

Tomiyama, A. (1998) Struggle with computational bubble dynamics. Invited lecture at the Third International Conference on Multiphase Flow, ICMF'98, Lyon, France, June 8th–12th 1998.

Tryggvason, G., Bunner, B., Esmaeeli, A., Juric, D., Al-Rawahi, N., Tauber, W., Han, J., Nas, S. and Jan, Y. J. (2001) A front tracking method for the computations of multiphase flow. *J. Comp. Phys.* **169**, 708–759.

Tryggvason, G., Bunner, B., Esmaeeli, A. and Al-Rawahi, N. (2002) Computations of multiphase flows. In E. van der Giessen and H. Aref (eds.) *Advances in Applied Mechanics*, Vol. 39, pp. 1–41. Academic Press.

Unverdi, S. O. and Tryggvason, G. (1992) A front-tracking method for viscous, incompressible, multi-fluid flows. *J. Comp. Phys.* **100**, 25–37.

Wang, L. P. and Maxey, M. (1993) Settling velocity and concentration distribution of heavy particles in homogeneous isotropic turbulence. *J. Fluid Mech.* **256**, 27–68.

Wolf, A. (1995) Film structure of vertical annular flow. Ph.D. Thesis, University of London (Imperial College).

Xue, M., Xu, H. B., Liu, Y. M. and Yue, D. K. P. (2001) Computations of fully nonlinear three-dimensional wave–wave and wave–body interactions. Part I: Dynamics of steep three-dimensional waves. *J. Fluid Mech.* **438**, 11–39.

Young, J. and Leeming, A. (1998) Particle deposition in turbulent pipe flow. *J. Fluid Mech.* **340**, 129–159.

Zaichik, L. I. (1991) Modelling of particle dynamics and heat transfer using equations for first and second moments of velocity and temperature fluctuations. In Eighth

International Symposium on Turbulent Shear Flows, Technical University of Munich, pp. 10-2-1 to 10-2-6.

Zaichik, L. I. and Alipchenkov, V. M. (2002) A kinetic model for predicting pair dispersion and preferential concentration of particles in isotropic turbulent flow, Tenth Workshop on Two-phase Flow Predictions, Merseberg, April 9–12.

Zhang, J., Miksis, M. J., Bankoff, S. G. and Tryggvason, G. (2002) Nonlinear dynamics of an interface in an inclined channel. *Phys. Fluids* **14(6)**, 1877–1885.

8

Guidelines and criteria for the use of turbulence models in complex flows

J. C. R. Hunt

University College London and Delft University of Technology

A. M. Savill

Cranfield University

Summary

This chapter begins with a review of the principles underlying general purpose turbulence models and the assumptions and procedures involved in applying them to calculate the kind of complex flows that are analysed in practical engineering and environmental problems. Secondly we develop, from considerations of basic mechanisms of turbulence and the different types of statistical turbulence model, a new guideline 'map' based on characteristic statistical parameters, which can be derived from standard models. This indicates in principle which types of turbulent flow can and cannot be approximately calculated with the current generation of 'CFD', one-point turbulence models, including those using k–ε and second order closure equations. No attempt is made to identify any one optimum model scheme. Thirdly, the proposed guidelines for the likely accuracy of turbulent modelling are tested by comparing them with the results of previous test-case studies for a range of complex turbulent flows, where standard models fail or need special adaptation. These include thermal convection, free stream turbulence, aeronautical flows and flows round bluff bodies. The relative merits of advanced models (e.g. involving two-point statistics) and numerical simulations are also discussed, but the CFD practitioner should note that the emphasis here is on why current models will not work in all circumstances. The technical level of this chapter is most suitable for readers with some formal training in fluid dynamics. These general guidelines are complementary to user guidelines for computational fluid dynamics codes.

1 Problems of modelling turbulent flows

The recent study programme on turbulence at the Isaac Newton Institute in Cambridge in 1999 began with extensive reviews of how computational models are

Prediction of Turbulent Flows, eds. G. F. Hewitt and J. C. Vassilicos.
Published by Cambridge University Press. © Cambridge University Press 2005.

Table 8.1 *Some critical flows for current general purpose turbulent models.*
The lower half are sensitive flows where models need special adaptation to
provide accurate answers.

	Rapidly/slowly changing turbulence	Spatial localness	Partially non-local	Significantly non-local
Shock b.l. interactions*	RCT	L		
Free stream turbulence outside boundary layers*	SCT		PNL	
Thermal convection*	SCT			NL
Trailing edge*	RCT			NL
Jets impinging onto bluff bodies	RCT			NL
Concentration fluctuations and combustion[†]	SCT			NL
Turbulent boundary layers as *Re* changes*	SCT	L →	PNL	
Wall jets*	SCT		PNL	
Separating flows*	RCT		PNL	

*k–ε (and adaptive stress) RSTM; [†]stochastic/pdf models.

used to calculate industrial and environmental flows (see Chapter 2) and included detailed consideration of all the current CFD methodologies in use (see Launder and Sandham, 2002). It was noted how there had been considerable progress in these models as practical techniques for industry. There is now general consensus that some standard models are sufficiently well specified that different users of the same codes, *provided the input data, boundary conditions and spatial/temporal resolution are properly specified*, will obtain the same results. However, even if the results are consistent, they may not be accurate. Furthermore, no systematic procedures are currently available to predict in advance the accuracy of any partic-ular model for a new type of flow, though previous authors have suggested some principles for indicating when standard models might fail.

A number of specific turbulent flow phenomena of critical importance in engi-neering and environmental systems were identified at the opening workshop of the INI programme where these modelling uncertainties are acute and need to be under-stood and hopefully reduced in the near future (see Table 8.1). At the same time a number of more general deficiencies in one-point turbulence models (see Section 2) were pointed out – in particular they do not adequately account for anisotropy and usually ignore non-local (strain history) effects where the flow changes so rapidly in space or time that the turbulence is not close to a state of local statistical equi-librium – the essential concept underlying most models. In other words, the finite length and time scales and particular features of eddy motion have to be taken into account.

More generally, the non-local relation between the velocity and pressure fields cannot be correctly represented in a single point closure model even for simple homogeneous turbulent flows in the presence of rapidly changing mean velocity gradients. This point can be substantiated, using linear theory, because in such cases the 'rapid' part of the pressure-strain rate dominates. That term is only approximately modelled in Reynolds stress transport model, RSTM, equations, but linear theory can be used to improve standard closure approximations, especially for rotating flows. Kassinos *et al.* (2001) used such an approach to incorporate the effects of length scale anisotropy directly in RSTM, the fundamental difficulties of which have been reviewed by Cambon in Launder and Sandham (2002) and in Oberlack and Busse (2002).

This chapter sets out to establish some possible principles for establishing such a priori estimates about the likely accuracy of models and simulations in different types of turbulent flows. We draw on earlier reviews by ourselves and others, notably Launder and Spalding (1972), Launder (1989), Lumley (1978), Hunt (1992 and 1995), Savill (1981, 1987 and 1996) and the basic text of Townsend (1976). The proposed principles are used to construct a guide based on the statistical properties of these flows, to enable practitioners to make a preliminary judgement about the likely suitability of any particular model for computing the particular type of flow. The present guide is more fundamental than, but complementary to the recent 'best practice' guidelines document for computational fluid dynamics (CFD) produced by ERCOFTAC (2000). This has provided CFD users with a well established blueprint for the selection, application and expected capabilities of models, so as to become a baseline reference for those involved in routine engineering CFD. The present chapter examines various turbulence model types and examines each for its key assumptions and how they may influence the accuracy of computations in various classes of flow. It is concluded that turbulent flows can generally be classified on the basis of rapidity of change in time and space (relative to the turbulence time and length scales) into broad groupings, where for practical purposes different sorts of models are most suitable. No one-point model is suitable for all cases. Most current turbulence model approaches have been calibrated against a small range of standard, ideal 'building-block' flows (involving at most two applied strain rates or other effects such as free stream turbulence and only rarely considering the full non-linear interplay of these). They have then been applied to real complex flows. We particularly consider here, as a result of industry input at the INI programme, important subsets of complex flows encountered in practice. Such 'screening' procedures are common in other areas of engineering and environmental fluid mechanics (e.g. for atmospheric dispersion).

In Section 2 we review some current basic ideas about turbulence and models, before proceeding in Section 3 to develop in detail principles, guideline algorithms and a 'map'. Together these sections provide the terminology and the concepts for our new guidelines. The map shows how the algorithms apply in practice to the main types of turbulent flow. In Section 4 we use the map to consider new approaches for calculating flows where current practical methods are not appropriate. In Section 5 we conclude with a further application of the 'route map', indicating the most appropriate prediction methodologies to adopt for various flow applications, based on measurable/computable parameters.

We conclude this introduction by reviewing the main features of turbulent flows. In all such flows the velocity fluctuates randomly with characteristic and repeated patterns over wide ranges of length and time scales. These different scale motions interact with each other, but they are not highly correlated.

There are various characteristic types of eddy motion at the different length scales. Those containing most energy are not universal, because they depend on the large scales, the nature of the flows and the initial and boundary conditions. However, small scale motions within any particular flow are weakly correlated with large scales, and therefore tend to have an approximately universal structure, although their overall energy and rate of decay are still related to the large scales via a 'cascade' mechanism.

These concepts explain why the main statistical properties of turbulence and the forms of the large eddy structures depend on the type of turbulent flow and why, in many material aspects, turbulence is not a universal state of nature. Nevertheless, it is of considerable practical importance that there are certain features of the small scale eddy motion, such as the energy spectra, that are independent of the large scales and are approximately universal.

Certain other features of the small scale motion, such as third moment spectra, are not independent and are not universal (e.g. Hunt *et al.*, 1988). Despite this broad conclusion that the statistical structure of turbulence is not universal, research has shown that the same qualitative mechanisms are found in all turbulent flows (for example, the generation of Reynolds stresses, production and dissipation of turbulent energy, diffusive transport of inhomogeneous turbulence, transfer of energy between the components of eddy motion with different direction and length/time scale) (e.g. Townsend, 1976). The development of these concepts has driven the systematic development of computational models for calculating the statistical properties of turbulence.

In most models used in practice only broad characteristics of turbulence are predicted, such as mixing length-Reynolds shear stress models at the simplest level, or k–ε models for the energy and dissipation rate (or effective velocity and length scales characterising the turbulent motion). However, to tackle many practical engineering and environmental problems, information is required on more detailed aspects of the statistical nature of turbulence. Some of these aspects can be predicted with more complex models, for example second order models for the time and spatial development of all the six covariances at one point (e.g. Launder *et al.*, 1975; Lumley, 1978) or if two-point statistics such as structure function and spatial cross correlations are needed (e.g. Mann, 1994), spectral models and numerical simulation methods are necessary. In some cases, but not all, more detailed predictions of the turbulence and unsteady statistics also help to make more accurate predictions of the mean velocity and pressure field. For a more detailed discussion of the full hierarchy of such methods and the interrelations of one-point, multi-point, probabilistic and quasi-linear closures, see Launder and Sandham (2002 – Chapter 9), Oberlack and Busse, 2002 – Chapter 4.

A wide range of 'validated' turbulence models are now available, varying in complexity and also in the types of turbulent flow for which they have been specifically developed (e.g. Pope, 2000). The uses of these models differ depending on their applications. For calculating engineering flows of different types, many organisations use the same advanced models for all applications. By comparison in environmental flows (e.g. atmospheric and ocean boundary layers and river flows) it is more common to use a variety of simpler models, or parameterisations, either for each type of flow or (as in meteorology) for the main types of eddy motion. Sometimes, where no data are available for a new problem, useful guidance about the reliability of predictions can be provided by comparing results of using several different models. The actual use of turbulence models (even within the same organisation) is generally decided pragmatically (e.g. for speed, convenience, familiarity) rather than being based on any systematic principles. This is one reason for more systematic procedures. However, judgements about models tend to be influenced by the outputs demanded of them in flow situations that may or may not be suitable for the model. We restrict ourselves to the more basic flow field quantities as guidelines; guidelines for the more demanding requirements for heat and mass transport need further consideration.

2 Overview of turbulence models

2.1 Statistical aspects and governing equations

In order to provide a basis for the guide in Section 3 as to the appropriate choice and use of turbulence models, in this section we first review the assumptions, objectives, methods of calculation, inputs and outputs of these models. It is also necessary nowadays to distinguish them from 'simulations' or real time predictions, which can provide information about turbulent flow fields as they evolve in time-space in arbitrary and in particular transient realisations of the flow. In the latter case unsteady forms of turbulence models can predict ensemble average statistics as a function of time. In the former case for truly random (eddy) simulations, whether direct Navier–Stokes (DNS) or large eddy simulations (LES), ensemble averaging over many integral time scales is required to obtain mean flow statistics. This type of modelling is reviewed by N. D. Sandham in Chapter 6 (see also Launder and Sandham, 2002; Sagaut, 2001; Hunt *et al.*, 2001).

Turbulence 'models', as they are known, are systems of differential equations for a finite set of statistical quantities of velocity, pressure and temperature, denoted by u_i^*, p^* and θ^* defined over many realisations, together with initial and boundary conditions relevant to their statistics. Most often in practice these quantities are moments of variables at a single point, up to second order. But practical models are also now being developed for higher order two-point statistical quantities even in complex turbulent flows; these include spectra (e.g. Mann, 1994; Hunt *et al.*, 1990; Parpais, 1997; Cambon, 2002) and probability density functions (Pope, 2000). For some problems, as we shall explain in Section 3, where models are not currently satisfactory, the techniques of 'simulation' and 'modelling' are being combined (e.g. Kenjeres and Hanjalić, 1999).

Using standard notation, the turbulence variables in given flows are expressed as the sum of the ensemble mean, denoted by an over-bar, and fluctuating quantities (i.e. in certain unsteady engineering and environmental flows, the mean has to be determined by boundary conditions so that convergent statistics can be evaluated from the set of realisations in the ensemble).

Thus for velocity, pressure, density, temperature and body force (per unit volume):

$$u_i^* = U + u_i, \quad \text{where } U \equiv \bar{u}_i,$$
$$p^* = \bar{p} + p,$$
$$\rho^* = \bar{\rho} + \rho,$$
$$\theta^* = \Theta + \theta,$$
$$b_i^* = \bar{b}_i + b_i. \tag{8.1}$$

The approximate equations used in turbulence models, denoted by

$$L^{(\mathrm{T})}\big(M^{(\mathrm{n})}(u)\big) = 0$$

for the *n*th moment, are derived from the exact (Navier–Stokes) governing equations of fluid flow denoted by $L^{\mathrm{NS}}(u_i^*) = 0$. This procedure involves both systematic procedures (e.g. for the linear terms) and also significant approximations for certain terms whose error is a priori unquantifiable (with our present state of knowledge of turbulence).

From the Navier–Stokes equations for momentum and continuity (for incompressible and approximately uniform density fluids):

$$L^{NS}(u_i^*) = \frac{\partial u_i^*}{\partial t} + u_j^* \frac{\partial u_i^*}{\partial x_j} + \frac{1}{\rho^*} \frac{\partial p^*}{\partial x_i} - b_i^* - \nu \frac{\partial^2 u_i^*}{\partial x_j^2} = 0, \tag{8.2}$$

$$\frac{\partial \rho^*}{\partial t} + \frac{\partial}{\partial x_i}(\rho^* u_i) = 0. \tag{8.3}$$

Where there are significant variations in the density ρ^*, it may be assumed that, in the 'low speed' flows considered here, their only dynamical effect in the presence of gravitational acceleration is to cause a buoyancy force $b_i = g_i \rho^*$. Otherwise ρ^* is taken as a constant value ρ_0. Where the inertial forces associated with large density fluctuations are dynamically significant, in combustion and in low speed variable density flows, the modelling and the physical understanding of their effects on turbulence (e.g. thickening or thinning shear layers) are arguably still quite limited (e.g. Eames and Hunt, 1997; Bray *et al.*, 2000). As a result of the Boussinesq approximation, the continuity equation reduces to

$$\frac{\partial u_i^*}{\partial x_i} = 0. \tag{8.4}$$

From the conservation of energy, for a homogeneous fluid, the temperature field is determined by

$$\frac{\partial \theta^*}{\partial t} + \frac{\partial}{\partial x_i}(u_i^* \theta^*) = \kappa_\theta \nabla^2 \theta^*, \tag{8.5}$$

where κ is the thermal diffusivity.

Note that the body force may also be produced by other effects such as electromagnetic forces or the action of fixed or moving resistive objects within the flow (e.g. spray particles), all of which influence the eddy motion and statistics of the turbulence (Hunt, 1995). Higher Mach numbers and compressibility effects are modelled using similar methods (e.g. Durbin and Zeman, 1992; Simone *et al.*, 1997).

The governing equations for the mean component follow by ensemble averaging (Reynolds, 1895) so that from (8.2) and (8.3):

$$\frac{\partial U_i}{\partial t} + U_j \frac{\partial U_i}{\partial x_j} = -\frac{1}{\rho_0} \frac{\partial \bar{p}}{\partial x_i} + \bar{b}_i - \frac{\partial \bar{\sigma}_{ij}}{\partial x_j} + \nu \frac{\partial^2 U_i}{\partial x_j^2}, \tag{8.6}$$

$$\frac{\partial \Theta}{\partial t} + U_j \frac{\partial \Theta}{\partial x_j} = -\frac{\partial F_{\theta j}}{\partial x_j} + \kappa_\theta \frac{\partial^2 \Theta}{\partial x_j^2}, \tag{8.7}$$

where the mean momentum and scalar fluxes are defined as $\sigma_{ij} = \overline{u_i u_j}$, $F_{\theta j} = \overline{\theta u_j}$. The magnitudes of the velocity fluctuations and Reynolds stresses (usually defined as $-\sigma_{ij}$), are normally much weaker than the typical changes in mean velocity and mean momentum fluxes (i.e. $\Delta|\overline{u_i u_j}| \ll \Delta|U_i\,U_j|$), where Δ denotes changes in the moments. Therefore the shear stress gradients (i.e. $\partial\sigma_{ij}/\partial x_j$) only have a significant effect on the mean flow when the mean accelerations are small and the turbulence is changing slowly along the mean streamlines. This, as we shall see, is why turbulence models work (i.e. computing mean flows) as well as they do!

The equations for the fluctuations (after subtracting the mean terms) are:

$$\frac{\partial u_i}{\partial t} + U_j\frac{\partial u_i}{\partial x_j} + u_j\frac{\partial U_i}{\partial x_j} + u_j\frac{\partial u_i}{\partial x_j} = -\frac{1}{\rho_0}\frac{\partial p}{\partial x_j} + b_i + \nu\frac{\partial^2 u_i}{\partial x_j^2},$$

$$\frac{\partial u_i}{\partial x_i} = 0,$$

$$\frac{\partial\theta}{\partial t} + U_j\frac{\partial\theta}{\partial x_j} + u_j\frac{\partial\theta}{\partial x_j} = \kappa_\theta\frac{\partial^2\theta}{\partial x_j^2}. \tag{8.8}$$

From (8.8), equations for the higher moments $M^{(n)}(u)$, $n \geq 2$ are derived. But simplifying assumptions and/or auxiliary equations for statistical quantities have to be introduced in order to solve the moment equations up to the highest moment that is to be considered. This is the 'closure' problem, as explained below. These heuristic steps involve choosing inputs from theoretical and numerical studies of turbulent structures and experimental data. Since there is no unique or correct approach a variety of model equations have been proposed. They tend to result in similar, but not exactly the same, approximate forms of soluble moment equations. However, the differences in certain circumstances have real practical consequences.

For the subsequent discussion of the approximation and limitations of solution methods, it is convenient to label terms in moment equations. Thus for the second moments σ_{ij} (following Launder et al. (1975), who brought together several approximate representations of different phenomena into one consistent model for second order moments), the following equation may be used:

$$\left(\frac{\partial}{\partial t} + U_j\frac{\partial}{\partial x_j}\right)\sigma_{ij} = I_{ij} + B_{ij} + ET_{ij} + V_{ij}. \tag{8.9}$$

Here the right hand side of the equation is made up of the following terms: inertial production made up of the Reynolds stress production and pressure-strain correlation so that

$$I_{ij} = R_{ij} + P_{ij}, \text{ where}$$

$$R_{ij} = -\left(\sigma_{ik}\frac{\partial U_j}{\partial x_k} + \sigma_{jk}\frac{\partial U_i}{\partial x_k}\right), \quad P_{ij} = \left(\overline{p\frac{\partial u_i}{\partial x_j}} + \overline{p\frac{\partial u_j}{\partial x_i}}\right); \tag{8.10}$$

body-force production,

$$B_{ij} = (\overline{u_i b_j} + \overline{u_j b_i}); \tag{8.11}$$

eddy transport (which includes pressure 'diffusion'),

$$ET_{ij} = -\left(\frac{\partial}{\partial x_j}(\overline{u_i p}) + \frac{\partial}{\partial x_i}(\overline{u_j p}) + \frac{\partial}{\partial x_k}\overline{u_k u_i u_j} \right); \tag{8.12}$$

and viscous 'production', made up of the energy dissipation rate and the viscous diffusion of stress, i.e

$$V_{ij} = -D_{ij} + V_{ij}^{(\sigma)}, \; D_{ij} = \nu \left(2\overline{\frac{\partial u_i}{\partial x_k}\frac{\partial u_j}{\partial x_k}} \right), \quad V_{ij}^{(\sigma)} = \nu \frac{\partial^2}{\partial x_k^2}\sigma_{ij}. \tag{8.13}$$

This set of equations (8.9) to (8.13), reduces to the equation for the eddy kinetic energy $k = \frac{1}{2}\sigma_{ii}$ when the three equations for normal stresses are summed. Then:

$$A = I + B_k + ET_k - \varepsilon, \tag{8.14}$$

where:

$$
\left.
\begin{aligned}
A &= \left(\frac{\partial}{\partial t} + U_j \frac{\partial}{\partial x_j} \right) k \\
I &= \frac{1}{2} R_{ii} = -\sigma_{ik} \frac{\partial U_i}{\partial x_k} \\
B_m &= \frac{1}{2} B_{ii} = \overline{u_i b_i}
\end{aligned}
\right\} \tag{8.15}
$$

$$ET_m = \frac{1}{2}ET_{ii} = -\left[\frac{\partial}{\partial x_m} \left(\overline{u_m p}\delta_{im} + \frac{1}{2}\overline{u_m u_i u_i} \right) \right], \tag{8.16}$$

and

$$\varepsilon = \frac{1}{2}D_{ii} - \frac{1}{2}V_{ii}^{(\sigma)}. \tag{8.17}$$

Auxiliary algebraic expressions and/or equations are needed to solve the equations for the second moments, in particular for estimating the two-point moments (e.g. the pressure-strain terms and dissipation) in terms of one-point moments (see Durbin and Pettersson-Reif, 2000). In some practical turbulence models (Lumley *et al.*, 1978; Wyngaard, 1979) equations for third moments are included in the set to be solved, especially in highly anisotropic and inhomogeneous flows. These equations have a similar form to (8.9)–(8.13) with production transport and dissipation terms (at least for low Reynolds number flows). Note that these equations for the moments involve the mean local strain $\partial U_i/\partial x_j$. But when this varies rapidly within the lifetime or length scale of eddies, as it does in some important practical flows, then new approaches are needed. These are considered in Section 4.

We next consider how these equations form the basis of approximate turbulence models.

2.2 Physical concepts and modelling approximations

Generally, given certain initial and boundary conditions, the first objective (or 'output') of all practical turbulence models when applied to a particular flow is to calculate the changes of mean Reynolds stress σ_{ij} and heat flux $F_{\theta i}$, and then the mean velocity and temperature fields. Achieving satisfactory accuracy of the latter predictions is usually the main criterion for the usefulness of the model. As the models become more elaborate, they are constructed so as to predict other statistical properties of the turbulence. Those we consider here are firstly 'mixing length models' (ML) for σ_{ij}, secondly energy dissipation (or k–ε) models for k and ε (or an approximately equivalent statistical quantity) and thirdly Reynolds stress transport models (RSTM) for σ_{ij} and ε. More complex and multi-point models exist (as already mentioned), are briefly discussed in Section 4. The 'input' data required by turbulence models vary according to the model, as we shall see, or the type of flow and the size and location of the volume within which the turbulence model is applied.

Because (as explained above) there are no exact or unique models for constructing equations for these quantities (one point, second moments) from the moment equations, the 'closure' hypotheses needed to make the equation soluble have to be based on physical concepts, theoretical and numerical studies, and comparison with experimental measurements of various 'reference' turbulent flows. Therefore, the models depend on the types of reference flow selected for this comparison. Indeed, variation in this selection causes significant differences between models. Nevertheless, a set of basic reference flows have been generally agreed and endorsed at international workshops (e.g. Kline *et al.*, 1981). As a result the approximate forms of computable moment equations that have emerged over the many years of turbulence research are broadly, but not exactly, similar to each other. However, the slight differences between the models do have practical consequences (e.g. ERCOFTAC, 2000).

We list first some of the principles and turbulent properties on which these practical models are based, and their physical justification (more details of the properties of turbulence are given in Chapter 2). Then we analyse why and where they fail, or are particularly sensitive to particular initial and/or boundary conditions.

All turbulence models are based on the fundamental concept that certain statistical properties of turbulence (to some level of approximation) do not change within a given flow, or for certain types of flow. It is unlikely that any model is valid for *all* types of turbulent flow because, except for certain properties of very small scale turbulence at very high Reynolds number, the statistical properties of turbulence are not universal. Nevertheless, within certain *types* of turbulent flow (and within certain parameter ranges) many features of the energy containing eddy motions and their statistical properties are similar. This is broadly the consensus of the views expressed in the literature and at the INI programme (see Chapter 2). However, two main qualitative features are present in *all* types of 'fully developed' turbulence, and they determine the assumptions made in all turbulence models, viz.:

(i) Turbulence occurs at high Reynolds number (Re), so that in the energy containing eddies with length scale L_x and velocity scale u_0 (where $Re = u_0 L_x / \nu \gg 1$), viscous stresses are negligible and dissipation occurs in eddies whose scale is much smaller than L_x and whose velocity fluctuations are much smaller than u_0. As the Reynolds number becomes larger the motions on these dissipation scales progressively tend to become more isotropic (Taylor, 1938a). Recent research indicates that in typical engineering and environmental flows at very high Re this is only an approximation because the higher order statistical moments of the smallest scale motions are not completely isotropic (see Chapter 2).

(ii) The random eddying motion of turbulence causes an exchange between random momentum in different directions and on different time scales, so that, provided the mean rate distortion of turbulence by the mean flow or external forces is not too large (relative to the natural time scale of the turbulence $T_L \sim L_x / u_0$), the energy spectrum $E(\hat{k})$ (and probability density function $pr_{(u)}(u)$) of fully developed turbulence generally have the same basic forms, i.e. a single maximum, where $E(\hat{k}) = E(\hat{k}_{mx})$ at the scale of the energy containing eddies ($\hat{k}_{mx} \sim L_x^{-1}$), and a single maximum in $p_{(u)}(u)$; though the symmetries and functional forms can vary significantly from one flow to another (e.g. Launder and Spalding, 1972). This provides a useful working definition of the term 'fully developed turbulence'. For example, in the near wake flow of a bluff body placed in a turbulent stream there are typically two maxima in $E(\hat{k})$ (e.g. Britter *et al.*, 1979); 'fully developed' turbulence only develops over a few diameters downstream (Fig. 8.1). (For a more detailed discussion of the 'canonical' form of $E(\hat{k})$ see Godeferd *et al.*, 2001.)

Most turbulent flows in practice are bounded by rigid surfaces (such as the walls of a pipe or an aircraft wing) or by fluid interfaces where the turbulence is damped or disappears (such as at the outer edge of a boundary layer – see Fig. 8.2). Both at the former type of boundary whose locations are well defined and the latter, which are randomly moving, the turbulence structure undergoes enormous changes so that it no longer has the characteristic form of fully developed quasi-homogeneous three-dimensional turbulence described above. In the former case of wall turbulence

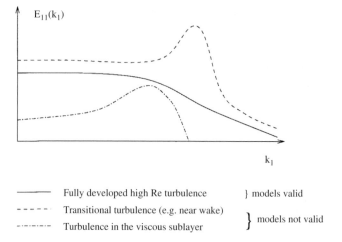

Fig. 8.1 Schematic diagram showing the typical forms of one-dimensional spectra $E_{11}(k_1)$ for which turbulence models are and are not valid.

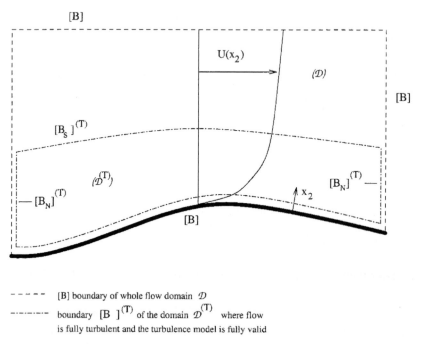

Fig. 8.2 Schematic diagram showing the regions within a typical flow domain where the turbulence models might be approximately valid. The nomenclature of the boundaries is also shown.

the Reynolds number decreases, the turbulence becomes highly anisotropic (nearly two-component) and the high frequency part of the spectrum is damped, all of which tend to reduce the exchange of energy between motions on different length scales. Some authors assume turbulence models are also valid in this limit and use this limit to define the parameters of their model (e.g. Craft and Launder, 1996). In the latter case of fluid interfaces between turbulent and non-turbulent motion the small scale rotational turbulence becomes intermittent and the probability density function of velocity fluctuations becomes skewed, with high flatness (Murlis *et al.*, 1982). However, within the fluctuating interface the turbulence structure is close to 'normal', so that the statistics of the turbulence at a given point depend greatly on how often that point is on the turbulent or non-turbulent side of the interface (Bisset *et al.*, 2002).

Therefore, in applying a model to a flow lying in a defined region of space, it is necessary to realise that the model is only valid within a 'conforming' part of that space. In particular, on both rigid and free boundaries there are thin 'non-conforming' layers where the turbulence departs from its 'fully developed' form. Then special 'matching' conditions are needed to describe the turbulence statistics inside these thin layers and relate them to the turbulence in the 'conforming' part of the flow. (See Sections 2.3 and 2.4.) The boundary conditions also have to be applied where the space adjoins upstream and downstream flows and have to be defined for each type of model. For the simplest models, dependent only on local properties of the mean velocity (e.g. mixing length models), these considerations may be formally irrelevant in deriving solutions, but they are relevant for interpreting the results of the computations.

2.3 Local (mixing length) models

Many studies have shown that mixing length models can provide a reasonable approximation for estimating Reynolds shear stress and hence the mean velocity in turbulent shear flows when the turbulence is changing slowly along the flow over the Lagrangian time scale (or T_L). But in more complex flows and for calculating heat transfer there are severe limitations to the use of these models. This limited applicability of simple models can be explained in terms of certain general properties of the turbulence in these sheared flows. It is found that the eddy structure and certain statistical properties have characteristic forms (e.g. elongated vortices, streaky jets, vortex sheets) in fully developed turbulent flows. Broadly, these eddy structures depend on the forms of the gradients of the mean velocity, which can be expressed as $\partial U_i / \partial x_j$ or $\nabla \mathbf{U}$. These forms are either swirling flows, shear flows or pure straining motion corresponding to values of the normalised strain parameter

$\tilde{S}$ derived from $\nabla \mathbf{U}$, namely

$$\tilde{S} = \frac{\Sigma^2 - \frac{1}{2}\Omega^2}{S^2}, \quad \text{where} \quad S^2 = \Sigma^2 + \frac{1}{2}\Omega^2, \tag{8.18a}$$

and Σ and Ω are the mean strain rate and rotation respectively. $\tilde{S}$ ranges from -1, to 0, to $+1$ (independent of the orientation of the axes). For these mean velocity gradients to control the eddy structure, their magnitude S (e.g. $S = |\partial U_1/\partial x_2|$ in a shear flow) must be at least comparable with or greater than those of the random straining motions of energy containing eddies, i.e.

$$S > u_0/L_x \sim T_L^{-1}. \tag{8.18b}$$

In most shear flows, where the mean velocity U_1 varies with x_2, the mean shear $S = |\mathrm{d}U_1/\mathrm{d}x_2|(\equiv U_1')$ is not constant. Typically, it varies in the x_2 direction over a distance of the order of or greater than the eddy integral scale. Thus $|U_1''| \leq S/L_x$.

Note that, in swirling flow (where $\tilde{S} \sim -1$) and straining flow (where $\tilde{S} \sim 1$) the turbulence structure and the energy of the turbulence are quite sensitive to the initial conditions and the history of the flow. However, in shear flows, the combination of shear and rotation leads to the slow and high speed 'streaks' form of the eddy structure that is quite insensitive to such non-local influences (e.g. Cambon and Scott, 1999). But note that the form of local straining is very significant. Even when the strain rate is of order $1/10$ of the mean shear, i.e. $|\tilde{S}| > 1/10$, the turbulence structure in boundary layers is changed from its form in pure shear where $\tilde{S} = 0$ (Bradshaw, 1971; Belcher *et al.*, 1991; Teixeira and Belcher, 2000).

These types of turbulent flow theory and observations show that the energy-containing three-dimensional eddies are both rotated and stretched out by the mean shear. In inhomogeneous flows especially, the eddies are propelled across the flow. Both mechanisms induce a mean Reynolds stress $\tau = -\sigma_{12}$, proportional to the mean shear. Here the coefficient of proportionality is the dimensional eddy viscosity ν_e. The first studies of turbulence suggested that profile curvature U_1'' rather than U_1' should be the main variable for defining τ (because of model calculations based on two-dimensional turbulence in the x_1-x_2 plane). However, for three-dimensional turbulence this effect is small because eddies being carried along and sheared by this mean flow at one level (say x_2') distort the mean vorticity gradient U'' in such a way that they have little effect on the turbulence at another level x_2. This distortion process determines the length scale L_x over which the vertical fluctuations and pressure fluctuations are correlated. It is found that, in non-uniform shear flows (Hunt and Durbin, 1999; Lumley *et al.*, 2000):

$$L_x \sim \frac{u_0}{S}. \tag{8.19}$$

If the mean shear is uniform and constant in time, as in certain laboratory and numerical studies, the length scale grows with time without limit and the formula (8.19) is not valid (e.g. Champagne *et al.*, 1970). This is why such experiments are not necessarily suitable 'test cases' for the use of turbulence models in practical flows where $U_1'' \neq 0$.

Consequently, in typical unidirectional non-uniform shear flows, provided the conditions (8.14)–(8.17) are satisfied, there is generally a local relation between the shear and mean gradient:

$$\tau = \nu_e \frac{\partial U_1}{\partial x_2}. \tag{8.20}$$

Exact idealised inviscid calculations of the development of velocity fluctuations (Townsend, 1976) or of the movement of fluid lumps in a mean shear flow (Hunt, 1987) (or indeed using the production term of the shear-stress transport equation as confirmed by experiments and simulation) show that the magnitude of ν_e is determined by the velocity fluctuations across the flow $(\overline{u_2^2} = \sigma_{22})$ and the time (t) of the development. An equivalent expression can be written in terms of $\sqrt{\overline{u_2^2}}$ and l_2, the approximate distance moved by fluid elements across the mean shear in a time t so that:

$$\nu_e \approx \lambda \overline{u_2^2} t \approx \lambda \sqrt{\overline{u_2^2}} l_2. \tag{8.21}$$

This result (8.21) should be applicable in a range of fully developed shear flows because the dimensionless coefficient λ is not very sensitive to the initial form of the turbulence and varies only slowly as the product St increases. Equation (8.21) is approximately equivalent in physical terms to Prandtl's (1925) estimations of the shear stress produced by fluid lumps being displaced over a certain 'braking' ('Bremsweg') distance l_2 before they are distorted and mixed with their surroundings. He also suggests that the r.m.s. velocity fluctuations $\sqrt{\overline{u_2^2}}$ could be estimated in terms of $\partial U_1/\partial x_2$, so that:

$$\sqrt{\overline{u_2^2}} \sim l_2 \frac{\partial U_1}{\partial x_2} \quad \text{and} \quad \nu_e \approx \lambda l_2^2 \left| \frac{\partial U_1}{\partial x_2} \right|. \tag{8.22}$$

Subsequently, Prandtl (1931) renamed the scale l_2 as the 'mixing length'. This is confusing because, erroneously, it suggests that momentum is transferred by a mixing process. (Hence the deprecating remarks on 'mixing length' theory by G. I. Taylor (1938b) and other authors. In fact momentum transfer, by the movement of 'lumps', is an inviscid mechanism, but its magnitude is limited by the action of viscosity.)

Note that the studies of shear flows with mean curvature (i.e. $U_1'' \neq 0$) already mentioned suggest that the displacement l_2 of fluid particles and the scale L_x of the energy containing eddies in these flows are limited more by the mean shear profile than by mixing processes. Thus l_2 is of the order of the length scale $L_x^{(2)}$ of the component of the velocity fluctuations directed across the shear flow (u_2). Note that this mechanism, quantified in (8.22), which also justifies (8.19), provides an inviscid (non-mixing) explanation of Prandtl's 'braking-mixing' scale. It also justifies some of the heuristic methods which have evolved for estimating l_2 in different types of shear flow.

For turbulent boundary layers, with thickness δ, the eddy motions of the normal velocity component are blocked by a rigid surface (or wall) at $x_2 = 0$ and, as we shall see, are reduced in scale by the increased shear $|\partial U_1/\partial x_2|$ near the surface. This suggests why the length scales of these motions are of the order of x_2, the distance from the surface. (This is only valid at high Reynolds number when there is a full spectrum of turbulence, Hunt and Carlotti (2001).)

Thus, if the other straining effects considered in (8.14)–(8.17) are negligible, then there are reasonable physical grounds for the experimental finding that, near the rigid surface ($x_2 \leq \delta/5$), $l_2 \approx \kappa x_2$, where $\kappa \approx 0.4$ is the von Karman dimensionless 'constant'. This is approximately equal to the length scale $L_x^{(2)}$ of the normal velocity fluctuations (at very high Reynolds number). Note that, although this length scale over which the turbulence is correlated decreases near the surface, at each level (x_2) it is comparable with the length scale (l_2) over which the velocity gradient varies. Therefore the turbulence structure is not exactly a local function of dU_1/dx_2, and is affected by the presence of the wall. This effect is considered in some turbulence models for σ_{ij} near the wall (see Launder et al., 1975; Durbin and Pettersson-Reif, 2000).

We now consider the boundary conditions at the boundaries of the 'conforming' part of the space where the model is valid, first considering free boundaries where the turbulence decays to zero. The space outside these random boundaries is 'non-conforming' – here the models do not strictly apply. Within the turbulent region, such as the upper part of turbulent boundary layers and the bulk of free shear flows, turbulent eddies grow with time by collision and coalescence. Here the largest eddy scales are comparable with the bulges of the interface, which are observed to be about 1/5 of the thickness δ of the shear layer. However, the energy containing scales that determine the 'mixing length' l_2 are only about 1/10 of the layer thickness.

At the outer region of these shear layers there is a zone where the random velocity fluctuations decay to zero. Within this zone, there is a very thin and approximately continuous interface at, say, $x_2 = x_{2I}$ (x_1, x_3, t) separating rotational turbulence within the interface ($x_2 \leq x_{2I}$) and irrotational turbulence outside it (where the average shear stress is zero). The large eddies displace this interface over distances of order $L_x^{(2)} \approx \delta/5$, but the thickness of this thin layer is much less than $L_x^{(2)}$.

At first sight, the application of the shear stress formula in a region where S is very small would appear to conflict with the assumption (8.18) for its validity. However, they can be applied meaningfully to *conditional* values of the mean velocity and turbulent shear stress at a given distance from the interface $(y - y_1)$ in the turbulent region (denoted by $\langle U_1 \rangle (x_2 - x_{21})$ and $\langle \sigma_{12} \rangle$). Recent numerical simulations by Bissett *et al.* (2002) indicate that the 'mixing length' scale in formula (8.22) is approximately valid for a conditional mixing length $\langle l_2 \rangle$ defined in the turbulent region. Furthermore, if the mean shear stress σ_{12} is then evaluated by averaging $\langle \sigma_{12} \rangle$ over the turbulent regions of the flow (i.e. $\sigma_{12} = \overline{\langle \sigma_{12} \rangle}$, noting that $\langle \sigma_{12} \rangle$ is zero outside these regions where $x_2 \geq x_{21}$), it is found that the 'mixing length' formula (8.22) can now be applied to the (unconditionally) average flow. Also the scale l_2 is approximately the same as the value $\langle \tilde{l}_2 \rangle$ determined by the eddy motion within the turbulent region (Bissett *et al.*, 2002). This provides some physical justification for the widespread use of the formula (8.22) even in the very inhomogeneous and intermittent turbulence at the edges of turbulent shear layers. Note that the values of σ_{12} and U in the intermittent transition zone are influenced as much by the probability density function (pdf) of the interface location $pr(x_{21})$ as by the local turbulence within the bulges of the interface. However, because the pdf is generally approximately Gaussian, the results are quite robust. This 'smoothing' by averaging out the sharp gradients of the actual turbulence leads to smooth and unique solutions of the mean momentum equation (8.6) where σ_{12} is determined by (8.20)–(8.22). In other words, the solutions do not depend on the numerical methods used (which is not true for certain other turbulence models, as we shall see later).

Although the mixing length approach can be useful, particularly for boundary layer flows, for many practical situations the approach cannot be applied. This is because the conditions (8.14)–(8.17) are not satisfied and the dependence of the mixing length scale l_2 either on the initial conditions and boundary conditions or on the mean flow profile is not generally known, so that other approaches are needed. Some authors still do not accept this limitation and assume that linear stress-strain or linear flux-gradient relationships are valid for any kind of strain, for example:

$$\sigma_{ij} = -\nu_e \left(\frac{\partial U_i}{\partial x_j} + \frac{\partial U_j}{\partial x_i} \right) \quad \text{or} \quad F_{\theta i} = -\kappa_e \frac{\partial \theta}{\partial x_i}, \qquad (8.23)$$

where ν_e and κ_e are eddy viscosity and diffusivity respectively, which are positive. A number of experimental and numerical studies of highly distorted or inhomogeneous turbulent flows (e.g. pipe bends or convective boundary layers) demonstrate the limitations of this assumption (Launder, 1989).

For those types of flow where the local models cannot be applied, because the turbulence does not satisfy the conditions necessary for the validity of the model, either some complex models should be used (to be discussed in Section 2.4) or physically based corrections to (8.22) can be introduced. This is sometimes computationally more convenient. One simplification occurs where, in certain regions of a flow, the changes of the mean velocity gradients and turbulence structure occur quite rapidly with distance along the mean streamlines. This means that the distortion time scale (T_D defined as $|\nabla \mathbf{U}| \Big/ \left(\mathbf{U} \dfrac{\partial}{\partial s} (\nabla \mathbf{U}) \right) \sim |\sigma_{ij}| \Big/ \left| \left(\mathbf{U} \dfrac{\partial}{\partial s} \sigma_{ij} \right) \right|$), is small in comparison with the adjustment or 'memory' time scale of the turbulence (or Lagrangian integral scale) T_L, so that the normalised turbulent time scale:

$$\tilde{T}_L = \frac{T_L}{T_D} > 1. \tag{8.24}$$

(In an unsteady flow T_D is defined in terms of the rate of change of the mean flow or of the turbulence, e.g. $T_D \sim |\sigma_{ij}|/|\partial\sigma_{ij}/\partial t|$.) In such a flow region, the expressions for σ_{ij} or $F_{\theta i}$ in (8.22) and (8.23) may be wrong in magnitude and even in sign (Craft and Launder, 1996; Savill, 1987). However, the effects of such errors on calculations of the mean flow and mean temperature are small because in rapidly changing flows the magnitude of the Reynolds stress term $\| \partial\sigma_{ij}/\partial x_j \|$ is much less than the term $|U_j \partial U_i/\partial x_j|$. In fact, neglecting the effects of turbulence on the stress can lead to a more accurate calculation of U, p than by including an erroneous estimate of σ_{ij} based on (8.22) (e.g. Belcher *et al.*, 1993). Note that the distortion time scale for scalar transport $T_{\theta D}$ (defined as $|F_{\theta i}|/(U \partial F_{\theta i}/\partial x_j)$) is not necessarily the same as T_D for the fluid flow, since the initial and boundary conditions for the scalar may differ. For example, as is quite usual, dispersion from a local source may occur rapidly in a slowly changing turbulent flow, i.e. $T_{\theta D}/T_L \ll 1$, where $T_D/T_L \sim 1$. Similarly, in some flows the turbulence changes rapidly as a result of fluctuating, rotational body forces b_i per unit mass (e.g. electromagnetic or buoyancy driven). Then, the time scale for distortion is $T_D \sim k/(|\overline{b_i u_i}|)$.

Where the shear stress is determined by the local mixing length model (8.22) (where $l_2(x_2)$ is specified) the initial and boundary conditions for turbulent flows are only applied to the mean velocity field, but they vary depending on the nature of the boundary. On an open boundary with $\mathbf{n}$ as the normal, $\mathbf{U} \cdot \mathbf{n}$ is specified subject to continuity conditions being satisfied over the whole boundary. Where the boundary is parallel and is a smooth surface at $x_2 = 0$ and parallel to $\mathbf{U}$, the model for turbulent shear stress (8.22)–(8.24) may not be valid very close to the boundary. There the turbulent eddies decay to zero at $x_2 = 0$ and the eddy structure ceases to be that of a 'fully developed' turbulence in a surface viscous layer. Therefore, for local modelling, a dynamically consistent boundary condition for U_1 is needed where the turbulent region meets the surface viscous layer (or the roughness elements if it is a rough surface). The total shear stress made up of turbulent plus viscous components and $U(x_2)$ must be continuous everywhere (as $x_2 \to 0$), and the boundary condition must be satisfied such that $U = 0$ when $x_2 = 0$.

In the flat surface case, the turbulence model is valid above a certain distance $x_{20} \sim (\nu/u^*)$ from the surface, and for $x_2 \ll x_{20}$ the mean velocity approaches a linear laminar form. Over a rough surface the whole turbulent field is displaced upwards by a distance d (of the order of $1/30$ of the height of roughness element at high Reynolds number) so that in (8.22), $l_2 = \kappa(x_2 - d)$ and the surface boundary condition for the turbulent region is $U_1 = 0$ when $x_2 - d = x_{20}$, where x_{20} is the roughness length. In both situations, the turbulence model is actually only valid a certain distance of order x_{20} above the false origin $x_2 = x_{20}$ or $x_2 = x_{20} + d$. The Karman–Prandtl solution to (8.6), (8.7) when $x_2 > x_{20}$ subject to (8.20)–(8.22) is:

$$U_1 = \frac{u_*}{\kappa} \left[\ln \left(\frac{x_2}{x_{20}} \right) + B \right] \quad \text{or} \quad U_1 = \frac{u_*}{\kappa} \ln \frac{(x_2 - d)}{x_{20}}, \qquad (8.25)$$

for smooth and rough walls respectively, where $u_*^2 = -\sigma_{12}$ (as $x_2 \to 0$) and the constant $B \simeq 5.24$. It is found that if the Reynolds number is high enough and the flow is developing slowly enough i.e. $Re \gg Re_{\text{crit}}$, (cf. (8.24)), the mean profile $U_1(x_2/x_{20})$ near a rigid surface has the same characteristic form for most types of turbulent flow (e.g. Schlichting, 1960). But if the turbulence approaching the wall is changing rapidly, i.e. $\tilde{T}_L \geq 1$, for example in a separated flow region (Dengel and Fernholz, 1990), or if the Reynolds number is low so that $Re \sim Re_{\text{crit}}$ near transition, the wall profiles differ significantly from the high Re form (8.25), as the more complex turbulence models demonstrate. Nevertheless (and without empirical or physical justification), (8.25) is often used as a convenient wall condition for the mean velocity in turbulence models. Note that (8.25) is also valid above or below horizontal fluid interfaces (at high enough Reynolds number) where U_1 is the difference between the mean velocities at the planes x_2 and at the interface located at $x_2 = 0$ (e.g. Lombardi *et al.*, 1996). In such flows the horizontal velocity fluctuations $\overline{u_1^2}, \overline{u_3^2}$ are not necessarily zero (relative to the interface).

As explained by Launder (Chapter 3) these wall functions change in the presence of strong body forces, which affects the structure of the turbulence, and its anisotropy.

2.4 Eddy viscosity and Reynolds stress transport models (EVM and RSTM)

2.4.1 General assumptions

There are many complex turbulent flows of practical importance where there are large variations in the mean velocity field $U_i\,(\underline{x},\,t)$ and in the turbulence kinetic energy k (e.g. the flow is highly curved or the streamlines are diverging or there are non-uniform body forces). In flows that are rapidly changing in space and/or time the variances $(\overline{u_i^2})$ and length scales e.g. $(L_x^{(i)})$ change at a dynamically significant rate along the mean streamlines over a distortion time scale T_D (see (8.24)) that is comparable with the turbulent time scale T_L (i.e. $\tilde{T}_L$ $(= T_L/T_D) \sim 1$). In spatially non-local flows the turbulence energy and/or the rate of strain (S) of the mean flow may have gradients on a length scale Λ *across* the streamlines that are dynamically significant. Non-local pressure fluctuations may then play a significant role (e.g. Durbin and Pettersson-Reif, 2000). This means that Λ is comparable with or less than the relevant length scale L_x of the turbulence, i.e. the normalised length scale:

$$\tilde{L}_x = L_x/\Lambda \geq 1, \tag{8.26}$$

where in a shear flow $L_x \equiv L_x^{(2)}$, and Λ is the lesser of

$$\Lambda^{(k)} = k/|\nabla k| \;\text{(which is the minimum of } k/\nabla k \text{ or } u_i^2/|\nabla u_i^2| \text{ for } i = 1, 2, 3\text{), or}$$
$$\tag{8.27a}$$

$$\Lambda^{(U)} = S/|\nabla S|. \tag{8.27b}$$

Examples of these variations, given in Fig. 8.3, show how in a turbulent boundary layer the normalised time scale $\tilde{T}_L$ varies from near zero at the wall to about unity at the outer edge (which is the typical value in most free shear flows, such as jet, wakes and shear layers). But if the layer is distorted very 'rapidly' as in a shock (and near the trailing edge of an aerofoil or in a bent tube), $\tilde{T}_L$ becomes significantly greater than unity. In separating flows and when turbulence is forced by buoyant convection, there are large gradients of turbulence across streamlines, such that $\tilde{L}_x \geq 1$. In these flows, the local variance, shear stress σ_{ij}, and fluxes $F_{\theta i}$ at any point x clearly depend on how the turbulence develops along and across the streamlines.

By drawing the analogy between molecular fluxes and turbulence fluxes (e.g. as Lumley, 1978) and by inspection of the moment equation, it is natural to model σ_{ij} terms of parabolic transport equations along the streamlines. Thus, schematically, the turbulent statistics at position x depend on the mean flow and turbulence dynamics over an upstream elongated region indicated in Fig. 8.4. The extent of the region depends on the type of modelling, especially as to whether this includes elliptic effects, which allow for the influence of the boundary and motions downwind. These act over distances of order L_x.

2.4.2 EVM models

In the simplest transport models (e.g. the widely used k–ε model) it is assumed that turbulence is changing slowly enough and homogeneously enough that, following

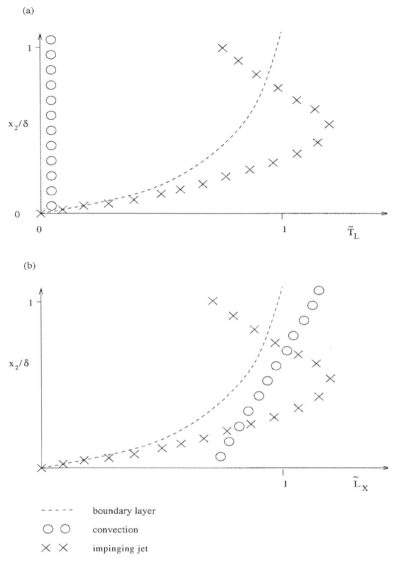

(a)

(b)

- - - - -   boundary layer

○ ○   convection

× ×   impinging jet

Fig. 8.3 Profiles perpendicular to the boundary of normalised time scales $\tilde{T}_{\mathrm{L}}$ and length scales $\tilde{L}_x$ in different kinds of turbulent flow.

(a) Profile of the normalised time scale of turbulence $\tilde{T}_{\mathrm{L}}$ (i.e. the ratio of eddy time scale to the distortion time in the flow).

(b) Profile of the normalised length scale of turbulence (i.e. the ratio of the energetic eddy size L_x to the distance Λ over which the turbulence or mean velocity changes).

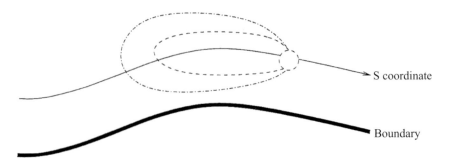

S coordinate

Boundary

$\left(\times\right)$ Zone of influence for M-L

$-----$ Zone of influence for k -ε-v_e

$-\cdot-\cdot-\cdot-$ Zone of influence for R.S.T.M.

Fig. 8.4 Typical mean streamline pattern in a complex turbulent flow showing the extent of the flow volume where turbulence (in the model) influences the computation of Reynolds stress at x.

Eqs. (8.24) and (8.26), $\tilde{T}_{\mathrm{L}} \ll 1$, $\tilde{L}_x \ll 1$. Then, especially in sheared flows, the fluxes of momentum σ_{ij} and scalar $F_{\theta i}$ are determined by:

(a) the mean gradients $\partial U_i / \partial x_j$ and $\partial \Theta / \partial x_i$ of mean velocity U_i and of a passive scalar Θ (for instance mean concentration of a contaminant or mean temperature);
(b) effective eddy viscosity v_e and eddy diffusivity κ_e, which are determined by the kinetic energy k and a length scale L_X, i.e., if v_e is an isotropic eddy viscosity (which is only a reasonable approximation for weakly strained turbulence, $ST_L \leq 1$);

with the result that

$$\sigma_{ij} = -v_e \left(\frac{\partial U_i}{\partial x_j} + \frac{\partial U_j}{\partial x_i} \right), \quad \text{where} \quad v_e = C_v k^{1/2} L_x, \qquad (8.28)$$

$$F_{\theta i} = -\kappa_e \frac{\partial \Theta}{\partial x_i}, \quad \text{where} \quad \kappa_e = C_\kappa k^{1/2} L_x, \qquad (8.29)$$

where C_v, $C_\kappa \sim 1$. The ratio of these coefficients depends on the eddy structure and tends to vary only slightly in any one type of turbulent flow. (More complex tensor models can also be constructed for a v_{ei} and κ_{ei} e.g. see Pope, 2000; Durbin and Pettersson-Reif, 2000.) Note that the appropriate value of k for computing v_e and κ_e in (8.28), (8.29) is not necessarily the actual kinetic energy, but rather some fraction of this corresponding to the 'energy containing' eddy motions that dominate the transport process (e.g. $\overline{u_2^2}$). This is discussed further in Section 4.

In some, 'k–L', models, k is calculated from an auxiliary equation while L_X is specified (e.g. as in Section 2.3 for the mixing length models). More usually, both k and L_X are computed from coupled k and ε equations, and L_X is estimated from the appropriate relation (for high Reynolds turbulence), e.g. $L_X \sim k^{3/2}\varepsilon^{-1}$.

To reduce the formal equations (8.14) and (8.17) in Section 2.1 to soluble sets of equations for U_i, p, k, ε and σ_{ij}, only involving second order moments, further approximations are required. These are usually based on assumptions about the localness of the turbulence structure (which is characterised here by $\tilde{T}_L$, $\tilde{L}_X$) and about the 'Gaussianity'.[1] As explained by Lumley (1978) and other authors, third order moments (e.g. $\overline{p\partial u_i/\partial x_j}$, $\overline{\partial u_j u_i^2/\partial x_{ij}}$) have a general relation to second order moments σ_{ij} only if the probability distribution of the velocity is close to that of a Gaussian variable. This is defined by the dimensionless ratio $\tilde{G}$ being small. If $\tilde{G}$ is of order unity these third order moments cannot be estimated without the use of higher order equations or particular assumptions. Here:

$$\tilde{G} = \max \left\{ \frac{\overline{u_i^4}}{\left(\overline{u_i^2}\right)^2} - 3, \quad \frac{\overline{u_i^3}}{\left(\overline{u_i^2}\right)^{3/2}} \right\} \quad \text{for } i = 1, 2, 3. \tag{8.30}$$

When $\tilde{G} \sim 1$ (e.g. in thermal convection or in a wake dominated by vortex shedding) the turbulence is significantly non-Gaussian, so that the coefficients defining its statistical properties may also differ from their values in near Gaussian turbulence (e.g. C_v, C_k) (e.g. see review by Hunt (1982) using data for scalar transport). In the equation for the kinetic energy the Reynolds stress production term $\sigma_{ij}\,\partial U_i/\partial x_j$ is expressed, using the local relations (8.28) and (8.29), in terms of k, ε and the mean velocity gradient, where:

$$P \approx C_1 k^2 \varepsilon^{-1} \frac{\partial U_i}{\partial x_j} \left(\frac{\partial U_i}{\partial x_j} + \frac{\partial U_j}{\partial x_i} \right), \tag{8.31}$$

an approximation that is only valid if $\tilde{T}_L \le 1$, $\tilde{L}_X \le 1$. Generally this local approximation is more accurate in shear flows than in non-shear flows (cf. Section 2.3), i.e. where:

$$\tilde{S} < \frac{1}{10}. \tag{8.32}$$

The transport terms can also be approximated in terms of local gradients if the turbulence is changing slowly, is not very inhomogeneous, and also is approximately

---

[1] This theoretical criterion for the correctness of closure approximations may not always be necessary from a practical point of view. Furthermore, it is not generally possible to obtain the required higher order moment information.

Gaussian, so that

$$ET_k \approx \frac{\partial}{\partial x_j} \left(\frac{C_T k^2}{\varepsilon} \frac{\partial k}{\partial x_j} \right),$$
(8.33)

$$\tilde{T}_L \ll 1, \quad \tilde{L}_X \ll 1, \quad \tilde{G} \le 1.$$
(8.34)

The dissipation term ε is computed from an even more heuristically derived equation (see Pope, 2000) and one which is quite questionable in highly strained flow where $\tilde{T}_L \gtrsim 1$. The forms of the terms have been constructed in terms of local gradients by rescaling the k equation so as to model ε in a number of practical types of flow and then adjusting the unknown coefficients appropriately. For a critical review and evaluation of the term for wall bounded flows, see Rodi and Mansour (1993). Criteria for the general validity of the equations might reasonably be proposed by considering the 'localness' of the turbulence dynamics in the particular 'types' of flow chosen for the construction of the model equation. Since these 'types' are typical shear flows, wakes, jets, boundary layers and slowly decaying homogeneous turbulence, the criteria for the wider validity of the flow are approximately the same as for the k equation, specifically (8.34). The ε equation is also not well adapted for flows in which the mean strain is mainly rotational or irrotational (i.e. where the criterion in (8.32) is not satisfied).

The rate of dissipation, which is determined by the viscous stresses acting on the smaller scales, depends on the latter's energy in relation to that of the large scales. Since this depends sensitively on the local Reynolds number and on the structure of the turbulence, for example near a wall or density interface or in the presence of body forces, corrections to the standard form for ε have been developed; especially for lower values of Re (see Launder and Sharma, 1974; ERCOFTAC, 2000).

One needs to consider for this more complicated transport model the boundary and matching conditions at the boundaries of the 'conforming domain' outside which the model is invalid. Defining these matching conditions requires special assumptions that, just as in the derivation of the model equations, are physically based. They cannot be uniquely derived from the governing equation. They depend on the nature of the turbulent flow, including the value of Re, and on the modelling. See Table 8.2. At rigid surfaces there are two approaches. In the 'wall function' approach it is assumed that in a surface layer $x_2 < l_s$ (where $l_s \sim \delta/10$) the profiles for U, k, ε have a universal 'wall' profile (based on (8.25)). Alternatively, closure models are used (e.g. k–ε) and boundary conditions are assumed as $x_2 \to 0$ for $U(x_2)$, $k(x_2)$, $\varepsilon(x_2)$ and their derivatives (which is necessary since the governing k–ε equations are differential equations). The solutions of the closure equations yield the 'structural' wall profiles for U, k, ε (as in (8.25) when $Re \gg 10^4$). A

Table 8.2 *Normalised length scales, integral time scales, Gaussianity parameter and straining parameter for different types of turbulent flow. These are denoted by* $\tilde{L}_x$, $\tilde{T}_L^*$, $\tilde{G}$, $\tilde{S}$, *where* $\tilde{L}_x = \max\{L_X/\Lambda^{(k)}, L_X/\Lambda^{(\Sigma)}\}$; $\tilde{T}_L = T_L/T_D$;

$$\tilde{G} = \max \left\{ \frac{\overline{u_i^4}}{\left(\overline{u_i^2}\right)^2} - 3, \quad \frac{\overline{u_i^3}}{\left(\overline{u_i^2}\right)^{3/2}} \right\} \; ; \tilde{S} = \frac{\Sigma^2 - \frac{1}{2}\Omega^2}{\Sigma^2 + \frac{1}{2}\Omega^2}; 0 \leq \tilde{L}_X < \infty;$$

$$0 \leq \tilde{T}_L \leq \infty; 0 < \tilde{G} < \infty; -1 < \tilde{S} < 1.$$

	Types of turbulent flow	$\tilde{L}_x$	$\tilde{T}_L^*$	$\tilde{G}$	$\tilde{S}$
(i)	Homogeneous stationary turbulence	$\ll 1$	$\lesssim 1$	$\ll 1$	—
(ii)	Homogeneous decaying turbulence	$\ll 1$	~ -1	0	—
(iii)	Strained turbulence	$\ll 1$	$\gtrsim 1$	0	$0 \leftrightarrow 1$
(iv)	Turbulent boundary layer				
	(near a wall)	$\sim 1/2$	$\ll 1$	0	$\ll 1$
	(near outer boundary)	~ 1	~ 1	$\lesssim 1$	$\lesssim 1$
(v)	Free shear flows and separated flows	~ 1	~ 1	$\lesssim 1$	$\lesssim 1$
(vi)	Large scale free stream turbulence interacting with a boundary layer	$\gtrsim 1$	~ 1	0	$\lesssim 1$
(vii)	Turbulence near a free surface	~ 1	$\ll 1$	0	—
(viii)	Natural convection (no mean flow)	~ 1	$\ll 1$	~ 1	—
(ix)	Shock boundary layer interaction	$\gtrsim 1$	$\gg 1$	$\ll 1$	$\lesssim 1$
(x)	Trailing edge boundary layer–wake interactions	$\gg 1$	~ 1	$\ll 1$	$\lesssim 1$
(xi)	Jet impaction	~ 1	~ 1	$\ll 1$	$\lesssim 1$
(xii)	Turbulent flow over small obstacles	$\gtrsim 1$	$\gtrsim 1$	$\ll 1$	$\lesssim 1$
(xiii)	Swirling shear flows	~ 1	~ 1	$\ll 1$	$\gtrsim -1$

noticeable feature of the latter approach is that the models have to be applied in a thin layer, where $0 \leq x_2 \leq x_{2w}$, even though this is a 'non-conforming' region where the models are not physically valid. Both approaches are effectively based on the assumption that the profiles of k, ε very near the wall are well understood and can be defined in any given flow situation.

However, as the Reynolds number of turbulence increases above about 10^4 there is a significant change in the structure of the near wall turbulence (Hunt and Morrison, 2000; Kim and Adrian, 1999). As suggested earlier by Townsend (1976) and Marusic and Perry (1995), k increases towards the wall in proportion to $k \sim u_*^2[-\ln(x_2/l_s) + A]$, where $A \sim 10$ and l_s is the depth of the surface layer ($\sim \delta/10$). This means that the maximum value of k, which occurs where $x_2 \sim x_{20} \sim \nu/u^*$, increases with Re (in proportion to $\ln Re$). Current transport-based turbulence models do not represent this trend. One way of avoiding the sensitivity of the ratio $k(x_2)/u_*^2$ to the value of Re is (for shear flows) to substitute $\overline{u_2^2}$ as the 'k' variable in

k–ε models. This is because near the 'wall' $\overline{u_2^2}/u_*^2$ is approximately constant outside the viscous layer as Re varies.

Other computational problems can arise at the fluctuating outer 'free' boundary of a shear layer, where the turbulent energy and dissipation tend towards zero. In numerical simulations of the k–ε equation it is customary to specify that outside the boundary where $x_2 > \delta$ the kinetic energy and dissipation rate (k, ε) are equal to small but finite values, i.e. $k = \lambda_k k_0$, $\varepsilon = \lambda_\varepsilon \varepsilon_0$, where λ_k, λ_ε are very small threshold values (e.g. $\lambda_k = \lambda_\varepsilon = 0.01$ and k_0, ε_0 are average values in the shear layer). Some modellers relate λ_k, λ_ε to molecular values. If $\lambda_k = \lambda_\varepsilon = 0$ the numerical solutions to the equations for the rate at which turbulence diffuses outwards into the non-turbulent region, i.e. the entrainment rate, would depend on the numerical diffusion (see Cazalbou *et al.*, 1994). Authors differ as to whether numerical assumptions about the outer boundary conditions cause differences in predictions about the entrainment rates of free shear flows, even when the same turbulence model is used (e.g. Kline *et al.*, 1981).

The general view about the practical value of EVM methods is that they lead to converged solutions which provide estimates for the principal Reynolds shear stresses. But they are often not very accurate. For example, in recirculating flows (e.g. in a rotating vessel or a separating flow) this is the simplest type of model. With mixing length methods the length scale l_2 has to be estimated by 'scaling' arguments, but formally the turbulence is not in a state of local equilibrium and the conditions for the model are generally not satisfied. When comparisons have been made between EVM methods and measurements (or more complex models) it is commonly found that, where the basic 'localness' assumptions are not satisfied, the results for shear stresses, for certain overall properties in non-equilibrium flows, and even the drag of the wavy surfaces, can be in error by as much as 100% (e.g. Belcher *et al.*, 1991; Huang and Leschziner, 1985).

Note that in low Re flows, additional turbulence damping models are needed. The Launder–Sharma (Launder and Sharma, 1974) version is the most widely used for a range of turbulent and transitional applications (Savill, 1996).

2.4.3 Reynolds stress transport models

A more advanced level of transport modelling is necessary when the turbulence is less local. Then eddies may develop fast enough and be transported perpendicular to the streamlines rapidly enough that the momentum and scalar fluxes σ_{ij}, $F_{\theta i}$ are not simply proportional to the local mean gradients. Now the full equations for the development of these moments (see (8.9)–(8.13)) are used. But as in the k–ε equations, simplifications have to be made to ensure a soluble set of equations. Second order closure models or Reynolds stress transport models involve the relations of two-point to one-point moments and third order to second order moments (Launder *et al.*, 1975; Craft and Launder, 1996).

Because the local flux assumptions are no longer applied, the equations can formally be used in rotating and straining flow, as well as sheared flow (cf. (8.32)). These models are particularly useful for curved shear flows and where there are maxima and minima in the mean velocity profiles. As the development of the turbulence is accounted for, the slow distortion assumption (8.34) can be partly relaxed, i.e. $\tilde{T}_L \lesssim 1$. Other limitations of weak local inhomogeneity can also be relaxed to some extent, so that shear stress can be 'transported' across streamlines; i.e. $\tilde{L}_X \lesssim 1$. The assumptions used in the various terms in the model can be summarised as follows (see Launder *et al.*, 1975; Hunt, 1992).

To model the pressure-strain term, $P_{ij} = \overline{u_i \partial p / \partial x_j}$, involves estimating p at $\underline{x}$ in terms of the velocity field at other points $\mathbf{x}'$ near $\mathbf{x}$. Thence P_{ij} can be expressed in terms of linear second order moments and non-linear third order moments. These are denoted by $P^{(l)}$ and $P^{(nl)}$, following Launder *et al.* (1975). The $P^{(l)}$ terms are usually termed 'rapid' because they can act on time scales faster than the 'slow' $P^{(nl)}$ terms; but note that in most shear/straining flows the linear terms continue to influence the flow as much as, and over the same period as, the non-linear terms. Hence we prefer the nomenclature 'linear', 'non-linear'. The $P^{(l)}$ term corresponds to products of fluctuations of velocity (denoted by $\overline{uu}$) and gradients of the mean velocity gradients (denoted by $\partial U / \partial x$) and the $P^{(nl)}$ term to products of moments of the fluctuating velocity field, the kinetic energy and the energy dissipation rate. Schematically:

$$P = P^{(l)} + P^{(nl)},$$

$$\text{where} \quad P^{(l)}(\mathbf{x}) \sim C^{(l)} \overline{uu} \frac{\partial U}{\partial x}(\mathbf{x}), \tag{8.35}$$

$$P^{(nl)}(\mathbf{x}) \sim C^{(nl)} \overline{uu} \varepsilon k^{-1}(\mathbf{x}). \tag{8.36}$$

In practice the coefficient $C^{(l)}$ for the linear term is determined by considering idealised turbulence undergoing a constant rate of deformation or strain. However, $C^{(l)}$ changes if the turbulence is highly anisotropic or if the mean deformation rate ($\propto \partial U / \partial x$) varies over the time scale T_L along the mean streamlines, as occurs in flows approaching bluff bodies or in trailing edge flows. In the latter type of flow the errors in the turbulence do not cause practically significant errors in the mean flow, as explained in Section 2.4.2. Several studies (e.g. Hunt and Carruthers, 1990) have shown that in shear flows the magnitude of $C^{(l)}$ is insensitive to the anisotropy of the turbulence, but in straining or rotating flows it is not. This needs to be remembered when considering errors and the results of different models. Also, the errors may be more significant near boundaries where the local homogeneity assumption is much less valid, because the scale of the eddies L_X is generally comparable with the distance x_2 over which the mean velocity gradient $(\partial U_1/\partial x_2)$ is varying. Any errors in the constants $C^{(1)}$ and $C^{(nl)}$ affect P and thence stresses (σ_{ij}).

To correct such wall errors by representing correctly the blocking of eddies at the boundary, either special wall correction functions have been proposed for $C^{(1)}$ and $C^{(nl)}$ (e.g. Gibson and Launder, 1978), or else auxiliary equations have been proposed to estimate P_{ij} which model more realistically the way that the blocking affects the pressure field (Durbin, 1993). An alternative approach is to describe this highly distorted turbulence in terms of an interpolation between three-dimensional turbulence and the limiting form of two-dimensional fluctuations near the wall. Making this idealised assumption avoids having to introduce any auxiliary equation (Craft and Launder, 1996). The differences between these models are most acute and affect the mean flow and heat transfer predictions when there are large mean accelerations and pressure fields near the boundary, for example where jets impinge onto obstacles (Durbin and Pettersson-Reif, 2000) or where shocks interact with boundary layers.

The non-linear 'pressure-strain' term $P^{(nl)}$ is finite even when there are no mean velocity gradients (i.e. $\partial U_i/\partial x_j = 0$). However, it is only significant (i.e. $P^{(nl)} \sim \varepsilon$) when the total and/or spectral distributions of energies of the three components of the turbulent velocity are not the same, i.e. the turbulence is anisotropic and is not in equilibrium.

Until recently the models did not reflect the form of the eddy structure and only showed how P was proportional to the anisotropy, as in:

$$P_{ij}^{(nl)} \propto C^{(nl)}(\overline{u_i u_j} - \overline{u_m u_m}\delta_{ij})k^{-1}\varepsilon, \qquad (8.37)$$

a model due to Rotta (1951) which effectively implies that the anisotropy decreases in proportion to its local magnitude. However, this general assumption (e.g. Pope, 2000) is inconsistent with many experimental studies which have shown that for certain types of anisotropic eddy (e.g. where elongated vortical tubes exist), the anisotropy does not decrease, and even increases (Townsend, 1976). More recently, this widely repeated error has been obviated by allowing the coefficient $C^{(nl)}$ to be a function of the statistical measure that defines the anisotropic eddy structure (see Craft and Launder, 1996). This approach is of practical value, for example to model the very weak rate of energy transfer between velocity components in highly distorted turbulence, such as occurs near rigid boundaries and in slowly decaying longitudinal vortices (Carlotti and Hunt, 2001).

The transport ET_{ij} terms in the equations for fluxes are modelled using the gradient diffusion mechanism as in the k equation (8.33). This is based on the assumption that the length scales are small, $\tilde{L}_X \lesssim 1$, and that the turbulence is approximately Gaussian, i.e. $\tilde{G} \leq 1$. By contrast, for turbulence with large scale

Table 8.3 *Boundary conditions and internal parameters for applying different types of model. The first four columns represent initial inflow boundary conditions on the boundary* $[B_N]^{(T)}$. *Note that the distance* X_s *along streamlines over which turbulence boundary conditions directly influence the turbulence is given by* $\hat{T}_D(X_s) \lesssim 1$, *where* $\hat{T}_D = T_D(X_s)/T_L(B_N)$; T_D *being the travel distortion time along the streamline and* $T_L(B_N)$ *being the integral time scale on the inflow boundary.*

	U	k	ε	σ_{ij}	Boundary conditions on boundaries parallel to the mean flow $[B_s]^{(T)}$	Internal parameters (for choosing the appropriate model)
Local	●				U, σ_{sn}	
Transport	●	●	●		$U, \sigma_{sn}, k, \varepsilon$	Re, Re_{crit}
k–ε–ν_e						
RSTM	●	●	●	●	$U, \sigma_{ij}, \varepsilon$	Re, Re_{crit}

eddies such as in a convective flow or in flow passing over an obstacle, the transport effects are determined by long range elliptical pressure fluctuations which act instantaneously across streamlines and do not diffuse. Other kinds of modelling methods are required in those conditions (see Wyngaard, 1979; Hunt *et al.*, 1990; Kenjeres and Hanjalić, 1999; Durbin, 1993).

The magnitude of the dissipation term D_{ij} for fluxes is the same order as that of the total dissipation term (ε) in the k equation, and in all one-point models the D_{ij} is proportional to ε. However, D_{ij} has to have an extra factor $\hat{D}_{ij}$ that depends on the variances of the different velocity components of the turbulence, on the form of the anisotropic spectra ($\Phi_{ij}(\boldsymbol{\kappa})$) (e.g. Kassinos *et al.*, 2001), and to some extent on the local Reynolds number of the turbulence, i.e. $k^{1/2}L_X/\nu$. Since these three effects vary rapidly near a rigid surface, modelling $\hat{D}_{ij}$ is quite uncertain in the surface layer of a boundary layer and in the shear layer where the flow separates, e.g. at trailing edges. Note that D_{ij} is only non-zero for the normal components of the momentum flux (i.e. $\hat{D}_{ij} \propto \delta_{ij}$).

In Reynolds stress transport models (RSTM) essentially the same equation for dissipation rate ε is used as in the k–ε models, but the extra information gained from the more accurate calculation of σ_{ij} should lead to a more accurate calculation of ε.

With the greater number (seven) of equations in RSTM for momentum flux and dissipation than for the k–ε equations in EVM, in principle a greater number of initial and boundary conditions are needed for the six independent components of σ_{ij} and for ε, as well as for the mean flow field $\underline{U}(\underline{x}, t)$ – see Table 8.3. In many practical calculations the turbulence quantities are not all known, or cannot be defined, for example where there is a transition from laminar to turbulent flow, or transition from fully turbulent flow to a highly distorted form at a resistive (whether rigid or flexible) boundary. However, the structure of the equations is such that, for most practical flows (e.g. containing significant shear and boundaries confining the flow), the calculations of σ_{ij} and ε are not very sensitive to the initial or *inflow* conditions (e.g. the flow within pipes entering a vessel). However, in unconfined flows with 'free boundaries' such as wakes, these transport models are consistent with experiments that show how the inflow turbulent boundary conditions affect the flow far from the boundary (e.g. Bevilaqua and Lykoudis, 1978; Launder, 1989). A simple guide to this sensitivity was given by Hunt (1992) in terms of the normalised time scale $\tilde{T}_L = T_L(x)/T_D(x)$, which is the ratio of the Lagrangian integral scale (or 'turn-over' time scale) to the mean time travelled by a fluid particle along the developing flow. For the former case of a confined flow where T_L is fixed and T_D increases, $\tilde{T}_L(x)$ becomes very small. In the latter case (as in *all* free shear flows) $\tilde{T}_L(x)$ tends to a constant value because the 'memory' time $T_L(x)$ increases at the same rate on $T_D(x)$. So the turbulence continues to be weakly influenced by its inflow conditions, which therefore have a significant influence on the whole flow. This also follows from the form of the EVM or RSTM equations because of the 'power-law' solutions for k, $\varepsilon(t)$ in developing flows, i.e. $k \propto k_0(T_D/T_L)^\beta$. Again for low Re regions similar damping treatments are required (Savill, 1996).

3 Proposed 'map' for the one-point models in complex flows

3.1 Outline concept

The previous reviews of one-point turbulence models show that their governing equations and their matching to boundary conditions contain significant approximations. Both these affect the accuracy of the predictions made by the models in any particular flow. The question to be addressed here is whether by analysing these approximations, local or overall (e.g. pressure drop) errors in such predictions can be estimated. It would be desirable to make such estimates for any particular type of complex flow without having to perform an elaborate calculation and to verify it experimentally. This would be of great value to those using these models in practice as well as stimulating new directions in the development of models. In the previous discussion we focused on the velocity field. But many of the same principles and criteria apply to models for scalar statistics. The analyses in Sections 2.3 and 2.4 have shown that in the one-point models the relations between two-point and one-point moments, and between third order and second order moments, are essentially local in space and time. This is because it is assumed that the turbulence is changing slowly over the integral time scale T_L and over the length scale L_X.

When this assumption of localisation is not valid, additional non-linear and non-equilibrium effects, such as eddy transport and inhomogeneous dissipation, are present. Then there is the strong possibility that the locally-based models are not valid and their predictions are erroneous. The departure from Gaussianity of the energy containing eddies in such complex flows can cause additional errors.

So the first objective here is to use this concept to produce a guide or map for assessing whether in certain types of flows, or regions of turbulent flows, or ranges of parameters this 'locality' or slow variation is satisfied. We particularly look at the flow types in Table 8.1.

Our second objective is to provide an indication of the size of the errors where the assumptions are not satisfied.

Finally, we discuss whether, even if the assumptions are not valid everywhere, there may be regions in the flow where local approximate analytical/computational methods may be suitable.

3.2 Proposed 'localness' guideline map

Our basic hypothesis, which has been developed in Sections 2.3 and 2.4, is that the standard one-point models are only strictly valid and likely to provide accurate predictions when the turbulence dynamics are local in time and space and when the turbulence characteristics are 'typical', e.g. with probability distribution close to Gaussian such that $\tilde{G} \leq 1$, and only moderate degrees of anisotropy of velocity and length scale (away from boundaries). In other words, we exclude highly intermittent and effectively two-dimensional turbulence, which rarely occur in engineering flows, though they have to be considered in environmental flows. Appropriate measures defined in Section 2.4 to define the localness property are the normalised time and length scales, viz.

$$\tilde{T}_L = \frac{T_L}{T_D} \qquad (8.38)$$

and

$$\tilde{L}_X = \max \left\{ \frac{L_X}{\Lambda^{(k)}}, \frac{L_X}{\Lambda^{(U)}} \right\}. \qquad (8.39)$$

The dynamical significance of these 'localness' parameters can be better understood by showing how they relate to the terms in the kinetic energy equation (8.14). The rate of change or advection term A is related to k and T_D. Let $A = \tilde{A}k/T_D$, where $\tilde{A}$ is the normalised coefficient $|\tilde{A}|$. For most spatially developing flows the magnitude of $\tilde{A}$ (namely $|\tilde{A}|$) is of order one. So that if $\tilde{A} > 0$ the kinetic energy of fluid particles is increasing as they move through the flow, and if $\tilde{A} < 0$, it is decreasing. For forced stationary turbulence $\tilde{A} = 0$. The rate of production of energy $I \propto k\nabla U \propto (kS)$, where S is the mean shear. The rate of dissipation $\varepsilon (\propto k^{3/2}/L \propto k/T_\mathrm{L})$ is also related to T_L, so that, from (8.14), the normalised time scale of turbulence $\tilde{T}_\mathrm{L}$ is related to the ratio of $(I+B)/\varepsilon$ by $\tilde{A}\tilde{T}_\mathrm{L} \approx (I+B)/\varepsilon - 1$. The eddy transport terms are comparable with the other terms. Thus if $\tilde{T}_\mathrm{L} \ll 1$, the turbulence is close to *local temporal equilibrium*, while if $\tilde{T}_\mathrm{L} \gtrsim 1$, it means that the turbulence is changing 'rapidly' with time as it moves along the mean streamlines or with time in a slowly moving/stationary flow.

The condition of local spatial equilibrium is satisfied if the rate of dissipation of energy (ε) is much greater than ET_k, the transport of energy by eddy movement and/or pressure fluctuations (see (8.16)). This only occurs if the gradients of turbulent energy (on a scale $\Lambda^{(k)} = k/\nabla k$) and/or of the mean strain (whether rotational or irrotational) (on a scale $\Lambda^{(U)}$), are small enough (see (8.27b)) that

$$\varepsilon \gtrsim |ET_k|$$

$$\text{or} \quad k^{3/2}/L_X \gtrsim k^{3/2}/\min\left[\Lambda^{(k)}, \Lambda^{(U)}\right]. \tag{8.40a}$$

Therefore the condition for non-local or inhomogeneous effects to significantly affect the local dynamics is that

$$\tilde{L}_X \lesssim 1, \tag{8.40b}$$

where $\tilde{L}_X$ is defined in (8.39).

In order to understand these temporal and spatial 'localness' properties of the turbulence, in different types of turbulent flows, which are discussed in Section 3.3, we briefly review the generic mechanisms and parameters such as Gaussianity $\tilde{G}$, and type of strain ($\nabla \tilde{S}$).

If it is 'rapidly' changing, turbulence is being distorted over periods T_D that are much less than T_L. This can occur when the mean strain is changing, whether by imposed unsteadiness (e.g. in an engine with moving parts) or by changing along mean streamlines (as when a turbulent flow approaches a bluff body). It can also occur by rapidly changing forces of the turbulence (e.g. in an electromagnetically driven flow), or by the boundary condition acting on the turbulent eddies changing rapidly either by a moving boundary (e.g. particles moving in turbulent flow) or as the flow moves past the boundary (e.g. the leading or trailing edges of aerofoils). In all these forms of rapidly changing flows, the response of the turbulence at time t to the distortion depends on its structure at an earlier time $t - T_\mathrm{L}$, e.g. before the distortion begins. The distortion may be varying rapidly over a time T_D which is less than T_L. The degree to which the 'memory' time is important depends on the type of strain, being shorter for shear flows (where $|\nabla \tilde{S}| \ll 1$) than for rotating/strained flows (where $|\nabla \tilde{S}| \sim 1$).

In spatially non-local turbulence, the statistical properties of the turbulence vary significantly over a distance $\Lambda^{(k)}$ that is comparable with or smaller than the length scale of the turbulence L_X. Although the random eddy motions of turbulence act to reduce these spatial variations, they are not completely effective because (unlike gas molecules) the eddies have a finite size and are themselves affected by the same factors that make the overall turbulence statistics inhomogeneous. These factors, which are listed below and have already been discussed in Section 2, do not act in isolation. As with those affecting the temporal non-localness, they also depend on the local structure of turbulence (e.g. defined by $\tilde{G}$ and Re) and the type of distortion (e.g. $\nabla \tilde{S}$).

In defining our 'map' we have to consider the domain of validity of the models and the boundary conditions that apply.

(i) At a rigid surface (or in the flow of less dense fluid near an inversion layer where there is a large density difference) the velocity at the surface tends to zero (the no slip condition). Therefore, as remarked in Section 2, the velocity fluctuations within a thin viscous layer do not have the same form as fully developed turbulence. On a smooth surface the typical thickness δ_w of this layer is of order $10v/u_*$, where u_* is the characteristic turbulent velocity just outside the layer. Therefore at high Reynolds number this layer is very small compared to L_X. The main effect of the boundary is on the normal component u_2 which is blocked by the surface. In general the surface conditions have markedly different effects on the profiles of the three velocity components, and on their one- and two-point moments depending on the Reynolds number. These profiles also depend sensitively on whether the magnitude of the mean shearing/straining flow parallel to the surface is larger or smaller than the level of velocity fluctuation above the surface (i.e. whether it is 'sheared' or 'shear-free') (see Chapter 2). In both cases kinematic analysis shows that the variance of the component normal to the interface, $\overline{u_2^2}$, decreases to zero as $x_2 \to 0$, for sheared and shear-free layers. In a shear-free turbulent boundary layer the mean velocity component parallel to the surface (x_1 direction) remains finite as $x_2 \to 0$, in contrast to the situation at a surface in a sheared boundary layer. There the boundary condition on the turbulent velocity components u_1, u_3 depends on the precise nature of the surface. At a rigid wall moving with the mean velocity $\overline{u_1^2}, \overline{u_3^2} \to 0$, but at the free surface of a liquid $\overline{u_1^2}, \overline{u_3^2} \neq 0$ (e.g. Thomas and Hancock, 1977).

Recent research shows that, broadly, the direct effect of the rigid surface on the eddy motions above the surface viscous layer is weakest when there are mean shear flows parallel to a rigid surface. The integral scale $L_X^{(2)}$ (for the normal component) is approximately equal to $x_2/2$, where x_2 is the distance to the surface. This is *less* than the length scale $\Lambda^{(U)} = |\nabla U| / |\nabla \nabla U|$ on which the mean shear varies, since from Eq. (8.38), $\Lambda^{(U)} \approx x_2$. It follows that in shear flows the normalised length scale

$\tilde{L}_X^{(2)} \approx 1/2$. However, when there is no mean straining, the surface only produces a blocking effect (assuming the Reynolds number is high enough). It is found that the anisotropy of turbulence $L_X^{(2)}$ is much greater than for a sheared boundary layer (for $x_2 > \delta_w$), with the ratio of the variances of the normal to parallel components $\overline{u_2^2}/\overline{u_1^2}$ varying approximately as $(x_2/L_X)^{2/3}$ (for $Re > 10^3$). Within a distance n from the surface of order $L_X^{(2)}$, $\Lambda^{(k)} = k/|\nabla k| \leq x_2$ (e.g. in convection and near a free surface where, from (8.27), $\Lambda^{(k)} = \overline{u_2^2}/|\partial \overline{u_2^2}/\partial x_2|$). The integral length scale $L_X^{(2)}$ is approximately equal to n (or greater) and therefore $\tilde{L}_X^{(2)} \gtrsim 1$. Thus the turbulence is non-local and the boundary condition has a much greater effect for these 'shear-free' flows near boundaries than in the previous case of sheared flows. In thermal convection, the fluctuating thermal conditions at the rigid surface also have a large effect on the turbulence (Hunt, 1998).

(ii) In transition zones between shear layers and external flows the turbulence and mean shear velocity vary over length scales (defined as $\Lambda^{(k)}$, $\Lambda^{(U)}$) that can be smaller than the length scales L_x of the turbulence, especially when there is relatively large scale turbulence outside the shear layer. Examples are free stream turbulence passing over aerofoil blades with thin boundary layers, or where external turbulence interacts with the narrow wakes of aerofoils or thin separated shear layers from bluff bodies. Although approximations for the transport terms of turbulence models (e.g. (8.12)) represent the tendency for the external turbulent energy to be transported by the eddies into the shear, they do not model correctly the dominant processes when turbulent flows with two distinct length scales interact as they do in these flows. For example, it is found that large scale environmental turbulence diffuses wakes of obstacles in cross flows much faster than one-point models predict. This is because in k–ε models the eddy viscosity or diffusivity, which is determined by the *largest* length scale L_X, is not then equal to the expression k^2/ε, which may be largely determined by eddies with length scales that are smaller. The same tendency but in reduced form is found in RSTM. Essentially this error is significant when $L_X/\Lambda^{(k)}$ is of order unity or greater. There is a second cause of error in applying turbulence models when the external turbulence has a larger length scale than the thickness of the shear layer, $\Lambda^{(U)}$, i.e. when $L_X/\Lambda^{(k)} \geq 1$. This is caused by the tendency of external eddies to be blocked by the mean vorticity in the thin shear layer (Wu et al., 1999). Therefore they diffuse into the layer much more slowly than given by the transport terms. This is very marked in the initial stages of developing free shear layers (e.g. in wakes of bluff bodies and aerofoils).

3.3 Mapping the characteristics of the main types of turbulent flows

In Section 3.2 we suggested guideline principles and a 'map' for the types of flow where one-point models are likely to be valid in principle, and reasonably accurate. Using the dimensionless parameters of the guidelines we now classify the main types of turbulent flow in Table 8.2. This shows where these 'types' should be

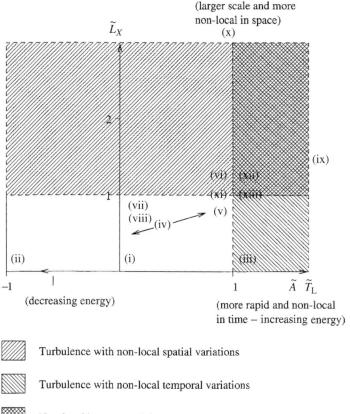

Fig. 8.5 Map for the localness of turbulence processes ($\tilde{T}_L = T_L/T_D$; $\tilde{A}$ is the sign of the advective term in the energy equation; $\tilde{L}_X = L_X/\Lambda$). The roman numerals on the graph refer to particular flow types described in Section 3. Λ is the appropriate inhomogeneity scale $\Lambda^{(k)}$, $\Lambda^{(U)}$.

placed on the 'localness' map in Fig. 8.5 corresponding to different ranges of the parameters $\tilde{T}_L$, $\tilde{L}_X$, which define temporal and spatial localness of the turbulence. The coefficient $\hat{A}$ defines whether the energy is increasing, decreasing or approximately stationary.

(i) Homogeneous turbulence with a *continual input of energy* (e.g. by body forces) is approximately stationary and close to local equilibrium because, if there are no boundaries, the integral length scale grows only slowly. Hence $\tilde{T}_L \sim 1$. Here $\tilde{A} = 0$. Note that even if initially the probability distribution of large scale turbulence is non-Gaussian (e.g. consists of discrete energetic eddies), it eventually tends to become near Gaussian. In general EVM and RSTM methods are valid for these flows but the former does not predict the normal stress anisotropy at all accurately.

(ii) Homogeneous *decaying* turbulence is one of the types of flow used to calibrate the EVM and RSTM models. Although the flow is local spatially, the turbulence decays with time over a period of order T_L, which is why $\tilde{T}_L \sim 1$, and $\tilde{A} < 0$. If the initial turbulence is non-isotropic, the turbulence may or may not tend to isotropy depending on how it is generated (Townsend, 1976). This dependence on the form of the spectrum is included in the latest version of RSTM (Craft and Launder, 1996).

(iii) Homogeneous *strained* turbulence (usually both pure shear and pure irrotational strain, i.e. $\tilde{S} = 0, 1$ respectively – see (8.18a)) is also used for calibrating the strain/shear terms in the k–ε form of the EVM and the RSTM models (e.g. Durbin and Pettersson-Reif, 2000; Pope, 2000). Usually the strain rate is low enough in the experiments and numerical simulations used for calibration that the turbulence dynamics are approximately local or slowly changing, i.e. $\tilde{T}_L \lesssim 1$. For higher strain rates, modified forms of RSTM have been developed (e.g. Jongen and Gatski, 1999). When the mean strain is purely rotational, i.e. $\tilde{S} \sim -1$, turbulence decay and its large eddy structure are affected by the formation for inertial waves (Cambon, 2002), the usual RSTM models are not appropriate.

(iv) In turbulent *boundary layers* the validity of one-point models is greatest near the 'wall', but outside the viscous/roughness wall layer. Here the turbulence structure is approximately local in space and time and is approximately Gaussian (at the large scales). Also the straining is dominated by shear. However, at the outer edge of the boundary layer the local assumptions are not strictly valid, and the turbulence is intermittent and non-Gaussian (as discussed in Section 2). Hence the one-point models are better near the wall than in the outer part. To rectify this deficiency, specific models for the eddy viscosity ν_e itself have been proposed for the outer regions of boundary layers and free shear layers (e.g. Spalart and Allmaras, 1994).

(v) In most *free shear flows and separated flows* the dynamics of turbulence are not quite local in space and time. Therefore models with spatial and temporal development are needed. However, if the mean shear is strong enough (in relation to the turbulence) locally or globally across the flow, i.e. from (8.18b) $S/(u_0/L_X) \gtrsim 1$, then mixing length or eddy viscosity models can provide a useful approximation for free shear flows. In such a flow eddy viscosity models (Section 2.4.2) are the simplest constant form of modelling for the mean strain and stresses, but do not predict the normal stress anisotropy at all accurately. Note that the mixing length model (8.22) requires some physical or numerical adaptation where the mean shear tends to zero.

In flows which are significantly inhomogeneous and in flows over rigid bodies where the flow separates and there are strong regions of pure strain (i.e. $\tilde{S} \sim 1$) and pure rotation (i.e. $\tilde{S} \sim -1$) the turbulence is not entirely determined by the local velocity gradient. Consequently non-local EVM or RSTM models have to be applied to obtain approximately correct predictions. Even then, because of the relatively large scales of the eddy turbulence (i.e. $\tilde{L}_x \geq 1$), the mean flow calculations of separated flows are less accurate (e.g. Murakami *et al.*, 1990).

(vi),(vii),(viii) In highly distorted and inhomogeneous shear-free turbulent flows near blocking surfaces, the turbulence structure does not, as it does in sheared flows near interfaces, have a universal form. For example, whether or not the large scale turbulence is approximately Gaussian near the boundaries largely depends on its structure away from the boundaries. Thus the Gaussian structure of large scale free stream turbulence differs considerably from the slightly intermittent and non-Gaussian structure of turbulence near a free surface in open channel flow and the highly skewed large eddy structure driven by natural convection (Wyngaard, 1979). In these flows the fluxes of momentum and heat are determined by whole patterns of large scale eddy motions and not by local gradients, which is why RSTM models are the simplest level of model to produce even qualitatively correct results (Gibson and Launder, 1978). For reasonable accuracy of calculations of all three turbulent components especially at high Re, multi-point models (e.g. Hunt, 1984) or numerical simulations are more reliable (e.g. Schmidt and Schumann, 1989). Note that unsteady RANS modelling can provide accurate enough predictions if any unsteadiness is well separated spectrally from turbulence motion (Yao *et al.*, 2002; Kenjeres and Hanjalić, 1999). An extreme example of a flow with high skewness (i.e. $\tilde{G} \sim 1$) and inhomogeneity ($\tilde{L}_x \sim 1$) occurs when there is low intensity turbulence outside a shear layer (e.g. Veeravalli and Warhaft, 1989). In this case it is found that one-point models can only describe the profiles of turbulent energy if the coefficient (C_T) in the model for eddy transport in (8.33) is multiplied by a large factor (~ 5) (Bertoglio, 1982).

(ix),(x),(xi),(xii) In another broad group of highly distorted flows there are rapid variations both of the strain (i.e. $\nabla \underline{U}$) and of the type of strain (i.e. $\tilde{S}$) along the mean streamlines, so that $\tilde{T}_L \gtrsim 1$. Also in many cases the turbulence is also spatially non-local so that $\tilde{L}_x \gtrsim 1$. Many studies (e.g. Murakami *et al.*, 1990) have shown how EVM models can over-estimate the effect of straining on the growth of k when $\tilde{L}_X \gtrsim 1$ near bluff bodies. RST models are better, but still are not able to predict the correct sensitivity to $\tilde{L}_X$ of the distorted turbulence. A number of modifications to one-point models have been proposed for such flows, including eddy viscosity models where ν_e depends non-linearly on the strain rate (e.g. Hunt *et al.*, 2001).

For these rapidly distorted inhomogeneous flows, upwind boundary conditions have to be provided even if only one-point statistics are required. This is because the upwind/initial anisotropy and form of the spectra can significantly affect the changes in the variances of the three components and in the length scales and dissipation rate (Durbin and Pettersson-Reif, 2000). These sensitivities are greatest when the mean velocity gradients are closer to pure strain, as they are here (where $\tilde{S} \sim 1$).

(xiii) Finally we mention shearing flows with strong swirl because of their widespread engineering interest; here the turbulence structure varies rapidly depending on the ratio of the strength of the swirl to the local mean shear (i.e. the values of the parameter $\tilde{S}$). For example, strong swirl ($\tilde{S} \sim -1$) can suppress the shear stress. Thus simple mixing length and EVM models are inappropriate, while RST models can be adapted to allow for how the highly distorted turbulence structure affects the production term (P_{ij}) and to a lesser extent the transport term ET_k.

4 Refined and alternative modelling approaches

4.1 Modifications on one-point model for 'difficult' flows

Different approaches are being adopted to model some of the more 'difficult' types of turbulent flow identified in the previous section, which are briefly reviewed here – see also Hunt *et al.* (2001). The first approach is the modification of one-point models (of whatever level of complexity) to particular flows or types of flows, by changing the coefficients in the models, or by introducing new terms or extra auxiliary equations. Metaphorically, the 'frontier' in Fig. 8.5 of the range of applicability of general one-point models is not being moved outwards, but at 'outposts' specific closure models are constructed for dealing with particular types of flow, incorporating appropriate aspects of non-local and non-equilibrium phenomena.

A current example of this approach is to modify the expression for the eddy viscosity. A limitation of the mixing length model (8.20)–(8.22) is that the shear stress is predicted to be zero where the mean velocity gradient is zero. In fact the stress may be finite, which can be estimated by evaluating σ_{ij} at each point and then taking an average value over a scale L_X (e.g. Hunt, 1992). This obviates some of the difficulties in computing stresses in jets and wall jets. Another way is to introduce an advective partial differential equation for eddy viscosity (v_e), which is particularly suitable for calculating the mean velocity profiles in boundary layers, and possibly thin shear layers (Spalart and Allmaras, 1994). Another example is the '$\overline{v^2} - f$' model of Durbin (1993) which modifies EVM and RST models so that the 'wall-echo' or blocking effects of a wall are incorporated via a non-local relaxation. (Here $\overline{v^2} = \overline{u_2^2}$ in our notation; f is an auxiliary function for pressure-strain.) This is particularly valuable when there is significant straining ($\tilde{S} \sim 1$), as with a jet impacting onto a wall and/or large spatial inhomogeneity (so that $\tilde{L}_X \sim 1$), but because of its computational merits in terms of minimal complexity and fast convergence, the EVM model with two auxiliary differential equations is more widely used than the RSTM models which may involve seven. However, some of the additional information conveyed by these extra equations can be used to adjust the coefficients used in mixing length or k–ε models, for example by noting the results of RST models on the effects of variations in the straining parameter S, and then simply adjusting the mixing length in (8.22) (e.g. Belcher *et al.*, 1991).

A more general and widely used approach for such flows is to use more complex 'algebraic' stress relations in place of the simple mixing length formula (8.21), in combination with the non-linear k–ε model for the energy and length scale of the turbulence. Because of the reduced number of auxiliary equations to be solved, this approach has many of the advantages and improvements of RSTM at far less computational cost. See, for example, Launder and Sandham (2002) and Oberlack and Busse (2002).

These models describe normal stress anisotropy and non-linear responses to local strain rates far better than any simple linear eddy viscosity scheme, but, like all classes of EVM and RSTM, they do not describe effects of non-local strain and boundary conditions. Following the same methodology, some quite simple modifications could be introduced to correct for the errors in these local or quasi-local models when the turbulence scale is so large that $\tilde{L}_X \gtrsim 1$, for example as occurs with very large free stream turbulence in a boundary layer. In such situations the only eddy motions that affect the dynamics of mean flow and the decay of turbulence are those with length scales $\hat{k}^{-1}$ which are less than or of the order of the inhomogeneous length scales $\Lambda^{(U)}$, $\Lambda^{(k)}$ of the shear and turbulent energy (see Eqs. (8.27)). These effects can be approximately described by pressure-diffusion and non-local pressure-strain correlation as well as the anisotropy of dissipation, D_{ij}. One must also allow for strain-history related lag-effects, which has so far been done principally at the mixing length level (e.g. by Johnson and King, 1984), an approach that has been subsequently adopted at the algebraic stress model (ASM) and RSTM level of closure (Savill, 1996). Other researchers have relaxed the assumption of a fixed algebraic relation between k and stress to make the associated structure constants $a_{ij} = \overline{u_i u_j}/k$ a function of total strain history, e.g. see Savill (1987) and especially the work of Gatski and co-workers: see Gatski and Rumsey (2002) and references therein.

An alternative approach to correcting for very large scales in the kinetic energy equation is to replace k with its 'dynamically' significant component k_Λ, where

$$\frac{k_\Lambda}{k} = \int_{\Lambda^{-1}}^{\infty} E(\hat{k})d\hat{k} \bigg/ \int_{0}^{\infty} E(\hat{k})d\hat{k}. \tag{8.41}$$

k_Λ/k is the fraction of the kinetic energy produced by eddy motions of the order of or less than the length scale Λ of the turbulence or mean flow inhomogeneities. Experimental measurements of spectra (e.g. Britter *et al.*, 1979; Thomas and Hancock, 1977; Gutmark *et al.*, 1978) show that this is a clear and physically meaningful distinction. The ratio (8.41) could be estimated approximately for practical calculations by, for instance, the Kolmogorov spectrum.

Also, for large scales, since the eddy viscosity ν_e or diffusivity κ_e are defined in relation to the gradient of the mean velocity and scalar profiles, so the relevant scale of the eddy motion should be Λ and not L_x where $L_x \gg \Lambda$. Therefore $\nu_e \sim k_\Lambda^2/\varepsilon$. For very high Reynolds number turbulence where $E(\hat{k}) \sim \varepsilon^{2/3}\hat{k}^{-5/3}$ (for $\hat{k} \gtrsim L_X^{-1}$) it follows from (8.41) that,

$$\nu_e \sim \varepsilon^{1/3} \Lambda^{4/3}. \tag{8.42}$$

The approach suggested here is a much simplified version of the method proposed by Hanjalić *et al.* (1980), which also involves dividing the spectrum into two or more ranges and then computing separately the evolution of the integrals of energy, and other variances, in these ranges (e.g. for large and small wave numbers, or large, intermediate and small). With the extensive experimental data, as well as theoretical two-point and spectral models (e.g. Parpais, 1997; Carlotti and Hunt, 2001; Schiestel, 1987) now available for studying how turbulence changes in these different parts of the spectrum, such 'separate sub-range' methods are becoming more practical.

The third way in which one-point modelling methods can be improved and speeded up is by developing more accurate and physically correct models of turbulence near solid boundaries or fluid interfaces. As explained in Section 2, in some models calculations cover the entire flow field right up to the boundary or interface – which requires including, in a model for fully developed turbulence, a representation of random motions near the boundary which are not fully turbulent. The other approach is to model separately the motions in these 'boundary zones'. It is now well understood that the 'wall-function' assumptions are not generally valid. This is because in complex flows the mean velocity or scalar profiles for $U(x_2)$ or $\Theta(x_2)$ do not follow the logarithmic–linear form for the turbulent viscous/molecular sub-layers in the 'wall' zone on rigid surfaces (or interfaces with a strong enough density difference, cf. (8.25)). The errors are greatest where $\tilde{L}_X \gtrsim 1$ or $\tilde{T}_L \gtrsim 1$, e.g. in flows with natural convection (Kenjereš and Hanjalić, 1999) or very energetic free stream turbulence, or impinging jets (e.g. Durbin and Pettersson-Reif, 2000).

To avoid this limitation many models now compute the entire unsteady flow field explicitly for enough realisations (or for long enough time) to derive reliable statistics. However, the considerable complexity and computational cost involved in such procedures is not necessary if more use is made of the increased understanding of the structure of complex turbulent flow near rigid surfaces and flexible interfaces on the boundary conditions, and how the external flow varies. For example, even small departures from a constant shear flow caused by accelerating and decelerating mean flows, or by unstable and stable buoyancy zones in horizontal flows, lead to large changes in the logarithmic profile (8.25) (Bradshaw, 1971). These effects can be expressed simply in terms of an empirical coefficient α as

$$\frac{dU}{dx_2} = \frac{u_*}{\kappa x_2}\left(1 + \frac{\alpha x_2}{L}\right), \quad \varepsilon \sim \frac{u_*^3}{\kappa x_2}\left(1 - (\alpha - 1)\frac{x_2}{L}\right), \qquad (8.43)$$

where L is the relevant length scale (which may be positive or negative) for these perturbations (e.g. $L \propto u_*^3/(du_*/dx_1)$ for accelerating/decelerating flows, and $L \approx u_*^3/(g F_\theta/\bar{\theta})$ for flows with a heat flux F_θ, Belcher *et al.*, 1991; Hunt *et al.*, 1988). This approach is currently used more in geophysical than engineering flows.

Now consider the wall boundary zones in the presence of energetic large scale turbulence where $\tilde{L}_X \gtrsim 1$. This occurs when the shear production at the wall characterised by u_* is relatively weaker than that generated by large scale eddying with velocity u_0, such as in horizontal shear flows with strong natural convection (Rao *et al.*, 1996), separated recirculating flows above a rigid surface (e.g. Wood and Bradshaw, 1982 and 1984) or shear flows on sloping surfaces or in sloping channels driven by buoyancy zones. Here, the turbulence, although it is 'blocked' by the surface, is largely driven by the buoyancy forces, and large scale motions in the local interior. Very close to the surface, the local fluctuating shear dominates in a thin layer which, at very high Reynolds number, is of thickness $L = u_*^3/\varepsilon_0$, where ε_0 is the dissipation rate in the interior of the flow. In such flows, where $\varepsilon_0 \sim u_0^3/L_X$, $L \ll L_X$. In some cases u_* is based on the *fluctuations* of shearing motions because the mean stream is very weak (Schmidt and Schumann, 1989). For $L_X > x_2 > L$ the eddies that determine the mean velocity and temperature profiles are determined by the distance (x_2) normal to the boundary so that (8.42) becomes

$$v_e \sim \varepsilon_0^{2/3} x_2^{4/3}. \tag{8.44}$$

Thence, from (8.25), since the fluxes u_*, F_θ tend to be constant near the boundary,

$$\left[\frac{dU}{dx_2}, \frac{d\Theta}{dx_2} \right] \sim \left[u_*, \frac{F_\theta}{u_*} \right] \varepsilon_0^{2/3} x_2^{-4/3}. \tag{8.45}$$

Within the surface shear layer where $x_2 < L$, the usual log-linear wall functions apply for $\underline{U}, \Theta$ (Rao *et al.*, 1996). Such profiles, which have been measured in engineering and environmental flows, are widespread and could be used more often as suitable boundary conditions for practical problems to improve the accuracy of one-point model predictions (especially in very high Reynolds number flows).

As mentioned in Section 3.3 and earlier in this section, some of the most difficult flows to model occur where the turbulence is highly non-local in space and time, i.e. $\tilde{L}_X \gtrsim 1$ and $\tilde{T}_L \gtrsim 1$. In some cases these non-localness parameters remain large even near a boundary and therefore have to be considered in matching the interior flow to the boundary 'zone'. This is essential where turbulent flows impinge onto a rigid surface, e.g. at the stagnation point of bluff bodies and aerofoils, impinging jets, or where separation streamlines attach to downstream surfaces. The heat transfer between the surface and the flow is very high in these regions, whereas the momentum flux is low. One-point models have produced quite a variety of results because of the sensitivity of the calculations to upstream boundary conditions and, for heat transfer to conditions at the surface (e.g. ERCOFTAC, 2000; Durbin and Pettersson-Reif, 2000). However, the turbulence structure is now better understood. Theoretical studies based on the non-linear models, computations and experiments of straining turbulence (Kevlahan and Hunt, 1997) show how the vorticity fluctuations and ε increase from the outer values (e.g. ε_0) towards the surface in these straining flows (i.e. $\varepsilon \propto x_2^{-1}$). But within a thin boundary layer of thickness δ, the mean strain decreases and the dissipation rate reaches an approximately constant high value $\varepsilon_\delta \gg \varepsilon_0$. Within an even thinner shear boundary layer of thickness $L \sim u_*^3/e_\delta$, the velocity fluctuations parallel to the wall decrease to zero.

When the interface bounding the region of turbulent flow consists of a stable horizontal layer where the density increases downwards (parallel to gravity), it acts partly like a rigid surface, by blocking the vertical fluctuations (if the density jump is great enough) and partly like a flexible sheet with waves formed on it. These waves tend to break down and generate intense local turbulence; some of their energy may be propagated away towards the stable non-turbulent flow on the other side of the interface. By modelling K, ε, U, Θ in these interface zones (especially the effects of wave-turbulent interactions) it is possible to provide boundary conditions for one-point models which can be applied in the turbulent interior (Uitenbogaard, 1994; Strang and Fernando, 2001). These boundary conditions are needed for both chemical engineering problems, if they involve 'stratified' flows with interfaces between gas–liquid and liquid–liquid flows, and also for atmospheric, hydraulic and oceanic flows (Lombardi *et al.*, 1996).

In concluding this brief review of modifications to one-point models for particular types of flow and types of boundary condition, we should note two points about how they can be applied.

(i) Within a given complex flow there may be various flow zones, such as boundary layers, separating shear layers, impinging jets, etc., each having somewhat different types of turbulence. Here the dimensionless parameters $\tilde{L}_X$, $\tilde{T}_L$ of non-localness may vary from being small to quite large. Therefore some modellers choose appropriate turbulence models for different regions of the flow as defined by their physical location and/or some physical (e.g. non-localness) parameters. This 'zonal modelling' approach, which is quite standard in meteorological modelling, was advocated for engineering flows by Kline *et al.* (1981), but has not been widely adopted because of their geometrical complexity (see Savill, 1981 and 1987 for further details). An alternative approach has been more successful in which the turbulence modelling is related to the variations in the computational mesh especially near boundaries, for example eddy viscosity mesh models near the wall, and Reynolds stress transport modelling away from the wall. In many cases the accuracy of CFD predictions is determined more by limitations of mesh resolution rather than turbulence modelling. For example, in moderate Reynolds number flow more advanced Reynolds stress transport modelling does not necessarily produce much better results than low *Re* k–ε modelling. One reason why simpler models are satisfactory for quite complex flows appears to be that multiple strain effects can often act together to minimise any large distortions of the turbulence.

(ii) Some of the physical processes and boundary conditions are quite sensitive to variations in Reynolds number. The form of sensitivity is different at lower Reynolds numbers; near transition, shear layers exhibit large scale intermittency, whereas at higher Reynolds number, with a fully developed inertial range spectrum, intermittency occurs at small scales. This necessitates different forms of the models (e.g. coefficients or even expressions) according to the flow regime, especially those near transition. At very high Reynolds number, where $Re_T \gtrsim 10^4$ and where there is a fully developed and self-similar spectrum, it is possible to estimate Reynolds number corrections by making an appropriate Kolmogorov cut-off in the energy spectrum where $\hat{k}_K = (v^3/\varepsilon)^{-1/4}$, so that the kinetic energy at finite Re is given in terms of k at very large Re by

$$
k(\text{for finite } Re) = k(Re \to \infty) \frac{\displaystyle\int_0^{\hat{k}_K} E(\hat{k})\widehat{dk}}{\displaystyle\int_0^{\infty} E(\hat{k})\widehat{dk}}. \tag{8.46}
$$

This is good reason for attempting to understand the very high Re limit for turbulent flows – a point made cogently by the late A. E. Perry.

4.2 Possible alternative approaches using multi-point models and simulations

The previous section, 4.1, has indicated how a deeper knowledge of the structure of complex turbulent flows can, if used appropriately, aid in the selection of one-point models and the assessment of their applicability for particular types of complex flow. For example, two-point linearised models (such as rapid distortion theory, RDT, models) can compute spectra of all the velocity components and pressure fluctuations most accurately in those types of flow where the one-point models are least accurate, where $\tilde{T}_L \gg 1$ and $\tilde{L}_X$ may have any value. Such calculations, because they are explicit and often analytical, indicate trends and error in one-point methods, and suggest possible corrections. But note that current RDT methods are not applicable for the very smallest scales of motion and for the dissipation rate ε, because these are controlled by non-linear processes. Non-linear models of the time dependent eddy structure provide useful concepts for improving the statistical models and their boundary conditions very close to rigid surfaces, especially for low Re turbulence (e.g. Holmes *et al.*, 1996).

With the recent development in models for two-point moments in complex flows (e.g. Parpais, 1997; Mann, 1994; Hunt and Carlotti, 2001), it should be possible to understand better and perhaps improve the approximate differential equation currently used for calculating ε, especially in flows at very high Reynolds number and where $\tilde{T}_L \gtrsim 1$ (see for example the remarks of Pope, 2000). These methods, combined with simulations of random Fourier modes, indicate how the non-linear processes change with strong stable stratification and rotation, which should also lead to improvement where current models are limited (e.g. Nicolleau and Vassilicos, 2000). Renormalisation group (RNG) and eddy damped quasi normal Markovian (EDQNM) models of turbulence spectra are being applied to homogeneous turbulence flows under the influence of various strains/body forces, etc. (e.g. Kraichman, 1991; Lesieur, 1990). It is not yet clear whether they can improve prediction of one-point statistics in complex, in homogeneous flows. There is another important type of limiting flow where one-point models also fail, but where approximate and readily computable simulation models can be applied. These occur where eddies are much larger (and therefore slower) than the small scales of motion defined by the mean velocity or gradients of turbulence, i.e. $\tilde{L}_X \gtrsim 1$, but where $\tilde{T}_L$ is arbitrary, for example where large scale free stream fluctuations impinge on a bluff body or aerofoil (a type of flow which is sometimes computed using gust models), or large eddy motions driven by natural convection extend throughout the depth of a container or boundary layer.

For certain situations where the large eddies have a well defined form, e.g. wake vortices impinging on a downwind aerofoil (Goldstein and Atassi, 1976; Saxena and Hunt, 1993) or convective eddies interact with simple boundaries (e.g. flat or vertical walls – Grossman and Lohse, 2000; Hunt, 1998) the eddy motions can be analysed or calculated quite simply, leading to interesting and useful results (some of which were used here in Section 4.1). However, more often the inflow and boundary conditions are quite complex, e.g. in convectively driven flows in electronic cooling systems, the turbulence statistics can be derived by numerical simulation of the large eddies, with the effect of the small eddies on the large scales being computed using one-point (or two-point) models (Kenjeres and Hanjalić, 1999).

So we conclude that there is a useful 'cascade' of ideas and methods from simulations and more elaborate models to help improve one-point closures in general, as well as to identify the deficiencies of specific EVM and RSTM schemes – see Fig. 8.4.

The evidence from meteorological and environmental modelling is that progress is made most rapidly by developing different specific models for the main types of complex flow (e.g. in different types of cloud) rather than attempting to have a single model to cover all the cases of interest (Hunt, 1999). One might describe this as establishing 'outposts' along trajectories from the origin of the 'map' rather than broadening the sphere of model influence about the origin – see Fig. 8.6. Although extending the overall applicability of models does not seem practical in general, it has been successfully achieved for some classes of low Re transition modelling (Savill, 1996). There are limitations to the alternative 'flow-class' approach for intermediate cases. It can be misleading to assume (as is often done) that these

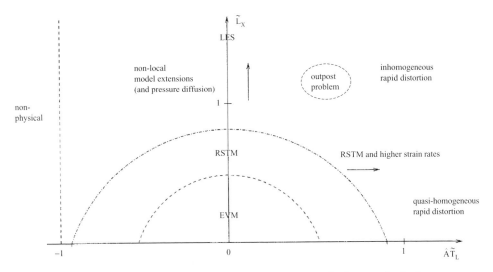

Fig. 8.6 Guideline map for choosing appropriate prediction methods for one-point turbulence statistics (and mean flows), in terms of the spatial and temporal localness parameters $\tilde{L}_X$, $\tilde{T}_L$. (Note that $\hat{A}$ is the sign of the advective term in the kinetic energy equation.) The line ------- is the 'frontier' of standard Reynolds stress transport models (RSTM), while ------ indicates a typical 'outpost' type of flow where flow specific modelling has been developed (e.g. for trailing edge regions of aerofoils).

are simply interpolations between 'known', well calibrated situations (e.g. in the transition from stably to unstably stratified boundary layers).

5 Remarks

5.1 Summary of 'guiding' principles

This review is not intended as a complete guide to the choice and accuracy level of calculations currently used for a broad range of practically significant turbulent flows. However, it has been shown that a useful test for the choice of method can be based on characteristic localness parameters, $\tilde{T}_L$, $\tilde{L}_X$ and, to a lesser extent, on the structural parameters, $\tilde{G}$ and $\tilde{S}$. Here $\tilde{L}_X$ is the ratio of the typical eddy length scale to a cross-stream length scale, determined by either the intensity or the rate of strain of the turbulence, and $\tilde{T}_L$ is the ratio of the 'memory' time scale of the turbulence to the imposed distortion time scale. The relationship between $\tilde{T}_L$, $\tilde{L}_X$ and the regions of applicability of the various models is shown in Fig. 8.6. In other flows, e.g. with strong buoyancy forces, anisotropy also has to be characterised.

Because these parameters have comparable values (or profiles) in flows of similar type they can be estimated (without detailed computations or measurements) by a brief inspection in any given flow or region of a flow, and by comparing with other similar flows. This is a quantitative demonstration of how the statistical properties of turbulent flows do not have universal forms, but do have strong similarities in flows of similar 'type' (convection, shear, rotating, etc.). In boundary layers, the type of turbulence depends on the range of Reynolds number of the flow. We have shown that in certain types of flow where the turbulence is close to a rate of local equilibrium and where shear is the dominant form of mean strain (i.e. $\tilde{S} \approx 0$), simple mixing length models are satisfactory; as the turbulence moves slightly away from local equilibrium, and the strain becomes purely irrotational or rotational, i.e. $|\tilde{S}| \sim 1$, then more complex models are needed. To calculate such flows with reasonable accuracy, progressively more non-local or 'history' effects need to be considered. When the turbulence dynamics are highly non-local, both in space and time, and especially if the probability distribution is non-Gaussian (i.e. $\tilde{G} \sim 1$), then the 'standard' one-point k–ε and RSTM models can be significantly in error. The physical reasons for this are given in Section 2 and examples of particular flows are discussed in Sections 4.1 and 4.2.

The sphere of influence on the 'localness map' of existing models is being expanded by introducing corrections to models to allow for non-local/non-equilibrium effects and other processes and by developing particular approximations for particular types of flow (e.g. as in meteorology) by making best use of increased understanding of flow physics.

Acknowledgements

It is a pleasure to record our thanks to Professor Geoff Hewitt for inviting us to write on this subject. We also both thank the other organisers of the Isaac Newton Institute Programme on Turbulence held in 1999, notably N. Sandham and J. C. Vassilicos, who encouraged this chapter to be written. The INI Programme was supported by Rolls-Royce plc, BAE Systems, British Gas Technology and British Energy as well as the Royal Academy of Engineering. The authors are grateful to these bodies, and to the referees for extensive and constructive comments. AMS also thanks his hosts at the University of Newcastle, New South Wales (particularly Professors N. W. Page and R. A. Antonia) where he worked on this review whilst a visiting research fellow there.

Appendix
Invariants of the strain rate tensors and the normalised form

For the mean velocity field $U_i(\underline{x}, t)$, at each point (and moment of time) the second moment (or quadratic) form of its gradient $\frac{\partial U_i}{\partial x_j}$ can be defined as

$$\Pi = \frac{\partial U_i}{\partial x_j} \frac{\partial U_j}{\partial x_i}.$$

This is invariant under rotation of axes. It can be expressed in terms of the symmetric strain (stretching) tensor $\Sigma_{ij} = \dfrac{1}{2}\left(\dfrac{\partial U_i}{\partial x_j} + \dfrac{\partial U_j}{\partial x_i}\right)$ and the vorticity $\Omega_l = \varepsilon_{lmn}\dfrac{\partial U_n}{\partial x_m}$, as

$$\Pi = \Sigma^2 - \frac{1}{2}\Omega^2,$$

$$\text{where} \quad \Sigma^2 = \Sigma_{ij}\Sigma_{ij} \quad \text{and} \quad \Omega^2 = \Omega_l\Omega_l.$$

It is convenient to normalise and rescale Π as the variable $\tilde{S}$, so that its range is limited. Then $\tilde{S} = \Pi/S$,

where $S = E^2 + \dfrac{1}{2}\Omega^2$.

(For further discussion and applications to turbulence modelling see Hunt and Vassilicos, 2000.)

Nomenclature

Nomenclature		*Super/Subscripts*	
A	Rate of change of eddy kinetic energy	$(\ )^*$	Dimensional variable
$[B_N]\,[B_S]$	Boundaries of the domain under consideration		
$B_i,\,B,\,B_{ij}$	Bodyforce vector, production rate, tensor	$(_-)$	Mean
C	Coefficients in (8.28), (8.29), (8.31), (8.35) and (8.36)	$(\ )_i$	Vector component
D_{ij}	Energy dissipation rate tensor	$(\ )^T$	Differential operator for turbulence model
d	displacement height for logarithmic profile		
$ET,\,ET_{ij}$	Eddy transport rate, tensor	$(\ )^{NS}$	Operator for Navier–Stokes equation
$E(\hat{k})$	Energy spectrum	$(\ )_{mx}$	Value at maximum
f	Auxiliary function for calculating stresses near wall	$(\ )^{(l)}$	Linear
$F_{\theta j}$	Turbulent heat flux vector	$(\ )^{(nl)}$	Non-linear
$\tilde{G}$	Measure of non-Gaussianity (higher moments)	$(\ )_L$	Lagrangian
$I,\,I_{ij}$	Inertial production rate, tensor	$(\ )_D,\,\theta D$	Distortion time for velocity, for scalar
$\hat{k}$	Wave number	$(\tilde{\ })$	Dimensionless ratios/coefficient
k	Kinetic energy	$\langle\ \rangle$	Conditional averages (e.g. relative to moving interface)
l	Eddy scale/action distance	$(\ )_w$	Wall level for defining thickness of non-conforming wall layer
L	Length scale for perturbed shear flows (e.g. Monin–Obukhov) (8.43)	$(\Lambda)^{(k),(U)}$	Inhomogeneity scales, energy and mean defined by turbulent velocity

$L_x, L_{x(i)}$ Integral length scale, for a particular component

$L^{()}()$ Differential operator, e.g. of turbulence model

$M^{(n)}()$ nth moment of variable

P_i, P_{ij}, P Pressure, pressure-strain tensor, magnitude

p Fluctuating pressure

$pr_{()}()$ Probability distribution for the variable ( )

Re Reynolds number of turbulence

R_{ij} Reynolds stress production tensor

$S, \tilde{S}$ Mean square 'total' strain, normalised strain

$T_{()}$ Integral time scales

t Time

U_i, u_i Velocity vector (mean, fluctuating)

U' $= [dU_1/dx_2]$ mean shear

U'' $[d^2U_1/dx_2{}^2]$ mean vorticity gradient

V_{ij} Viscous production tensor

$V_{ij}^{(\sigma)}$ Viscous diffusion of stress rate tensor

X_s distance along near stream lines

x_i, or x, y, z Spatial coordinate (x_1, x_2 in flow, normal directions of a shear flow or x, y)

x_{20} Roughness length

x_{2I} Interface height

$(k)_\Lambda$ Kinetic energy associated with scales Λ

$()_k$ Reference to processes in kinetic energy equation

$()_s$ Surface layer

Mathematical

$\nabla()$ Denotes gradient operator

Greek

α Coefficient for mean shear in perturbed flows

β Exponent in developing flows

δ Thickness of boundary layer

δ_{ij} Kronecker delta ($=1$ if $i=j$ $=0$ if $i\neq j$)

$\Delta()$ Change in a variable (e.g. over a certain distance)

ε Energy dissipation rate

κ_e Thermal diffusivity

κ Von Karman contstant

Θ, θ Temperature (mean/fluctuating)

$\lambda, \lambda_K, \lambda_\varepsilon$ Coefficient for estimating eddy viscosity (8.21), threshold values of energy and dissipation

Λ Inhomogenety length scales

ν, ν_e Viscosity, eddy viscosity

Π Invariant for strain moments

ρ density

$-\sigma_{ij}$ Reynolds stress tensor

Σ^2 Strain rate of the mean flow (squared)

τ Reynolds shear stress

$\Phi_{ij}(\hat{k})$ Energy spectrum tensor

Ω^2 Vorticity of the mean flow (squared)

References

Belcher, S. E., W. S. Weng and J. C. R. Hunt (1991) Structure of turbulent boundary layers perturbed over short length scales. *Eighth Symp. on Turb. Shear Flows*, paper 12–2 Tech. Univ. of Munich.

Belcher, S. E., T. M. J. Newley and J. C. R. Hunt (1993) The drag on an undulating surface induced by the flow of a turbulent boundary layer. *J. Fluid Mech.* **249** 557–596.

Bertoglio, J. P. (1982) A model of three-dimensional transfer in non-isotropic homogeneous turbulence. *Turbulent Shear Flows* 3 253–261.

Bevilaqua, P. M. and P. S. Lykoudis (1978) Turbulence memory in self-preserving wakes. *J. Fluid Mech.* **89** 589–606.

Bisset, D. K., J. C. R. Hunt and M. M. Rogers (2002) The turbulent/non-turbulent interface bounding a far wake. *J. Fluid Mech.* **451** 383–410.

Bradshaw, P. (1971) Variations on a theme of Prandtl. *AGARD Conference Proceedings on Turbulent Shear Flows*, p. C-1.

Bray, K. N. C., R. S. Cant and A. M. Savill (2000) The modelling of obstructed explosions. In *Proc. Third Int. Symp. on Scale Modelling*, Nagoya.

Britter, R. E., J. C. R. Hunt and J. C. Mumford (1979) The distortion of turbulence by a circular cylinder. *J. Fluid Mech.* **92** 269–301.

Cambon, C. (2002) From rapid distortion theory to statistical closure theories of anisotropic turbulence. In *Theories of Turbulence* (M. Oberlack and F. Busse, eds.), Springer.

Cambon, C. and J. F. Scott (1999) Linear and non-linear models of anisotropic turbulence. *Annual Review of Fluid Mechanics* **31** 1–54.

Carlotti, P. and J. C. R. Hunt (2001) Developments in turbulence for aeronautical and other applications. In *Festschrift for Pierre Perrier* (J. Periaux and P. Thomas, eds.).

Cazalbou, J. B., P. R. Spalart and P. Bradshaw (1994) On the behaviour of 2-equation models at the edge of a turbulent region. *Phys. Fluids* **6**(5) 1797–1804.

Champagne, F. H., V. G. Harris and S. Corrsin (1970) Experiments on nearly homogeneous turbulent shear flow. *J. Fluid Mech.* **41** 81–139.

Craft, T. J. and B. E. Launder (1996) A Reynolds stress closure designed for complex geometries. *Int. J. Heat and Fluid Flow* **17** 245–254.

Dengel, P. and H. H. Fernholz (1990) An experimental investigation of an incompressible turbulent boundary layer in the vicinity of separation. *J. Fluid Mech.* **212** 615–636.

Durbin, P. A. (1993) A Reynolds stress model for near wall turbulence. *J. Fluid Mech.* **249** 465–498.

Durbin, P. A. and B. A. Pettersson-Reif (2000) *Statistical Theory of Modelling for Turbulent Flow*, John Wiley.

Durbin, P. A. and O. Zeman (1992) Rapid distortion for homogeneous compressed turbulence with application to modelling. *J. Fluid Mech.* **242** 349–370.

Eames, I. and J. C. R. Hunt (1997) Inviscid flow around bodies moving in weak density gradients without buoyancy effects. *J. Fluid Mech.* **353** 331–355.

ERCOFTAC (2000) Best practice guidelines for industrial computational fluid dynamics. c/o A. G. Hutton, DERA, aghutton@dera.gov.uk.

Gatski, T. B. and C. L. Rumsey (2002) Linear and non linear eddy viscosity models. In *Closure Strategies for Turbulent and Transitional Flows* (Brian Launder and Neil Sandham, eds.), Cambridge University Press.

Gibson, M. M. and B. E. Launder (1978) Ground effects on pressure fluctuations in the atmospheric boundary layer. *J. Fluid Mech.* **86** 491–511.

Godeferd, F. S., C. Cambon and J. F. Scott (2001) Two-point closures and their applications: report on a workshop. *J. Fluid Mech.* **436** 393–407.

Goldstein, M. E. and H. Atassi (1976) A complete second-order theory for the unsteady flow about an airfoil due to a periodic gust. *J. Fluid Mech.* **74** 741–765.

Grossmann, S. and D. Lohse (2000) Scaling in thermal convection: a unifying theory. *J. Fluid Mech.* **407** 27–56.

Gutmark, E., M. Wolfshtein and I. J. Wygnanski (1978) The plane turbulent impinging jet. *J. Comp. Fluid Dyn.* **12**(1) 67–97.

Hanjalić, K., B. E. Launder and R. Schiestel (1980) Multiple time scale concepts in turbulent transport modelling. In *Turbulent Shear Flow* (L. J. S. Bradbury, ed.), Volume 2, Springer-Verlag.

Holmes, P., J. L. Lumley and G. Berkooz (1996) *Turbulence, Coherent Structure, Dynamical Systems and Symmetry*, Cambridge University Press.

Huang, P. G. and M. A. Leschziner (1985) Stabilization of recirculating flow computations performed with second-moment closure and third order discretization. *Proceedings of the Fifth Symposium on Turbulent Shear Flows* 20.7–21.2.

Hunt, J. C. R. (1982) Diffusion in the stable boundary layer. In *Atmospheric Turbulence and Air Pollution Modelling* (F. T. M. Nieuwstadt and H. van Dop, eds.), pp. 231–274, Reidel.

(1984) Turbulence structure in thermal convection and shear-free boundary layers. *J. Fluid Mech.* **138** 161–184.

(1987) Vorticity and vortex dynamics in complex turbulent flows. *Trans. Can. Mech. Eng.* **11** 21–35.

(1992) Developments in computational modelling of turbulent flows. In *ERCOFTAC Workshop on Numerical Simulation of Unsteady Flows* (O. Pirroneau, W. Rodi, I. L. Rhyming, A. M. Savill and T. V. Truong, eds.), pp. 1–76, Cambridge University Press.

(1995) Practical and fundamental developments in the computational modelling of fluid flows. *J. Mech. Eng. Sci. (Proceedings Part C)* **209** 297–314.

(1998) Eddy dynamics and kinematics in convective turbulence. In *Buoyant Convection in Geophysical Flows* (E. J. Plate, E. Fedorovich, D. X. Viegas and J. C. Wyngaard, eds.), pp. 41–82, Kluwer.

(1999) Environmental forecasting and modelling turbulence. In *Proceedings of Los Alamos Conference on Predictability* (M. Lesieur, ed.), Volume 133 of *Physica D*, pp. 270–295.

Hunt, J. C. R. and P. Carlotti (2001) Statistical structure at the wall of the high Reynolds number turbulent boundary layer. *Flow Turbulence and Combustion* **66**(4) 453, 475.

Hunt, J. C. R. and D. J. Carruthers (1990) Rapid distortion theory and the 'problems' of turbulence. *J. Fluid Mech.* **212** 497–532.

Hunt, J. C. R. and P. A. Durbin (1999) Perturbed vortical layers and shear sheltering. *Fluid Dynamics Research* **24** 375–404.

Hunt, J. C. R. and J. F. Morrison (2000) Eddy structure in turbulent boundary layers. *European Journal of Mechanics* **19** 673–692.

Hunt, J. C. R. and J. C. Vassilicos (2000) *Turbulence Structure and Vortex Dynamics*, Cambridge University Press.

Hunt, J. C. R., S. Leibovich and K. J. Richards (1988a) Turbulent shear flows over hills. *Quart. J. Roy. Meteorol. Soc.* **114** 1435–1470.

Hunt, J. C. R., D. D. Stretch and R. E. Britter (1988b) Length scales in stably stratified turbulent flows and their use in turbulence models. *Proc. I. M. A. Conf. on 'Stably Stratified Flow and Dense Gas Dispersion'* (J. S. Puttock, ed.), pp. 285–322, Clarendon Press.

Hunt, J. C. R., H. Kawai, S. R. Ramsey, G. Pedrizetti and R. J. Perkins (1990) A review of velocity and pressure fluctuations in turbulent flows around bluff bodies. *J. Wind Eng. and Ind. Aero.* **35** 49–85.

Hunt, J. C. R., N. Sandham, J. C. Vassilicos, B. E. Launder, P. A. Mokewitz and G. F. Hewitt (2001) Developments in turbulence research: a review based on the 1999 programme of the Isaac Newton Institute, Cambridge. *J. Fluid Mech.* **436** 393–407.

Johnson, D. A. and L. S. King (1984) A new turbulence closure model for boundary layer flow with strong adverse pressure gradients and separation. *AIAA* paper 84-0175.

Jongen, T. and T. B. Gatski (1999) A unified analysis of planar homogeneous turbulence using single point closure equations. *J. Fluid Mech.* **399** 117–150.

Kassinos, S. C., W. C. Reynolds and M. M. Rogers (2001) One point turbulence structure tensors. *J. Fluid Mech.* **428** 213–248.

Kenjeres, S. and K. Hanjalić (1999) Transient analysis of Rayleigh Bénard convection with a RANS model. *Int. J. Heat & Fluid Flow* **20** 329–340.

Kevlahan, N. and J. C. R. Hunt (1997) Nonlinear interactions in turbulence with strong irrotational straining. *J. Fluid Mech.* **337** 333–364.

Kim, K. C. and R. J. Adrian (1999) Very large-scale motion in the outer layer. *Physics of Fluids* **11**(2) 417–422.

Kline, S., B. Cantwell and G. K. Lilley (1981) Comparison of computation and experiments. *Proc. 1980–81 AFOSR-HTTM Stanford Conf. on Complex Turb. Flows*, Stanford.

Kraichman, R. H. (1991) Stochastic modelling of isotropic turbulence. *New Perspectives in Turbulence* (L. Sirovich, ed.), pp. 1–54, Springer-Verlag.

Launder, B. E. (1989) Second moment closure: present and future. *Int. J. Heat and Fluid Flow* **10**(4) 282–299.

Launder, B. E. and N. D. Sandham (eds.) (2002) *Closure Strategies for Turbulent and Transitional Flows*, Cambridge University Press.

Launder, B. E. and B. I. Sharma (1974) Application of the energy dissipation model of turbulence to the calculation of flow near a spinning disc. *Letters in Heat and Mass Transfer* **1** 131–138.

Launder, B. E. and B. Spalding (1972) *Mathematical Models of Turbulence*, Academic Press.

Launder, B. E., G. J. Reece and W. Rodi (1975) Progress in the development of a Reynolds stress turbulence closure. *J. Fluid Mech.* **68** 537–566.

Lesieur, M. (1990) *Turbulence in Fluids* (Second edn.), Kluwer.

Lesieur, M., O. Metais and R. Rogallo (1989) Large-eddy simulation of turbulent-diffusion. *Comptes de l'Academie des Sciences Serie II* **308**(16) 1395–1400.

Lombardi, P., V. DeAngelis and S. Bannerjee (1996) Direct numerical simulation of near-interface turbulence in coupled gas-liquid flow. *Phys. Fluids* **8**(6) 1643–1665.

Lumley, J. L. (1978) Computational modelling of turbulent flows. *Adv. Appl. Mech.* **26** 183–309.

Lumley, J. L., O. Zeman and J. Siess (1978) The influence of buoyancy on turbulent transport. *J. Fluid Mech.* **84** 581–597.

Lumley, J. L., Z. Yang, and T. H. Shih (2000) A length-scale equation. *Flow, Turbulence, Combustion* **63** 1–21.

Mann, J. (1994) The spatial structure of neutral atmospheric surface layer turbulence. *J. Fluid Mech.* **273** 141–168.

Marusic, I. and A. E. Perry (1995) A wall-wake model for turbulence structure of boundary layer – part 2: further experimental support. *J. Fluid Mech.* **298** 389–407.

Murakami, S., A. Mochida and Y. Hayashi (1990) Examining the $k–\varepsilon$ model by means of a wind tunnel test and large eddy simulation of the turbulence structure around a cube. *J. Wind Eng. and Ind. Aero.* **35** 87–100.

Murlis, J., H. M. Tsai and P. Bradshaw (1982) The structure of turbulent boundary layers at low Reynolds numbers. *J. Fluid Mech.* **122** 13–56.

Nicolleau, F. and J. C. Vassilicos (2000) Turbulent diffusion in stratified non-decaying turbulence. *J. Fluid Mech.* **410** 123–146.

Oberlack, M. and F. Busse (eds.) (2002) *Theories of Turbulence*, Springer-Verlag.

Parpais, S. (1997) *Développement d'un modéle spectral pour la turbulence inhomogéne; resolution par une méthode d'éléments finis.* Ph.D. Thesis, Ecole Centrale de Lyon.

Pope, S. B. (2000) *Turbulent Flows*, Cambridge University Press.

Prandtl, L. (1925) Bericht über Untersuchungen zer ausgebildeten Turbulenz. *Zs. Angew. Math. Mech.* **5** 136–139.

   (1931) *Abrißder Strömungslehre*, Vieweg (see *Fluid Dynamics*, Blackie, 1956).

Rao, K. G., R. Narasimha and A. Prabhu (1996) Estimation of drag coefficient at low wind speeds over the monsoon trough land region during montblex-90. *Geophys. Res. Lett.* **23**(19) 2617–2620.

Reynolds, O. (1895) On the dynamical theory of turbulent incompressible viscous fluids and the determination of the criterion. *Phil. Trans. Roy. Soc. London A* **186** 123–164.

Rodi, W. and N. N. Mansour (1993) Low Reynolds number $k–\varepsilon$ modelling with the aid of direct simulation data. *J. Fluid Mech.* **250** 509–529.

Rotta, J. (1951) Statistiche Theorie nichthomogener Turbulenz. *Z. Physik* **129** 547.

Sagaut, P. (2001) *Large Eddy Simulation for Incompressible Flows* (second edn.), Scientific Computing Series, Springer.

Savill, A. M. (1981) Zonal modelling? *Proc. AFSOR-HTTM-Stanford Conf. on Complex Turbulent Shear Flows 2*, pp. 999–1004, Stanford University Press.

   (1987) Recent developments in rapid-distortion theory. *Ann. Rev. Fluid Mech.* **19** 521–531.

   (1996) One point closures applied to transition. In *Turbulence and Transition Modelling* (M. Hallback, D. S. Henningson, A. V. Johnsson and P. H. Alfredsson, eds.), Kluwer Academic Press.

Saxena, V. and J. C. R. Hunt (1993) A second order solution for the distortion of large-scale turbulence around a 2-d obstacle. *Applied Scientific Research*, **51** 457–461. Also in *Advances in Turbulence* (F. T. M. Niewstadt, ed.), Volume 4, Kluwer.

Schiestel, R. (1987) Multiple timescale modelling of turbulent flows in one-point closures. *Phys. Fluids* **30** 722–731.

Schlichting, H. (1960) *Boundary Layer Theory* (fourth edn.), McGraw-Hill.

Schmidt, H. and U. Schumann (1989) Coherent structure of the convective boundary layer derived from large-eddy simulations. *J. Fluid Mech.* **200** 511–562.

Simone, A., G. N. Coleman and C. Cambon (1997) The effect of compressibility on turbulent shear flow: a rapid-distortion theory and direct-numerical-simulation study. *J. Fluid Mech.* **330** 307–338.

Spalart, P. R. and S. R. Allmaras (1994) A one-equation turbulence model for aerodynamic flows. *Recherche Aerospatiale* **1** 5–21.

Strang, E. J. and H. J. S. Fernando (2001) Entrainment and mixing in stratified shear flows. *J. Fluid Mech.* **428** 349–386.

Taylor, G. I. (1938a) The production and dissipation of vorticity in a turbulent fluid. *Proc. Roy. Soc. A* **164** 15–23.

   (1938b) Turbulence. In *Modern Developments in Fluid Dynamics* (S. Goldstain, ed.), Clarendon Press.

Teixeira, M. A. C. and S. E. Belcher (2000) Dissipation of shear-free turbulence near boundaries. *J. Fluid Mech.* **422** 167–191.

Thomas, N. H. and P. E. Hancock (1977) Grid turbulence near a moving wall. *J. Fluid Mech.* **82** 481–496.

Townsend, A. A. (1976) *The Structure of Turbulent Shear Flow*, Cambridge University Press.

Uitenbogaard, R. (1994) In preprint of *Stably Stratified Flow Conference*, Grenoble.

Veeravalli, S. and Z. Warhaft (1989) The shearless turbulent mixing layer. *J. Fluid Mech.* **207** 191–229.

Wood, D. H. and P. Bradshaw (1982) A turbulent mixing layer constrained by a solid surface. Part 1: measurements before reaching the surface. *J. Fluid Mech.* **122** 57–89.
  (1984) A turbulent mixing layer constrained by a solid surface. Part 2: measurements in the wall bounded flow. *J. Fluid Mech.* **139** 347–361.

Wu, X., R. G. Jacobs, J. C. R. Hunt and P. A. Durbin (1999) *CTR Summer School Abstracts*, Stanford.

Wyngaard, J. C. (1979) The atmospheric boundary layer – modelling and measurement. In *Turbulent Shear Flows*, vol. 2 (L. J. S. Bradbury, ed.), Springer-Verlag.

Yao, Y. F., A. M. Savill, N. D. Sandham and W. N. Dawes (2002) Simulation and modelling of turbulent training-edge flow. *J. Flow Turbulence and Combustion* **68** 313–333.